M. Nachtegael, D. Van der Weken, D. Van De Ville, E. E. Kerre (Eds.)

Fuzzy Filters for Image Processing

Dear MyCopy Customer,

This Springer book is a monochrome print version of the eBook to which your library gives you access via SpringerLink. It is available to you at a subsidized price since your library subscribes to at least one Springer eBook subject collection.

Please note that MyCopy books are only offered to library patrons with access to at least one Springer eBook subject collection. MyCopy books are strictly for individual use only.

You may cite this book by referencing the bibliographic data and/or the DOI (Digital Object Identifier) found in the front matter. This book is an exact but monochrome copy of the print version of the eBook on SpringerLink.

Springer-Verlag Berlin Heidelberg GmbH

Studies in Fuzziness and Soft Computing, Volume 122
http://www.springer.de/cgi-bin/search_book.pl?series=2941

Editor-in-chief
Prof. Janusz Kacprzyk
Systems Research Institute
Polish Academy of Sciences
ul. Newelska 6
01-447 Warsaw
Poland
E-mail: kacprzyk@ibspan.waw.pl

Further volumes of this series can be found at our homepage

Vol. 103. N. Barnes and Z.-Q. Liu
Knowledge-Based Vision-Guided Robots, 2002
ISBN 3-7908-1494-6

Vol. 104. F. Rothlauf
Representations for Genetic and Evolutionary Algorithms", 2002
ISBN 3-7908-1496-2

Vol. 105. J. Segovia, P.S. Szczepaniak and M. Niedzwiedzinski (Eds.)
E-Commerce and Intelligent Methods, 2002
ISBN 3-7908-1499-7

Vol. 106. P. Matsakis and L.M. Sztandera (Eds.)
Applying Soft Computing in Defining Spatial Relations", 2002
ISBN 3-7908-1504-7

Vol. 107. V. Dimitrov and B. Hodge
Social Fuzziology, 2002
ISBN 3-7908-1506-3

Vol. 108. L.M. Sztandera and C. Pastore (Eds.)
Soft Computing in Textile Sciences, 2003
ISBN 3-7908-1512-8

Vol. 109. R.J. Duro, J. Santos and M. Graña (Eds.)
Biologically Inspired Robot Behavior Engineering, 2003
ISBN 3-7908-1513-6

Vol. 110. E. Fink l. 112. Y. Jin
Advanced Fuzzy Systems Design and Applications, 2003
ISBN 3-7908-1523-3

Vol. 111. P.S. Szcepaniak, J. Segovia, J. Kacprzyk and L.A. Zadeh (Eds.)
Intelligent Exploration of the Web, 2003
ISBN 3-7908-1529-2

Vol. 112. Y. Jin
Advanced Fuzzy Systems Design and Applications, 2003
ISBN 3-7908-1537-3

Vol. 113. A. Abraham, L.C. Jain and J. Kacprzyk (Eds.)
Recent Advances in Intelligent Paradigms and Applications", 2003
ISBN 3-7908-1538-1

Vol. 114. M. Fitting and E. Orowska (Eds.)
Beyond Two: Theory and Applications of Multiple Valued Logic, 2003
ISBN 3-7908-1541-1

Vol. 115. J.J. Buckley
Fuzzy Probabilities, 2003
ISBN 3-7908-1542-X

Vol. 116. C. Zhou, D. Maravall and D. Ruan (Eds.)
Autonomous Robotic Systems, 2003
ISBN 3-7908-1546-2

Vol 117. O. Castillo, P. Melin
Soft Computing and Fractal Theory for Intelligent Manufacturing, 2003
ISBN 3-7908-1547-0

Vol. 118. M. Wygralak
Cardinalities of Fuzzy Sets, 2003
ISBN 3-540-00337-1

Vol. 119. Karmeshu (Ed.)
Entropy Measures, Maximum Entropy Principle and Emerging Applications, 2003
ISBN 3-540-00242-1

Vol. 120. H.M. Cartwright, L.M. Sztandera (Eds.)
Soft Computing Approaches in Chemistry, 2003
ISBN 3-540-00245-6

Vol. 121. J. Lee (Ed.)
Software Engineering with Computational Intelligence, 2003
ISBN 3-540-00472-6

Mike Nachtegael
Dietrich Van der Weken
Dimitri Van De Ville
Etienne E. Kerre (Eds.)

Fuzzy Filters for Image Processing

Springer

Dr. Mike Nachtegael
e-mail: mike.nachtegael@rug.ac.be
Drs. Dietrich Van der Weken
e-mail: dietrich.vanderweken@rug.ac.be
Prof. Dr. Etienne E. Kerre
e-mail: etienne.kerre@rug.ac.be

Ghent University
Dept. of Applied Mathematics
and Computer Science
Krijgslaan 281 (S9)
9000 Gent, Belgium

Dr. Dimitri Van De Ville
Swiss Federal Institute
of Technology Lausanne
Biomedical Imaging Group
EPFL/STI/LIB
Bat. BM4.140
1015 Lausanne EPFL, Switzerland

DOI 10.1007/978-3-540-36420-7

Library of Congress Cataloging-in-Publication-Data applied for

A catalog record for this book is available from the Library of Congress.

Bibliographic information published by Die Deutsche Bibliothek
Die Deutsche Bibliothek lists this publication in the Deutsche Nationalbibliographie; detailed bibliographic data is available in the internet at <http://dnb.ddb.de>.

http://www.springer.de

Originally published by Springer-Verlag Berlin Heidelberg New York in 2003
MyCopy version of the original edition 2003

Typesetting: camera-ready-pages delived by editors
Cover design: E. Kirchner, Springer-Verlag, Heidelberg
Printed on acid free paper 62/3020/M - 5 4 3 2 1 0
www.springer.com/mycopy

Preface

The ongoing increase in scale of integration of electronics makes storage and computational power affordable to many applications. Also image processing systems can benefit from this trend. A variety of algorithms for image processing tasks becomes close at hand. From the whole range of possible approaches, those based on fuzzy logic are the ones this book focusses on. A particular useful property of fuzzy logic techniques is their ability to represent knowledge in a way which is comprehensible to human interpretation.

The theory of fuzzy sets and fuzzy logic was initiated in 1965 by Zadeh, and is one of the most developed models to treat imprecision and uncertainty. Instead of the classical approach that an object belongs or does not belong to a set, the concept of a fuzzy set allows a gradual transition from membership to nonmembership, providing partial degrees of membership. Fuzzy techniques are often complementary to existing techniques and can contribute to the development of better and more robust methods, as has already been illustrated in numerous scientific branches.

The present book resulted from the workshop "Fuzzy Filters for Image Processing" which was organized at the 10th FUZZ-IEEE Conference in Melbourne, Australia. At this event several speakers have given an overview of the current state-of-the-art of fuzzy filters for image processing. Afterwards, the book has been completed with contributions of other international researchers.

Fuzzy Filters for Image Processing covers a wide range of both theoretical and practical applications of this exciting topic. The chapters have been grouped into four parts: *Fuzzy Filters for Noise Reduction* (Chapters 1 to 4), *Fuzzy Filters for Edge Detection and Segmentation* (Chapters 5 to 9), *Fuzzy Filters for Image Enhancement* (Chapters 10 and 11) and *Specific Applications of Fuzzy Filters* (Chapters 12 to 15).

The first part deals with fuzzy filters for the reduction of noise in images. In Chapters 1 and 2, several fuzzy filters for gray-scale images are presented, and their performance is illustrated with numerous figures and evaluated w.r.t. some well-known error measures. The third chapter is devoted to the construction of a fuzzy filter based on the evaluation of fuzzy similarities between pixels in a local image neighbourhood. It turns out that this filter is very suitable for fast hardware implementation! Finally, Chapter 4 covers three fuzzy filters for color images, based on local image models and using statistical indices. It should be noted that color figures are reproduced in black and white throughout the book, but that the original coloured images are included in an appendix at the end.

Another very important application of fuzzy filters is situated in the area of edge detection and segmentation. Five chapters of this book are directly related to this topic. In Chapter 5 several recently developed fuzzy-based techniques for image segmentation are introduced and reviewed. Also, a robust fuzzy integral based image segmentation algorithm is extensively discussed and investigated. The following chapter provides a comprehensive discussion of several thresholding techniques that employ the concept of the measure of fuzziness and shows interesting applications, both to gray-scale and to color images. Chapter 7 presents yet another color image segmentation method. The method is based on an analysis of the color histogram and involves the use of fuzzy morphological filters. Chapter 8 addresses two very practical issues, namely the fact that several applications require that the edge detection algorithms are both fast and robust, and secondly the fact that edge detection algorithms should also be robust when noise is present in the images. The final chapter of this part investigates fuzzy data fusion techniques for multiple cue image and video segmentation.

Image enhancement, which is the topic of the third part of the book, also is a well-known image (pre-)processing task. In Chapter 10 a very efficient method for image enhancement, based on logaritmic models, is introduced. The experimental results show that these logaritmic models can be very usefull, both for gray-scale images and color images. Chapter 11 addresses the problem that image (quality) evaluation by humans is very subjective by nature. When images are enhanced – to support the visual perception of human experts – the quality assessement of the images depends directly on the image enhancement procedures. Therefore, this chapter extensively discusses an observer-dependent system for subjective image enhancement evaluation.

Finally, Part 4 of the book offers the reader some specific applications of fuzzy filters in image processing. Chapter 12 presents applications of several fuzzy operators in image processing, e.g. fuzzy location and scale estimators, the fuzzy cell Hough transform, and generalized fuzzy mathematical morphology. The next chapter discusses two specific adaptive fuzzy filters and their application to online maneuvering target tracking. The performance of the fuzzy filters is compared with the results of classical Kalman based techniques. Chapter 14 deals with lossy image compression. It not only introduces an image compression method based on fuzzy relational equations, but also proposes a new image reconstruction algorithm. The final chapter is devoted to avoidance of highlights, which is a very important practical problem in computer vision systems. Highlights can be considered as a type of noise, caused by incidence of light on a specular surface. A solution for the avoidance of highlights, based on the fuzzy integral as image fusion operator, is presented.

We are convinced that the reading of this book will stimulate your interests in this challenging research, and will give you an indication of important keypoints for future research. For your convenience the index has numerous entries, which makes finding a specific topic very easy. The book also contains hundreds of references to specialized scientific literature, which makes it a great source of information.

Gent,
November 2002

Mike Nachtegael[1]
Dietrich Van der Weken
Dimitri Van De Ville
Etienne E. Kerre

[1] Editors' contact address: Ghent University, Department of Applied Mathematics and Computer Science, Fuzziness and Uncertainty Modelling Research Unit, Krijgslaan 281-S9, B-9000 Gent, Belgium. Email: mike.nachtegael@rug.ac.be, dietrich.vanderweken@rug.ac.be, dimitri.vandeville@epfl.ch, etienne.kerre@rug.ac.be

Contents

Part IV. Specific Applications of Fuzzy Filters

List of Contributors

Yannis Avrithis
National Technical University of Athens
Greece

Otman Basir
University of Waterloo
Canada

Claudine Botte-Lecocq
Université Sciences & Technologies
Lille 1
France

Vasile Buzuloiu
Politehnica University of Bucureşti
Romania

Vassilios Chatzis
Aristotle University of Thessaloniki
Greece

Leonardo Javier Delgado-Rangel
Universidad Nacional de Colombia
Colombia

Manuel Guillermo Forero-Vargas
Universidad Nacional de Colombia
Colombia

Antony Fotinos
University of Patras
Greece

Spiros Fotopoulos
University of Patras
Greece

Aymeric Gillet
Université Sciences & Technologies
Lille 1
France

Kaoru Hirota
Tokyo Institute of Technology
Japan

Spyros Ioannou
National Technical University of Athens
Greece

Ivan Kalaykov
Örebro University
Sweden

Fakhri Karray
University of Waterloo
Canada

Stefanos Kollias
National Technical University of Athens
Greece

Hon Keung Kwan
University of Windsor
Canada

Ludovic Macaire
Université Sciences & Technologies
Lille 1
France

Socrates Makrogiannis
University of Patras
Greece

Mohammad B. Menhaj
AmirKabir University
Iran

Hajime Nobuhara
Tokyo Institute of Technology
Japan

Witold Pedrycz
University of Alberta
Canada

Vasile Pătraşcu
Politehnica University of Bucureşti
Romania

Ioannis Pitas
Aristotle University of Thessaloniki
Greece

Jack-Gerard Postaire
Université Sciences & Technologies
Lille 1
France

Aureli Soria-Frisch
Fraunhofer IPK
Germany

Giorgos Stamou
National Technical University of Athens
Greece

Yasufumi Takama
Tokyo Institute of Technology
Japan

Hamid R. Tizhoosh
University of Waterloo
Canada

Gustav Tolt
Örebro University
Sweden

Constantin Vertan
Politehnica University of Bucureşti
Romania

Hongwei Zhu
University of Waterloo
Canada

Part I

Fuzzy Filters for Noise Reduction

Chapter 1

Fuzzy Filters for Noise Removal

Manuel Guillermo Forero-Vargas and Leonardo Javier Delgado-Rangel

Universidad Nacional de Colombia
System Engineering Departament
OHWAHA Research Group
Ciudad Universitaria, Bogotá, Colombia
email: mforero@ing.unal.edu.co, ldelgado7@eudoramail.com

Summary. An image may be subject to noise from several sources. The presence of noise in an image can affect the accuracy of the results considerably. Because of its wide applicability to image filtering, several fuzzy filter methods have been proposed. In this chapter, a survey of different design techniques for fuzzy filters is presented. Six filters are investigated: multipass fuzzy, fuzzy multilevel median, histogram adaptive, fuzzy vector rank, fuzzy vector rational median, and fuzzy credibility color filters. An effort is made to evaluate the performance of the filters using criteria such as: mean average error (MAE), mean square error (MSE), normalized mean square error (NMSE), signal to noise error ratio (SNR) and mean chromaticity error (MCRE). The evaluation is based on some real world images.

1 Introduction

The presence of noise affects the accuracy of image processing. In order to reduce the noise, nonlinear filtering techniques are often employed because they provide better results than linear methods for noise removal and do not degrade the edges and the details of the image. Recent progress in fuzzy logic allows different possibilities for developing new image noise reduction techniques. Using the nonlinear methods based on fuzzy logic reasoning, good performance can be obtained. Different new techniques can be mentioned. In spite of the proliferation of fuzzy filters, very little formal studies among the different types of techniques have been attempted. The purpose of this chapter is to investigate six filters: multipass fuzzy [9], fuzzy multilevel median [4], fuzzy vector rank [1], fuzzy vector median rational [5], histogram adaptive [3] and fuzzy credibility color [10]. An effort is made to evaluate the performance of this filters using criteria such as: mean average error (MAE), mean square error (MSE), normalized mean square error (NMSE), signal to noise error ratio (SNR) and mean chromaticity error (MCRE). The evaluation is based on two real-world images.

The following notation is adopted in the discussion below. Let (x, y) be the spatial coordinates of each pixel in an image Q of size $N \times M$, $q(x, y)$ be the grey level or intensity of the pixel in the position (x, y), where $0 \leq q \leq L-1$ and L is the maximum number of grey levels in the image. The intensity is normalized to the interval $[0, 1]$, in order to fuzzify the image employing the fuzzy inference choice.

A multichannel or color image $\mathbf{Q}$ is represented by a vector:

$$\mathbf{q}(x, y) = q_{\bar{1}}(x, y), \ldots, q_{\bar{n}}(x, y),$$

where n is the dimension of the vector given by the number of channels of the image.

The pixels, situated in a window W of size $l \times l$ centered in (x, y), are represented by:

$$q_1(x, y), q_2(x, y), \ldots, q_{l^2}(x, y) = q_i(x, y) \text{ for } i = 1, \ldots, l^2,$$

where the central pixel is noted simply $q(x, y)$.

The set of neighboring pixels of $q(x, y)$ inside a operation window W of size $l \times l$ are noted $w_1, \ldots, w_{l^2-1}$, except when otherwise specified. When $l = 3$, the pixels are labelled as follows:

$$\begin{bmatrix} w_1 & w_2 & w_3 \\ w_8 & q & w_4 \\ w_7 & w_6 & w_5 \end{bmatrix}.$$

2 Fuzzy filters

According to the fuzzy theory, the belonging of element x to a set A is gradual and characterized by a membership or belonging function $\mu_A(x)$. Let X be the universal set of elements x, then a fuzzy set A in X is defined by:

$$A = \{(x, \mu_A(x)) \, | x \in X\},$$

where $\mu_A(x)$ is called the membership function of the fuzzy set A.

The filters used in the image processing that employs fuzzy logic model the characteristics of the image as a fuzzy set. Two types of uncertainty have been identified: vagueness and ambiguity [6].

An important element of the fuzzy theory is related to the rules of inference which are applied directly to the pixels. These rules make it possible to determine the outputs of the fuzzy inferential filters. They are composed of a group of logic connectors and IF-THEN sentences, also known as fuzzy conditional statements. The form of the rules of inference is antecedent or premise-consequence or conclusion. The output of these filters depends on the fuzzy rules and on the defuzzifying process, which combines the effects of the established rules.

The vector median filters are used in order to reduce the noise present in a multichannel image. These filters minimize the distance in the vector space between the image vectors as an error criterion. The fuzzy ranked order filters are based on order statistics of different distances functions between a pixel and its neighbors. Fuzzy membership functions are used to determine the weights of the ordered statistic filter. There are other filters in which the input vector valued signal at a pixel is replaced by a linear combination of the ordered input vectors in the neighborhood of the pixel. The output of the vector rational filters are the result of a vector rational function. The hybrid filters combines fuzzy and non-fuzzy components.

2.1 Multipass fuzzy filter

In order to reduce the noise present in an image, Fabrizio Russo introduced a multipass fuzzy filter (MFF) [9]. This filter consists of three cascaded blocks (Fig. 1). Each block is based on a fuzzy operator which attempts to cancel the noise while preserving the image structure, so the filter can be repeatedly applied to the image without increasing the blur. Each operator is developed using fuzzy rules according to the specific kind of noise that it is designed to eliminate.

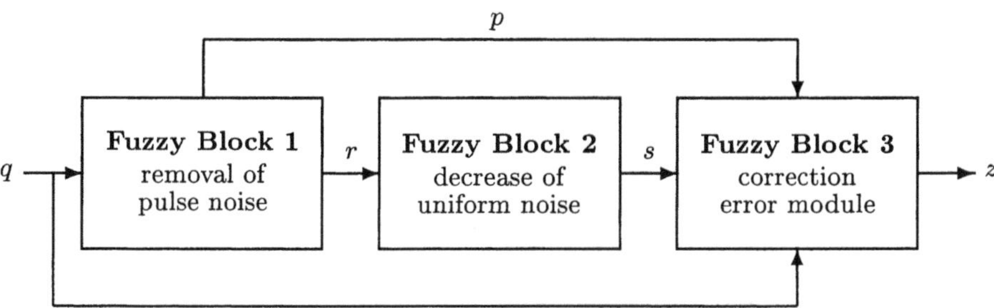

Fig. 1. Structure of the multipass fuzzy filter [9].

Given the input image, which is corrupted by mixed impulse and uniform noise. It is understood that an input pixel $q(x, y)$ examined for noise reduction is centered in a 3×3 window.

The first operator is a nonlinear recursive filter devoted to eliminating the noise pulses to obtain the output $r(x, y)$. The process of detecting the pulse noise is made by applying the filter windows $W_1, \ldots, W_{12}$ shown in Fig. 2.

The input of the operator, denominated $\Delta w_{ij}[q(x, y)]$, is given by the differences in the luminance between each one of the dark neighbor pixels in each window and $q(x, y)$. It is obtained as follows:

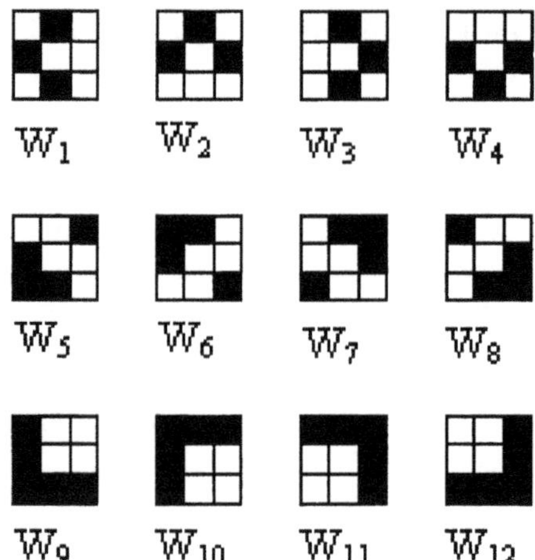

Fig. 2. Windows configurations for eliminating impulse noise.

$$\Delta w_{ij}[q(x,y)] = \{w_i \in W_j[q(x,y)]\} - q(x,y) \qquad \text{for } i = 1, \ldots, 8; j = 1, \ldots, 12.$$

These differences are used to calculate a correction term $\Delta r(x,y)$ as follows:

$$\Delta r(x,y) = (L-1)\left\{\max_{j=1}^{12}\left(\min_{i=1}^{8}\left[\mu_{LA}\left(\Delta w_{ij}[q(x,y)]\right)\right]\right) - \max_{j=1}^{12}\left(\min_{i=1}^{8}\left[\mu_{LA}\left(-\Delta w_{ij}[q(x,y)]\right)\right]\right)\right\}.$$

$\Delta r(x,y)$ is composed of two subterms. The first one is used to analyze the positive differences in the luminance and the second one the negative differences. $\Delta r(x,y)$ is added to the input to obtain the operator output:

$$r(x,y) = q(x,y) + \Delta r(x,y).$$

Notice that the fuzzy sets are defined in the interval $[-L+1, L-1]$.

μ_{LA}, shown in Fig. 3, denotes the membership function of a fuzzy set denoted Large given by:

$$\mu_{LA}(u) = \begin{cases} 0 & \text{if } -L+1 \le u < a \\ \frac{b(u-a)}{(L-1)(b-a)} & \text{if } a \le u < b \\ \frac{u}{L-1} & \text{if } b \le u \le L-1 \end{cases}.$$

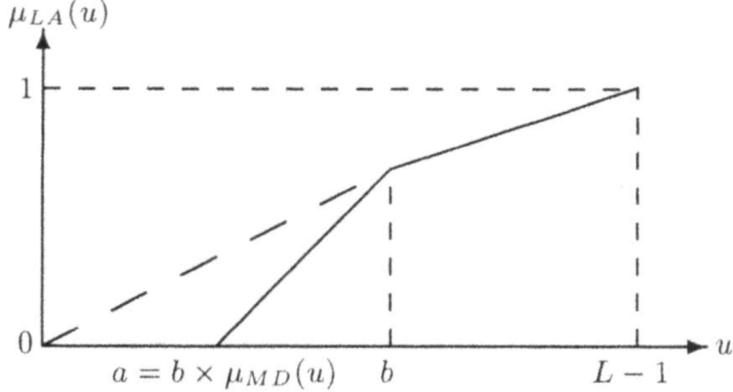

Fig. 3. Membership function of the fuzzy set LARGE [9].

The fuzzy set Large is designed to reduce the impulse noise, without affecting the details of the image. According to the amplitude of the input pulse noise, the output of the operator changes, from no correction (0) to maximal correction given by $u/(L-1)$. It is considered that if the amplitude is small, there is uncertainty about whether or not it is really noise. As a consequence, there is no filter action. To avoid this effect, Russo suggests a second fuzzy set denominated Medium (MD) that is directly applied on $q(x,y)$. Its membership function μ_{MD}, shown in Fig. 4, is specified by two parameters a_1, a_2 with $0 \leq a_1 \leq (L-1)/2$ and $0 \leq a_2 \leq a_1$ as follows:

$$\mu_{MD}(u) = \begin{cases} 0 & \text{if } 0 \leq u < \frac{L-1}{2} - a_1 \\ \frac{u+a_1-\frac{L-1}{2}}{a_2-a_1} & \text{if } \frac{L-1}{2} - a_1 \leq u < \frac{L-1}{2} - a_2 \\ 1 & \text{if } \frac{L-1}{2} - a_2 \leq u < \frac{L-1}{2} + a_2 \\ \frac{u-a_1-\frac{L-1}{2}}{a_1-a_2} & \text{if } \frac{L-1}{2} + a_2 \leq u < \frac{L-1}{2} + a_1 \\ 0 & \text{if } \frac{L-1}{2} + a_1 \leq u \leq L-1 \end{cases} .$$

μ_{MD} is employed to calculated a, by using the equation: $a = b * \mu_{MD}(x)$.

The flexibility of the form of the membership function of the Medium set makes it possible to modify it to maximize the maintenance of the details of the image.

The details are best preserved when the luminance of the pixel to be processed is medium. The pixels processed by this fuzzy operator belong to directional patterns.

$p(x,y) = |\Delta[r(x,y)]|$ can be understood as an estimate of the amplitude of a possible noise pulse in the position (x,y) and will be employed by the third operator.

The second block is designed to cancel the small amplitude noise characteristic of uniform noise and its output image $s(x,y)$ is supposed to be a good estimate of the original uncorrupted image.

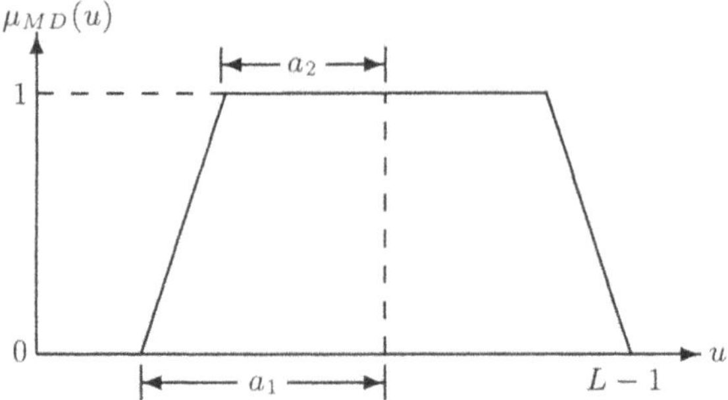

Fig. 4. Membership function of the fuzzy set MEDIUM [9].

This operator consists of a cross window size 5×5 centered in (x, y) and not including $r(x, y)$ given by $W_0[s(x, y)] = \{w_i[s(x, y)]$ for $i = 1, \ldots, 20\}$. Like the first operator, its input values are given by the differences of luminance of the image input: $\Delta w_i[s(x, y)] = w_i[s(x, y)] - r(x, y)$ for $i = 1, \ldots, 20$.

The differences $\Delta w_i[s(x, y)]$ obtained are used to calculate a correction term $\Delta s(x, y)$, that is added to $r(x, y)$ in order to produce the second operator output $s(x, y)$. $\Delta s(x, y)$ is obtained as follows:

$$\Delta s(x, y) = \frac{c}{20}\left[\sum_{i=1}^{20} \mu_{SM}(\Delta w_i) - \sum_{i=1}^{20} \mu_{SM}(-\Delta w_i)\right],$$

where μ_{SM} is the membership function of the fuzzy set SMALL (SM), shown in Fig. 5, given by:

$$\mu_{SM}(u) = \begin{cases} 0 & \text{if } -L+1 \leq u < -c \\ \frac{u+c}{2c} & \text{if } -c \leq u < c \\ \frac{3c-u}{2c} & \text{if } c \leq u < 3c \\ 0 & \text{if } 3c \leq u \leq L-1 \end{cases},$$

c is fixed to adjust the filter's edge preserving behavior. Then, the second filter output is calculated as follows:

$$s(x, y) = r(x, y) + \Delta s(x, y).$$

In order to improve the noise elimination, a third block, an error correction module, is adjoined in order to obtain the final image Z. The second operator reduce the amount of uniform noise, so its output $s(x, y)$ can be employed to improve the impulse noise reduction of $q(x, y)$. The third operator is very similar to the first one, because it employs the Large and Medium fuzzy

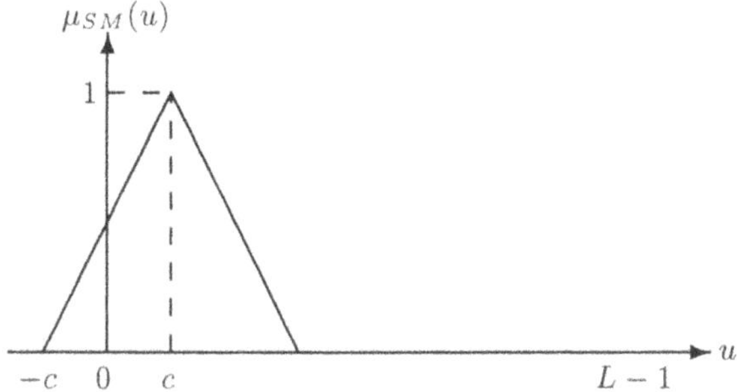

Fig. 5. Membership function of the fuzzy set SMALL [9].

sets, using as input the differences of luminance to obtain a correction term $\Delta z(x, y)$:

$$\Delta w_{ij}[z(x,y)] = \{w_i \in W_j[s(x,y)]\} - q(x,y) \qquad \text{for } i = 1,\ldots,8; j = 1,\ldots,12.$$

The estimate of the impulse noise amplitude $p(x, y)$ is used to define the operator output:

$$z(x,y) = \begin{cases} q(x,y) + \Delta z(x,y) & \text{if } p(x,y) > 0, \\ s(x,y) & \text{if } p(x,y) = 0. \end{cases}$$

2.2 Fuzzy multilevel median filter

The fuzzy multilevel median filter (FMMF) introduced by Jiu [4], is based on a multilevel median filter which is expanded to include fuzzy rules in order to improve impulse noise elimination. The multilevel median filter is designed as follows.

Let $W(x, y)$ be a square window of size $(2l + 1) \times (2l + 1)$ centered in the position (x, y). Four subwindows are defined as follows:

$$\begin{aligned} W_1(x,y) &= q(x,y+j), & -l \le j \le l \\ W_2(x,y) &= q(x+j,y+j), & -l \le j \le l \\ W_3(x,y) &= q(x+j,y), & -l \le j \le l \\ W_4(x,y) &= q(x+j,y-j), & -l \le j \le l. \end{aligned}$$

Let $\text{med}_i(x, y)$, for $i = 1, 2, 3, 4$, be the median values of each subwindow. Then, two variables can be defined:

$$\text{med}_{\max}(x,y) = \max_{i=1}^{4} \text{med}_i(x,y),$$

$$\mathrm{med}_{\min}(x,y) = \min_{i=1}^{4} \mathrm{med}_i(x,y).$$

The output of this filter is:

$$r(x,y) = \mathrm{med}(\mathrm{med}_{\max}(x,y), \mathrm{med}_{\min}(x,y), q(x,y)).$$

This filter is improved by using fuzzy reasoning. Two additional terms are calculated and adjoined in order to calculate $r(x,y)$ and a median fuzzy credibility function C is defined. To define the credibility of the median of each subwindow, the difference between the median and each one of the pixels in each of the same subwindows is studied.

Given the elements of each subwindow, w_{ij} for $i = 1, \ldots, 4; -l \leq j \leq l$, the following criterion is used to define this credibility function:

- If at least one of the differences is very high (higher than a threshold t) then the credibility of med for that subwindow should be very low. Jiu suggests using $30 \leq t \leq 40$ for $L = 256$ [4].
- Equally, if at least one of the differences is close to zero (less than 5 for $L = 256$ [4]) or null, then the credibility of med for that subwindow should be very low.
- If all differences are medium, then the credibility should be very high.

Different forms can be established to define the fuzzy credibility function. Jiu proposed the functions shown in Fig. 6. These fuzzy functions are employed to fuzzify the credibility of the median of each subwindow.

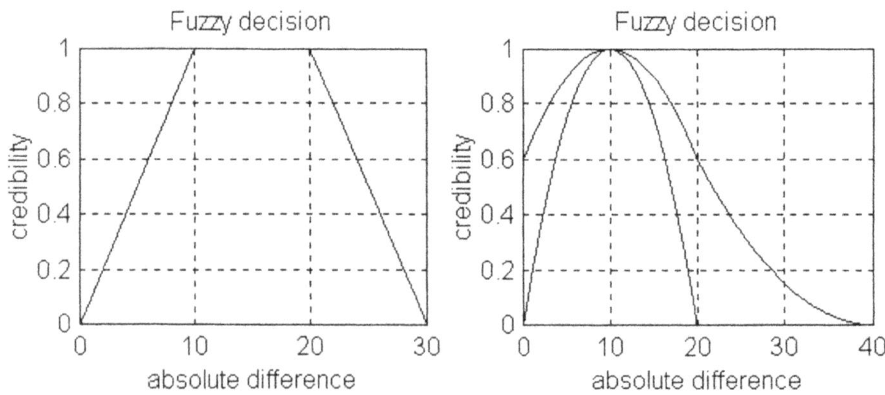

Fig. 6. Credibility functions [4].

Then a function $s_i(x, y)$ is obtained for each subwindow. $s_i(x, y)$ is defined by:

$$s_i(x, y) = \sum_{j=l-1}^{l+1} C_{\text{med}}(|\text{med}_i - w_{ij}|) \qquad \text{for } i = 1, \ldots, 4.$$

The output fuzzy filter is given by:

$$r(x, y) = \text{med}[\text{med}_{\text{max}}(x, y), \text{med}_{\text{min}}(x, y), q(x, y), s_{\text{max1}}(x, y), s_{\text{max2}}(x, y)],$$

where $s_{\text{max1}}(x, y)$ and $s_{\text{max2}}(x, y)$ correspond to the two largest values of $s_i(x, y)$.

2.3 Histogram adaptive filter

The histogram adaptive filter (HAF) introduced by Wang and Chu [12] belongs to a class of filters that employs the histogram for reducing noise.

Let the gray level fuzzy sets be dark (D), medium (M) and light (Li), a membership function associated with each gray level of the histogram must be determined. A generalized bell member function, based on the Cauchy distribution is a good option [3]:

$$\mu_j[q_i(x, y)] = \frac{1}{1 + \frac{|q_i - m_j|^{2b_j}}{d_j}}, \qquad \text{for } i = 0, \ldots, 9; j = D, M, Li.$$

This function is specified by three parameters, d_j, m_j and b_j, that determine the behavior of the membership function. Wang and Chu recommend making $b_j = 15$ for $j = D, M, Li$ [12]. The gray level intensity is normalized to the interval $[0, 1]$.

The fuzzy inference rules employed to define the membership of the gray level for each fuzzy set are given by:

- Rule 1:
  IF $q_i(x, y) \in D \; \forall \, i = 1, \ldots, 9$ THEN $r(x, y) \in D$.
- Rule 2:
  IF $q_i(x, y) \in M \; \forall \, i = 1, \ldots, 9$ THEN $r(x, y) \in M$.
- Rule 3:
  IF $q_i(x, y) \in Li \; \forall \, i = 1, \ldots, 9$ THEN $r(x, y) \in Li$.

These rules form the knowledge base of the filter where $r(x, y)$ is the output variable of the consequent.

Let N_{imp} be the set of the pixels in the window that are the most likely to be noisy. A pixel $q_i(x, y) \in N_{imp}$ if:

- $q_i(x,y) = \min W(x,y)$.
- $q_i(x,y) = \max W(x,y)$.
- $q(x,y) \leq t$.
- $q(x,y) \geq 1-t$.

where t is a value defined as the maximum quantity of noise permitted on the filtered image. Good results are obtained through $0.05 \leq t \leq 0.2$.

The normalized membership degree of each pixel is given by:

$$\varpi_j[q_i(x,y)] = \frac{\mu_j[q_i(x,y)]}{\sum_{i=1}^{9} \mu_j[q_i(x,y)]}, \qquad \text{for } i=1,\ldots,9;\ j=D,M,Li.$$

A free-noise estimate of $q(x,y)$ is obtained before the defuzzification process:

$$\bar{q}(x,y) = \frac{\sum_{i=1}^{9} \hat{q}_i(x,y)}{\sum_{i=1}^{9} I_i},$$

where:

$$\hat{q}_i(x,y) = \begin{cases} 0 \wedge I_i = 0, & \text{if } q_i(x,y) \in N_{imp} \\ q_i(x,y) \wedge I_i = 1, & \text{otherwise} \end{cases}.$$

To obtain the output of the filter defuzzification must be done. This is given by:

$$z(x,y) = \begin{cases} s_j(x,y) \mid \min_{j=D,M,Li} |E_j(x,y)| & \text{if } q(x,y) \in N_{imp}, \\ \bar{q}(x,y) & \text{otherwise} \end{cases}$$

where:

$$s_j(x,y) = \sum_{i=1}^{9} \varpi_j q_i(x,y),$$

and:

$$E_j(x,y) = \bar{q}(x,y) - s_j(x,y).$$

The normalized histogram $h_n(q)$ can be viewed as an estimate of the probability density function of the gray levels on the image. Let $h_{imp}(q)$ be a histogram composed of the pixels in the image most likely to be noisy. An estimate free-noise histogram $\bar{h}_n(q)$ can be obtained, given the expression:

$$\bar{h_n}(q) = \frac{h(q) - h_{imp}(q)}{\sum_{q=0}^{L-1} [h(q) - h_{imp}(q)]}.$$

To determine the parameters d_j and m_j, $h(q)$ is divided in three equal regions. For each region, the mass and the mean are obtained as follows:

$$\text{mass}_j = \frac{\sum_{q \in j} \bar{h_n}(q)}{\sum_{q=0}^{L-1} \bar{h_n}(q)}, \qquad \text{for } j = D, M, Li,$$

$$\text{mean}_j = \frac{\sum_{q \in j} q\bar{h_n}(q)}{\sum_{q \in j} \bar{h_n}(q)} \qquad \text{for } j = D, M, Li,$$

where mean_j and mass_j are employed by Wang et Chu [12] as initial values to find the parameters d_j and m_j. Then, the following algorithm is applied:

- While $\text{mass}_D \geq \text{mean}_D - t$ then $\text{mean}_D = \text{mass}_D = \sqrt{(\text{mean}_D - t)\,\text{mass}_D}$.
- While $\text{mass}_{Li} \geq (1 - t) - \text{mean}_{Li}$ then $\text{mean}_{Li} = \text{mass}_{Li} = \sqrt{[(1 - t) - \text{mean}_{Li}]\,\text{mass}_{Li}}$.
- Given $T = t$ or $T = t - 1$, while $\text{mass}_M \geq |\text{mean}_M - T|$ then $\text{mass}_M = |\text{mean}_M - T|$.
- Finally, $m_j = \text{mean}_j$ and $d_j = \text{mass}_j$ for $j = D, M, Li$.

2.4 Fuzzy vector rank filters

Androutsos et al. [1] proposed a new class of filters, denoted fuzzy vector rank filters (FVRF), based on a combination of different distance measures, fuzzy membership values and α-trimmed functions. Fuzzy membership functions are adopted to determine the weights of the ordered statistics filter.

Taking the pixels situated in a window W centered in (x, y), then an order statistic filter will replace the input vector $\mathbf{q}(x, y)$ by a linear combination of all the ordered input vectors in the window, that is:

$$\mathbf{r}(x, y) = \sum_{i=1}^{l^2} \omega_i \mathbf{q}_{(i)}(x, y),$$

where ω_i are constant weights and $\mathbf{q}_{(i)}(x, y)$, named the $i-th$ order statistic, represents the pixels in the window ranked according to some criterion in ascending order of magnitude such as:

$$\mathbf{q}_{(1)}(x, y) \leq \mathbf{q}_{(2)}(x, y) \leq \cdots \leq \mathbf{q}_{(l^2)}(x, y).$$

According the value of the weights, this filter could be a vector median or a linear mean filter.

The fuzzy weights ω_i are found by employing a fuzzy membership function defined by a distance criterion chosen to rank the data [7].

Androutsos et al. [1] proposed as distance criterion the minimum angle value that is defined as the sum of angles made up of an input vector and the remaining input vectors in the window. Hence, the sum a for the central vector $\mathbf{q}(x, y)$ is given by:

$$a(x,y) = \sum_{i=1}^{l^2-1} \theta[\mathbf{q}(x,y), \mathbf{w}_i(x,y)],$$

where:

$$\theta(\mathbf{AB}) = \arccos\left(\frac{\mathbf{AB}}{|\mathbf{A}||\mathbf{B}|}\right), \qquad 0 \leq \theta(\mathbf{AB}) \leq \pi,$$

with $\mathbf{A}$ and $\mathbf{B}$ two vectors.

Because the angle criterion is chosen, a sigmoid function is adopted to determine the weights:

$$\omega_i = \frac{1}{1 + e^{a_i(x,y)^b}}, \qquad \text{for } i = 0, \ldots, l^2,$$

where $a_i(x, y)$ denotes the value of a for all the $\mathbf{q}_i(x, y)$ and b is a smoothing positive parameter defined according to the problem to be considered. Then the weights are organized in ascending order of magnitude as:

$$\omega_{(1)}(x,y) \leq \omega_{(2)}(x,y) \leq \cdots \leq \omega_{(l^2)}(x,y),$$

and the input vectors are also ordered according to the weights.

The output of the filter is given by a weighted average of the ranked vector inputs inside the window:

$$\mathbf{r}(x,y) = \frac{\sum_{i=1}^{k} \omega_{(i)} \mathbf{q}_{(i)}(x,y)}{\sum_{i=1}^{k} \omega_{(i)}},$$

where the parameter k corresponds to the number of vector inputs $\mathbf{q}_{(i)}(x, y)$ in the window, whose fuzzy weights correspond to the maximum fuzzy values. The value of k can be a constant or be fixed for each window according to a criterion. Androutsos et al. [1] proposed using the number of vectors of which the fuzzy weights are larger than $1/l^2$ as criterion.

Another distance criterion that can be used to rank the input vector is the sum of Minkowski's measures between one input vector and the remaining input vectors in the window. This sum for the central vector $\mathbf{q}(x,y)$ is given by:

$$d[\mathbf{q}(x,y)] = \left\{ \sum_{i=1}^{l^2} \sum_{j=1}^{n} |(q_j(x,y) - q_{ij}(x,y)|^{\delta} \right\}^{\frac{1}{\delta}} .$$

When $\delta = 1$ Minkowski's measure becomes the Hamming distance, and for when $\delta = 2$ it become the Euclidean one. If this distance criterion is used, Androutsos et al. [1] suggests the following fuzzy transformation:

$$\omega_i = \exp[-d[\mathbf{q}_i(x,y)]^b] \qquad \text{for } i = 1, \ldots, l^2.$$

2.5 Fuzzy vector rational median filter

Khriji and Gabbouj [5] have proposed a multichannel filter, denoted fuzzy vector rational median filter (FVRMF), that combines fuzzy rational and median functions. This filter preserves the edges and the chromaticity of the image.

The output filter is the result of combining different fuzzy functions. Let r_1 and r_3 be the output of two fuzzy vector rank filters and r_2 the output of a fuzzy center weighted vector median filter.

In Sec. 2.4 the fuzzy vector rank filter was studied. Two windows W_1 and W_3, shown in Fig. 7, are used instead of the original window W to obtain the output of the filters r_1 and r_3. The transfer functions for r_1 and r_3 are defined by looking for a good compromise between noise elimination and detail and chromaticity preservation [5].

The output of the filter is the result of a vector rational function given by:

$$s(x,y) = r_2(x,y) + \frac{\sum_{i=1}^{3} \alpha_i r_i(x,y)}{h + k\,||r_1(x,y) - r_3(x,y)||_2},$$

where $||.||_2$ represent an L_2 norm vector, h and k are two positives parameters where k is used to control the quantity of the nonlinear effect [8]. $\alpha = [\alpha_1, \alpha_2, \alpha_3]$ is a constant vector coeficient. Khriji and Gabbouj [5] employ as criterion $\sum_{i=1}^{3} \alpha_i = 0$, chosing $\alpha = [1, -2, 1]^T$ [5].

The behavior of the filter change depends on the value of the parameter k:

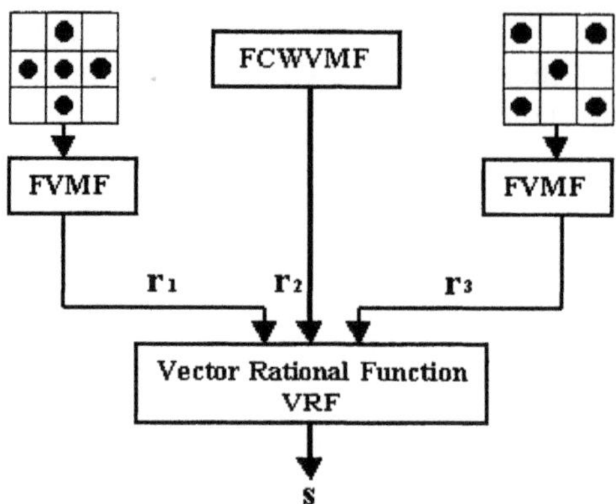

Fig. 7. Structure of the fuzzy vector rational median filter.

- If $k \to 0$, the filter can be seen as a linear combination of the three non linear subfunctions:

$$s(x,y) = c_1 r_1(x,y) + c_2 r_2(x,y) + c_3 r_3(x,y),$$

  where c_1, c_2 and c_3 are constants.
- If $k \to \infty$ the second term is cancelled and the output is equal to the output of the second filter:

$$s(x,y) = r_2(x,y).$$

- For other values of k, the $||r_1(x,y) - r_3(x,y)||_2$ term serves to detect the presence of an edge in the image and in consequence, the smoothing effect of the filter is reduced.

$r_2(x,y)$ is the output of a vector filter defined by:

$$r(x,y) = q_j(x,y) \mid j = \arg\{\min[d[\mathbf{q}_i(x,y)]]\}, \text{ for } i = 1, \ldots, l^2,$$

$$d[\mathbf{q}(x,y)] = \left\{ \sum_{i=1}^{l^2} \sum_{j=1}^{n} |(q_j(x,y) - q_{ij}(x,y)|^{\delta} \right\}^{\frac{1}{\delta}}.$$

2.6 Fuzzy credibility color filter

The fuzzy credibility color filter (FCCF) was introduced by Vertan et al. [10]. The colors are modelled in the CIELAB color space. The CIELAB model is used for its capability of representing perceived color difference through Euclidean distances [11].

They define each color c_k of the color space as a fuzzy color set C_i. Therefore, each color c has a membership function $\mu_{C_k}(c)$ associated given by a gaussian function:

$$\mu_{C_i}(c) = \begin{cases} 1 & \text{if } d \leq \epsilon \\ \exp\left[-\frac{\left(\frac{d}{\epsilon - 1}\right)^2}{2\alpha^2}\right] & \text{otherwise} \end{cases},$$

where $\epsilon = 2.3$ is the minimum distance between two distinguishable colors and $d = ||\mathbf{c}_i - \mathbf{c}||$.

The authors also suggest another membership function in order to improve the computation time:

$$\mu_{C_i}(c) = \begin{cases} 1 & \text{if } d \leq \epsilon \\ 0 & \text{if } d \geq (\beta + 1)\epsilon \\ 1 - \frac{1}{\beta}\left(\frac{d}{\epsilon} - 1\right) & \text{otherwise} \end{cases}.$$

Two output results should be given by the filter, correction or no correction. Then, the fuzzy inference rules must be established to calculate the credibility of color values $s_i(x, y)$ in window W for each pixel $\mathbf{q}_i(x, y)$ with respect to the set of remaining pixels in the window C_r. Between of the possible T-conorms available, they suggest:

$$\begin{aligned}
\sigma(a, b) &= \max(a, b) \\
\sigma(a, b) &= a + b - ab \\
\sigma(a, b) &= \min(1, a + b) \\
\sigma(a, b) &= \frac{a + b - 2ab}{1 - ab} \\
\sigma(a, b) &= \frac{a + b}{1 + ab}
\end{aligned}$$

where a and b are two fuzzy values.

The fuzzy inference is calculated as follows:

$$s_i(x, y) = \sigma\{\mu_{C_r}(\mathbf{q}_1(x, y)), \sigma[\mu_{C_r}(\mathbf{q}_2(x, y)), \ldots, \sigma[\mu_{C_k}(\mathbf{q}_{l^2-1}(x, y)), \mu_{C_k}(\mathbf{q}_{l^2}(x, y))]]\}$$

for $i = 1, \ldots, l^2$.

The output of the filter is the most credible color among the input colors in the window:

$$r(x, y) = q_j(x, y) \mid j = \text{argmax}[s_i(x, y)] \text{ for } i = 1, \ldots, l^2$$

3 Extension to color of gray level fuzzy filters

The MFF, FMMF and the HAF were initially designed for gray level images. A first attempt to extend these filters to color image noise reduction consists of applying each method separately for each color channel. The results could be improved by applying a measure of distance in order to avoid the loss of chromaticity. This concept was applied in all the color filters seen above.

4 Applications

In this section, the fuzzy techniques described earlier are applied to two natural images, "lena" size 256 × 256 and "bear" size 771 × 460, shown in Fig. 8.a, 11.a. and 12.a. The statistical values determined from the results are shown in Tables 1 to 3. The investigation was limited to how the filters behave according to impulse noise and how they reduce it. Uniform noise was not included since only one filter was designed for that and therefore, it was not considered important enough to study. The performance of these filters is compared analytically using mean average error (MAE), mean square error (MSE), normalized mean square error (NMSE), signal to noise error ratio (SNR), for gray level images, given by:

$$MAE = \frac{\sum_{x=1}^{N}\sum_{y=1}^{M}|q_f(x,y) - q_o(x,y)|}{NM},$$

$$MSE = \frac{\sqrt{\sum_{x=1}^{N}\sum_{y=1}^{M}|q_f(x,y) - q_o(x,y)|^2}}{NM},$$

$$NMSE = \frac{\sqrt{\sum_{x=1}^{N}\sum_{y=1}^{M}|q_f(x,y) - q_o(x,y)|^2}}{\sqrt{\sum_{x=1}^{N}\sum_{y=1}^{M}q_o(x,y)^2}},$$

$$SNR = \frac{\sum_{x=1}^{N}\sum_{y=1}^{M}|q_f(x,y) - q_o(x,y)|^2}{\sum_{x=1}^{N}\sum_{y=1}^{M}|q_n(x,y) - q_o(x,y)|^2},$$

and signal to noise error ratio (SNR) and mean chromaticity error (MCRE), for color images, given by:

$$SNR = \frac{\sum_{RGB}\sum_{x=1}^{N}\sum_{y=1}^{M}|q_f(x,y) - q_o(x,y)|^2}{\sum_{RGB}\sum_{x=1}^{N}\sum_{y=1}^{M}|q_n(x,y) - q_o(x,y)|^2},$$

$$MCRE = \frac{\sum_{RGB} \sum_{x=1}^{N} \sum_{y=1}^{M} |q_f(x,y) - q_o(x,y)|}{3NM},$$

where $q_f(x,y)$ is the output image of the filter, q_o is the non corrupted original image and q_n is the corrupted image.

The images obtained by the methods considered are shown in Figs. 8 to 14. The parameters employed were:

- MFF : $a_1 = 100$, $a_2 = 70$, $b = 60$, $c = 30$ (optimal values obtained experimentally).
- FVMF: $b = 1$.
- FVRMF: $b = 1$.

The color implementation of the MFF, FMMF and the HAF filters was limited to applying the same filter to each color channel.

Filter	MAE	MSE	NMSE	SNR
Noise	4.5	449	0.024	
MFF	3.9	75	0.004	0.17
FMMF	2.6	63	0.003	0.14
HAF	3.6	139	0.007	0.31
FVRF	4.0	62	0.003	0.14
FVRMF	3.8	60	0.003	0.13
FCCF	3.8	60	0.003	0.26

Table 1. Statistical results for gray level image "Lena".

Filter	MCRE	SNR
Noise	2.26	
MFF	1.66	0.10
FMMF	1.19	0.12
HAF	2.15	0.14
FVRF	1.63	0.09
FVRMF	1.65	0.08
FCCF	1.63	0.26

Table 2. Statistical results for color image "Lena".

According to the above test, all the filters have very good results on gray and color images. A subjective opinion will be given for the detail and cromaticity preservation. For color images, the FCCF has the best performance.

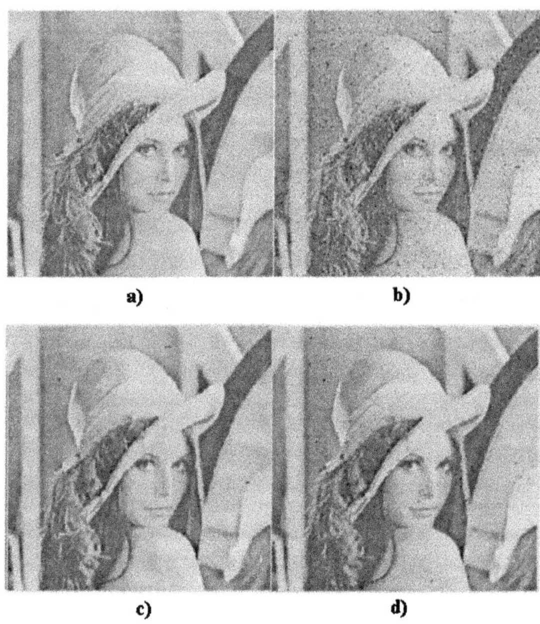

Fig. 8. a) Original "Lena" image. b) 5% impulse noise degrade image. c)MFF result. d) FMMF result.

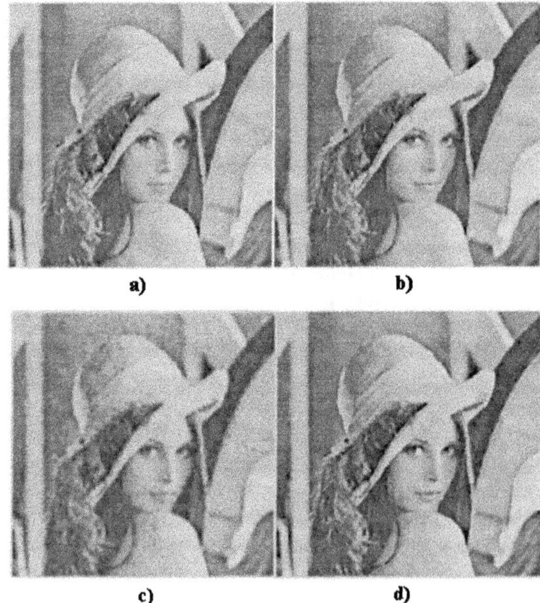

Fig. 9. a) FVRF result. b) FVRMF result. c) HAF result. d) FCCF result.

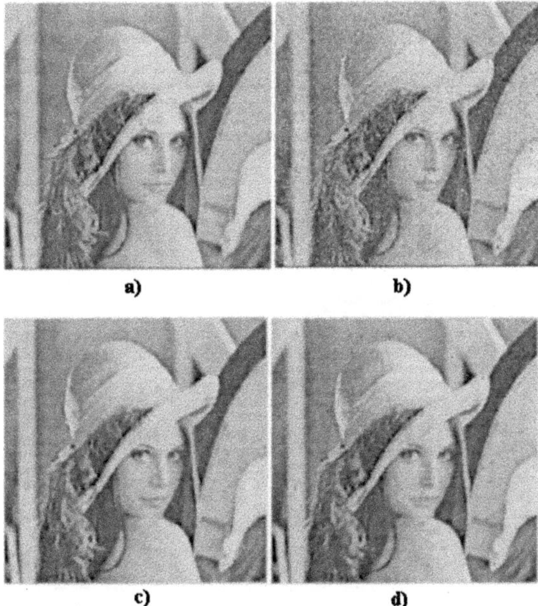

Fig. 10. a) Original "Lena" image. b) 8% impulse noise degrade image. c)MFF result. d) FMMF result.

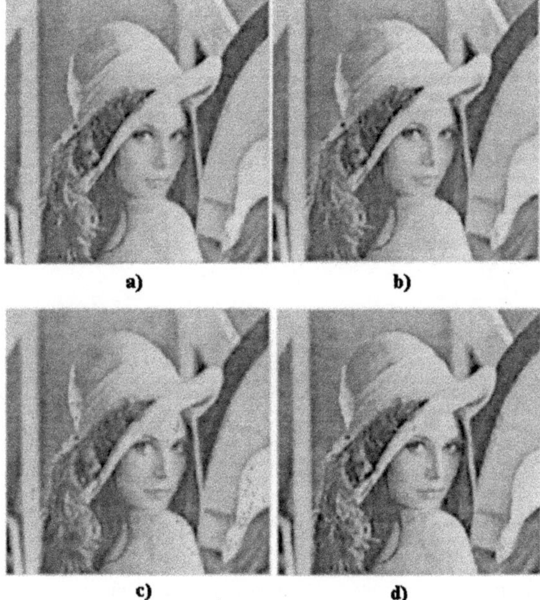

Fig. 11. a) FVRF result. b) FVRMF result. c) HAF result. d) FCCF result.

Filter	MAE	MSE	NMSE	SNR
Noise	64.52	12222	1.19	
MFF 1 Iteration	28.05	4846	0.47	0.250
MFF 6 Iterations	6.82	317	0.03	0.026
HAF	6.26	268	0.026	0.022

Table 3. Statistical results for image "Bear".

Fig. 12. a) Original "Bear" image. b) 50% impulse noise degrade image.

The HAF has the best performance when the image is highly corrupted with impulse noise, but this performance is affected when other kinds of noise are also present.

5 Conclusion

In summary, this chapter presented a survey of various fuzzy filtering methods. All the methods described here are intended to reduce the impulse noise but only MFF aims to reduce uniform noise. Attempts also have been made to evaluate the performance of each filter on noisy images. From this study,

Fig. 13. MFF result. a)After 1 iteration. b) After 6 iterations.

Fig. 14. HAF result.

it is concluded that for the set of test images all of the methods provide reasonably good noise reduction.

References

1. Androutsos D., Plataniotis K.N., Venetsanopoulos A.N. *Colour image processing using vector rank filters*, International conference on digital signal processing, Vol.2, pp. 614-619, 1995.
2. A.M. Eskicioglo, Fisher P.S., Chen S., *Image quality measures and their performance*,IEEE Trans. on Communication, Vol. 43, pp. 2959-2965, 1995.
3. Jang J.S.R., Sun C.T., Mizutani E., *Fuzzy sets*, Neuro-fuzzy and soft computing, pp. 13-46, 1997.
4. Jiu J.Y., *Multilevel median filter based on fuzzy decision*, DSP IC Design Lab E.E. NTU., 1996.
5. L.Khriji, M.Gabbouj, *A New Class of Multichannel Image Processing Filters: Vector Median Rational Hybrid Filters*, IEICE Transactions on Information and Systems, Vol. E82-D, No.12, pp. 1589-1596, 1999.
6. Lin C.T., Lee G., *Fuzzy measures*, Neural fuzzy systems: A neuro-fuzzy synergism to intelligent systems, pp. 63-88, 1996.
7. Paulus D., Hornegger J.,*Applied Pattern Recognition:a Practical introduction to image and speech processing in C++*,2.ed., Vieweg, Braunschweig, Wiesbaden, pp. 237, 1998.
8. Roberts R T., Mullis C. T. *Digital Signal Processing.* Addison Wesley Publishing Co. USA, 1987.
9. Russo F., Ramponi G., *A noise smoother using cascaded FIRE filters*, in: Proceedings of FUZZ-IEEE'95- 4th IEEE Int. Conf. on Fuzzy Systems, Vol. 1, pp. 351-358, 1995.
10. Vertan N.C.,*A Fuzzy Color Credibility Approach To Color Image Filtering*, http://citeseer.nj.nec.com/299826.html.
11. Vertan C., Buzuloiu V.,*Fuzzy nonlinear filtering of color images: A survey*, in: Fuzzy techniques in image processing,Kerre E., Nachtegael M, (ed.), Heidelberg, Physica Verlag, pp. 248-264, 2000.
12. Wang J.H. y Chiu H.C., *HAF: an adaptive fuzzy filter for restoring highly corrupted images by histogram estimation*, Proc. Natl. Sci. ROC(A), Vol. 23, No. 5 pp. 630-643, 1999.

Chapter 2

Fuzzy Filters for Noise Reduction in Images

Hon Keung Kwan

University of Windsor
Department of Electrical and Computer Engineering
401 Sunset Avenue
Windsor, Ontario, Canada N9B 3P4
email: Kwan1@uwindsor.ca

Summary. In this chapter, seven fuzzy filters for noise reduction in images are introduced. These seven fuzzy filters include the Gaussian fuzzy filter with median center (GMED), the symmetrical triangular fuzzy filter with median center (TMED), the asymmetrical triangular fuzzy filter with median center (ATMED), the Gaussian fuzzy filter with moving average center (GMAV), the symmetrical triangular fuzzy filter with moving average center (TMAV), the asymmetrical triangular fuzzy filter with moving average center (ATMAV), and the decreasing weight fuzzy filter with moving average center (DWMAV). Each of these fuzzy filters, applies a weighted membership function to an image within a window to determine the center pixel, is easy and fast to implement. Simulation results on the filtering performance of these seven fuzzy filters and the standard median filter (MED) and moving average filter (MAV) on images contaminated with low, medium, high impulse and random noises are presented. Results indicate that these seven fuzzy filters achieve varying successes in noise reduction in images as compared to the standard MED and MAV filters.

1 Introduction

The two common types of noise in images are impulse (or salt and pepper) noise, and random (or Gaussian) noise. Impulse noise is defined by noise density. Random noise is expressed in terms of its mean and variance values. Noise can be generated during image capture, transmission, storage, as well as during image copying, scanning, and display. For examples, impulse noise can be generated through TV broadcasting and due to information losses; and random noise can be generated during film exposure and development. Noise reduction in images has been one of the common tasks in image processing. For the case of impulse noise, most part of an original image is unaltered, and the image is characterized by some corrupted samples that vary drastically. Compared to impulse noise, random noise is a more challenging type of noise, it is important to be able to reduce random noise effectively in images.

In image processing, various linear and nonlinear filtering methods have been proposed. Linear filtering techniques used for noise reduction in images are characterized by mathematical simplicity and can effectively reduce noise with spectral components that do not overlap with those of an image. However, linear filters cannot effectively reduce impulse noise and have a tendency to blur the edges of an image. In such situations, median filters [1-3], which are nonlinear filters, provide an effective solution. Median filters have good edge preserving ability, can eliminate impulse noise, and have moderate noise attenuation ability in the flat regions of an image. The operations of a classical median (MED) filter involve the application of a window to move over an image and to replace the value at the center pixel with the median of all the pixel values within the window. In so doing, a pixel with a distinct intensity (in the case of an impulse) as compared to those of its predefined neighbors will be eliminated. The implementation of a standard median filter is simple and the filter can process an image in a fast manner. The performance of a median filter is average for filtering random noise in an image. This difficulty can be overcome with some success by another nonlinear filtering technique using moving average (MAV) filters [1-3]. Moving average filters can smooth random noise, but they cannot suppress impulse noise and cannot preserve sharp edges of an image. The idea of a standard moving average filter is to replace its center pixel by the average value of its predefined neighboring pixels, which can be easily implemented.

The ability to filter unwanted impulse noise and random noise while preserving the edges and details of an image is a non-trivial task. Various nonlinear filters based on classical and/or fuzzy techniques [1-8] have emerged in the past few years for this challenging task. Review on fuzzy-type of filters can be found in [5-7] and a comparison study has been reported in [8]. Depending on their filtering strategies, these filters can be classified as classical filters, classical-fuzzy filters, and fuzzy filters. We shall briefly go through a few representative filters in each of these three classes. The classical filters include the standard median filters for reduction of impulse noise, the standard moving average filters and the adaptive wiener filters (AWF) for reduction in Gaussian noise. We have already described the standard MED filters and the standard MAV filters in the previous paragraph. In the adaptive wiener filter [9], the center pixel of a window is replaced by the sum of the mean value within the window and a fractional of the difference between the center pixel and the mean. The fraction is defined as the normalized difference between the local and global variances of the noise.

Under the classical-fuzzy filters, we have the fuzzy median (FM) filter, the weighted fuzzy mean (WFM) filter, the Type-1 adaptive weighted fuzzy mean (AWFM1) filter, and the Type-2 adaptive weighted fuzzy mean (AWFM2) filter. The FM filter [10-11] is a modification of the classical median filter and is designed for reducing impulse noise. Using fuzzy rules, the FM filter determines the degree (as a weight) that a center pixel is not a noisy pixel. The weight is 1 when the center pixel is not noisy. If the center pixel is not a noisy pixel, the center pixel will remain unchanged. Otherwise, the center pixel will be replaced by a sum of the mean value within the window and the weighted difference between the center pixel and the mean value. The WFM filters adopt one or more fuzzy systems to determine the weights of a weighted linear filter to give the filtered output. In the WFM filter described in [12-13], it adopts three local features to estimate the weights. These local features incorporate the variances of the data and the additive Gaussian noise, the difference between a center pixel and the median value of its neighbors, and the normalized distance between any pixel and the center pixel, as the input of a fuzzy system. Another example of the WFM filters can be found in [14-15], in which triangular membership functions describing dark, medium, and bright are used to determine the weight of a pixel within a 3x3 neighborhood in order to calculate the normalized weighted pixel value. Among the three normalized weighted pixel values, the one closest to the estimated output by fuzzy interval is chosen as the final output. Given an original image and its noisy version, optimal membership functions can be adaptively calculated for each type of images. These optimal or adaptive membership functions are used in the AWFM1 filters [14]. The AWFM2 filters [15] are designed to filter medium-tailed and long-tailed impulse noises. For a given noisy image, the difference between a pixel and its AWFM1 output is computed. Based on the difference value obtained, four classes of fuzzy subspaces are determined by some fuzzy detectors. A dynamic selector then uses four corresponding fuzzy decision rules to determine the filtered output.

Under the fuzzy filters, we have the Fuzzy Inference Ruled by Else-action (FIRE) filters [16-19], the iterative fuzzy control (IFC) filter [20], and the GOA filter [21]. The FIRE filters are a family of nonlinear operators that adopt fuzzy rules to remove impulse noise from images. A FIRE filter evaluates the information in the neighborhood of the pixel by considering the luminance differences between this pixel and its neighbors. The fuzzy rules aim at evaluating a correction term that should cancel the noise. If no rule is satisfied, the central pixel remains unchanged. The FIRE filter [16] is based on two-step fuzzy reasoning and is designed to remove impulse noise from images. Firstly, fuzzy rules are applied to different patterns of the 3x3 neighborhood of a center pixel in order to determine a correction term. Secondly, a small correction is regarded as insignificant and will be further reduced to preserve fine details. The Dual Step FIRE (DS-FIRE) filter [17] makes use of more information from its neighborhood and therefore able to preserve the quality of fine details and textures while removing impulse noise from images. In the piecewise linear FIRE (PWL-FIRE) filter [18], two piecewise linear fuzzy sets are used for removing impulse noise from images which allows the neutralization of small corrections to be carried out in an implicit way depending on the starting and ending points of the two piecewise linear fuzzy sets. A FIRE filter [19] adopting a hierarchical rule base was designed for reducing mixed impulsive and Gaussian noise. The IFC filter [20] is designed for reducing both impulse noise and Gaussian noise. The idea of the

IFC filter is that: If the differences between a center pixel and its neighbors belong to a certain class, then the correction term of the center pixel should also belong to that class. The GOA filter [21] is designed for reducing Gaussian-like noise. The idea is to average a pixel using its neighborhood pixels, while simultaneously taking care of important image details such as edges. To achieve this, the filter estimates a fuzzy gradient in each direction so as to distinguish local variations due to noise from those due to image structure. Moreover, the membership functions are adapted according to the noise level in performing fuzzy smoothing.

In [22], we have described median filtering using fuzzy concept. Symmetrical and asymmetrical triangular membership functions with median center are used for filtering impulse, random, and mixed noises of a periodic rectangular pulse. At the same time, fuzzy filters consisting of symmetrical and asymmetrical triangular membership functions with median center and moving average center have been applied to filtering of images contaminated with impulse, random and mixed noises. The latter work has recently been reported in [23]. In this chapter, we present a summary of our earlier study on 2-dimensional fuzzy filters for noise reduction in images. Seven fuzzy filters are defined and their filtering performance on impulse noise and random noise are presented. This chapter is divided into six sections. Section 1 gives the introduction. Section 2 defines the seven fuzzy filters. Section 3 describes the simulations and results, and Section 4 concludes the chapter with some remarks.

2 Definitions of fuzzy filters

Let $x(i,j)$ be the input of a 2-dimensional fuzzy filter, the output of the fuzzy filter is defined as:

$$y(i,j) = \frac{\sum\limits_{(r,s)\in A} F[x(i+r,j+s)] \cdot x(i+r,j+s)}{\sum\limits_{(r,s)\in A} F[x(i+r,j+s)]} \tag{1}$$

where $F[x(i,j)]$ is the general window function and A is the area of the window. For a square window of dimensions $N \times N$, the range of r and s are: $-R \le r \le R$ and $-S \le s \le S$, where $N = 2R+1 = 2S+1$. With the definitions of different window functions, seven fuzzy filters can be obtained, which we shall call the Gaussian fuzzy filter with median center (GMED), the symmetrical triangular fuzzy filter with median center (TMED), the asymmetrical triangular fuzzy filter with median center (ATMED), the Gaussian fuzzy filter with moving average center (GMAV), the symmetrical triangular fuzzy filter with moving average center (TMAV), the asymmetrical triangular fuzzy filter with moving average center (ATMAV), and the decreasing weight fuzzy filter with moving average center (DWMAV). The standard median filter (MED) and the standard moving average filter (MAV) can be considered as special cases of these fuzzy filters. The definitions of all these nine filters are given in the following paragraphs.

2.1 Median filter (MED)

In the case of a standard median filter, the window function is defined as:

$$F_{med}[x(i+r,j+s)] = \begin{cases} 1 & \text{for } x(i+r,j+s) = x_{med}(i,j) \\ 0 & \text{otherwise} \end{cases} \quad (2)$$

such that the output value at the center of a window $y(i,j)$ is replaced by the median value $x_{med}(i,j)$ among all the input values $x(i+r,j+s)$ for $r,s \in A$ in the window A at discrete indexes (i,j).

2.2 Moving average filter (MAV)

In a standard moving average filter, the window function is defined as:

$$F_{mav}[x(i+r,j+s)] = 1 \text{ for } r,s \in A \quad (3)$$

The moving average filter is equivalent to a 2-dimensional rectangular-shape fuzzy filter covering all the input values $x(i+r,j+s)$ for $r,s \in A$ in the window A.

2.3 Gaussian fuzzy filter with median center (GMED)

The Gaussian fuzzy filter with the median value within a window chosen as the center value is defined as:

$$F_{gmed}[x(i+r,j+s)] = e^{-\frac{1}{2}\left[\frac{x(i+r,j+s)-x_{med}(i,j)}{\sigma(i,j)}\right]^2} \text{ for } r,s \in A \quad (4)$$

where $x_{med}(i,j)$ and $\sigma(i,j)$ represent, respectively, the median value and the variance value of all the input values $x(i+r,j+s)$ for $r,s \in A$ in the window A at discrete indexes (i,j).

2.4 Symmetrical triangle fuzzy filter with median center (TMED)

The symmetrical triangular fuzzy filter with the median value within a window chosen as the center value is defined as:

$$F_{tmed}[x(i+r,j+s)] = \left\{ \begin{array}{l} 1 - \frac{|x(i+r,j+s)-x_{med}(i,j)|}{x_{mm}(i,j)} \\ \quad \text{for } |x(i+r,j+s)-x_{med}(i,j)| \leq x_{mm}(i,j) \\ 1 \\ \quad \text{for } x_{mm} = 0 \end{array} \right\} \quad (5)$$

where

$$x_{mm}(i,j) = \max[x_{max}(i,j) - x_{med}(i,j), x_{med}(i,j) - x_{min}(i,j)]$$

$x_{max}(i,j)$, $x_{min}(i,j)$ and $x_{med}(i,j)$ are, respectively, the maximum value, the minimum value, and the median value of all the input values $x(i+r, j+s)$ for $r, s \in A$ within the window A at discrete indexes (i,j).

2.5 Asymmetrical triangle fuzzy filter with median center (ATMED)

The asymmetrical triangular fuzzy filter with the median value within a window chosen as the center value is defined as:

$$F_{atmed}[x(i+r,j+s)] = \begin{cases} 1 - \frac{x_{med}(i,j) - x(i+r,j+s)}{x_{med}(i,j) - x_{min}(i,j)} \\ \quad \text{for } x_{min}(i,j) \leq x(i+r,j+s) \leq x_{med}(i,j) \\ 1 - \frac{x(i+r,j+s) - x_{med}(i,j)}{x_{max}(i,j) - x_{med}(i,j)} \\ \quad \text{for } x_{med}(i,j) \leq x(i+r,j+s) \leq x_{max}(i,j) \\ 1 \\ \quad \text{for } x_{med}(i,j) - x_{min}(i,j) = 0 \\ \quad \text{or } x_{max}(i,j) - x_{med}(i,j) = 0 \end{cases} \quad (6)$$

Unlike Equation 5, the triangle window function in Equation 6 is asymmetrical. The degree of asymmetry depends of the difference between $x_{med}(i,j) - x_{min}(i,j)$ and $x_{max}(i,j) - x_{med}(i,j)$. $x_{max}(i,j), x_{min}(i,j)$ and $x_{med}(i,j)$ are, respectively, the maximum value, the minimum value, and the median value of all the input values $x(i+r, j+s)$ for $r, s \in A$ within the window A at discrete indexes (i,j).

2.6 Gaussian fuzzy filter with moving average center (GMAV)

The Gaussian fuzzy filter with the moving average value within a window chosen as the center value is defined as:

$$F_{gmav}[x(i+r,j+s)] = e^{-\frac{1}{2}\left[\frac{x(i+r,j+s) - x_{mav}(i,j)}{\sigma(i,j)}\right]^2} \quad \text{for } r, s \in A \quad (7)$$

where $x_{mav}(i,j)$ and $\sigma(i,j)$ represent, respectively, the moving average value and the variance value of all the input values $x(i+r, j+s)$ for $r, s \in A$ in the window A at discrete indexes (i,j).

2.7 Symmetrical triangle fuzzy filter with average center (TMAV)

The symmetrical triangular fuzzy filter with the moving average value within a window chosen as the center value is defined as:

$$F_{tmav}[x(i+r,j+s)] = \left\{ \begin{array}{l} 1 - \frac{|x(i+r,j+s) - x_{mav}(i,j)|}{x_{mv}(i,j)} \\ \quad \text{for } |x(i+r,j+s) - x_{mav}(i,j)| \leq x_{mv}(i,j) \\ 1 \\ \quad \text{for } x_{mv} = 0 \end{array} \right\} \tag{8}$$

where

$$x_{mv}(i,j) = \max[x_{max}(i,j) - x_{mav}(i,j), x_{mav}(i,j) - x_{min}(i,j)]$$

$x_{max}(i,j)$, $x_{min}(i,j)$ and $x_{mav}(i,j)$ represent, respectively, the maximum value, the minimum value, and the moving average value of $x(i+r,j+s)$ within the window A at discrete indexes (i,j).

2.8 Asymmetrical triangle fuzzy filter with moving average center (ATMAV)

The asymmetrical triangular fuzzy filter with the moving average value within a window chosen as the center value is defined as:

$$F_{atmav}[x(i+r,j+s)] = \left\{ \begin{array}{l} 1 - \frac{x(i+r,j+s) - x_{mav}(i,j)}{x_{max}(i,j) - x_{mav}(i,j)} \\ \quad \text{for } x_{mav}(i,j) \leq x(i+r,j+s) \leq x_{max}(i,j) \\ 1 - \frac{x_{mav}(i,j) - x(i+r,j+s)}{x_{mav}(i,j) - x_{min}(i,j)} \\ \quad \text{for } x_{min}(i,j) \leq x(i+r,j+s) \leq x_{mav}(i,j) \\ 1 \\ \quad \text{for } x_{max}(i,j) - x_{mav}(i,j) = 0 \\ \quad \text{or } x_{mav}(i,j) - x_{min}(i,j) = 0 \end{array} \right\} \tag{9}$$

The degree of asymmetry depends of the difference between $x_{mav}(i,j) - x_{min}(i,j)$ and $x_{max}(i,j) - x_{mav}(i,j)$. $x_{max}(i,j)$, $x_{min}(i,j)$ and $x_{mav}(i,j)$ represent, respectively, the maximum value, the minimum value, and the moving average value of $x(i+r,j+s)$ within the window A at discrete indexes (i,j).

2.9 Decreasing weight fuzzy filter with moving average center (DWMAV)

The decreasing weight fuzzy filter with the moving average value within a window chosen as the center value is defined as:

$$F_{dwmav}[x(i+r,j+s)] = 1 - \frac{\max(|r|,|s|)}{\max(|R|,|S|) + t} \tag{10}$$

where

$$-R \leq r \leq R \text{ and } -S \leq s \leq S, \text{ and } 2R+1 = 2S+1 = N$$

N is the width of a square window of dimensions $N \times N$. t is the threshold value that determines the height of the decreasing triangular-shape weighted function when $|r| = R$ and/or $|s| = S$. In general, $t = 1, 2$, and 3 gives a varying degree of filtering performance. For ease of explanation, we shall call the DWMAV filters with $t = 1, 2$, and 3 as DWMAV1, DWMAV2, and DWMAV3 respectively.

3 Simulations and results

In all the simulations, three 8-bit mono images of dimensions $M1 \times M2$ ($= 256 \times 256$) pixels are used. In each of the images, the pixels $s(i,j)$ for $1 \leq i \leq M1$ and $1 \leq j \leq M2$, are corrupted by adding two kinds of noise, namely, impulse (or salt and pepper noise) noise $n_i(i,j)$, and random (or Gaussian) noise $n_g(i,j)$. These three images are the Slope, Peppers, and Lena images as shown in Figs. 1-3. Each of these images represents a slightly different class of image. Low, medium, and high levels of impulse noise, with respective density values of 0.03, 0.15, and 0.3 are added to each of these three images as shown in Figs. 4-6 for Slope image, Figs. 22-24 for Peppers image, and Figs. 40-42 for Lena image. Also, low, medium, and high levels of random noise, each has a mean value of 0.0 and a respective variance value of 0.0052, 0.021, and 0.106 is added to each of the three images as shown in Figs. 7-9 for Slope image, Figs. 25-27 for Peppers image, and Figs. 43-45 for Lena image. The two input noisy images $x_i(i,j)$ for $i = 1$ to 2 can be expressed as:

$$x_1(i,j) = s(i,j) + n_i(i,j) \tag{11}$$

$$x_2(i,j) = s(i,j) + n_r(i,j) \tag{12}$$

In all the simulations, square windows of dimensions $N \times N$ pixels and with different values of width N (= 3, 5, 7) are used. The mean squared error (MSE) is used to compare the relative filtering performance of various filters. The MSE between the filtered output image $y(i,j)$ and the original image $s(i,j)$ of dimensions $M1 \times M2$ pixels is defined as:

$$MSE = \frac{\sum_{i}^{M} 1 \sum_{j}^{M} 2[y(i,j) - s(i,j)]^2}{M1 \cdot M2} \tag{13}$$

The MSE of the original and filtered noisy Slope, Peppers, and Lena images for the 3 levels of impulse noise and the 3 levels of random noise for $N = 3, 5, 7$

are respectively summarized in Tables 1-2, Tables 3-4, and Tables 5-6. As seen from Tables 1-6, the MSEs of the impulse and random noise filtered images share some similar properties. As the window width N increases, nearly all the MSEs increase for low-level noises while majority of the MSEs decrease for high-level noises, and there is a combination of MSEs increase and decrease for medium-level noises. In general, for reduced MSE performance, a narrower window width is appropriate for low-level noises, and a wider window width is appropriate for high-level noises. It should be noted that the edges and details of an image become blur as the window width N increases. From the filtered images, it is observed that edges and details are well preserved for $N = 3$ in all the seven filters. To have a closer look at the relative filtering performance, all the seven filters (in which the DWMAV has three sub-filters) are ranked according to their MSE values for $N = 3$. The filter with the minimum MSE value will be ranked first and so on. As a result, six ranking tables, Tables 7-12, are obtained from the corresponding Tables 1-6. From the 6 ranking tables, the top three filters for low, medium, and high levels of impulse and random noises are listed in Tables 13-14. The MED filter is a standard filter for impulse noise filtering and the MAV filter is a standard filter for random noise filtering. For comparisons, the filtered images ($N = 3$) of the best filters out of the seven fuzzy filters are placed side-by-side (a) with the MED filter for low, medium, and high impulse noise filtering as shown in Figs. 10-15 for Slope image, Figs. 28-33 for Peppers image, and Figs. 46-51 for Lena image; and (b) with the MAV filter for low, medium, and high random noise filtering as shown in Figs. 16-21 for Slope image, Figs. 34-39 for Peppers image, and Figs. 52-57 for Lena image.

4 Concluding remarks

In this chapter, a study of seven fuzzy filters and their filtering performance has been presented. Each of these fuzzy filters applies a weighted membership function to an image within a window to compute the value of the center pixel, is easy and fast to implement and can suppress low, medium, and high levels of impulse noise and random noise with a varying degree of success. Depend on the features of an image, the performance of each of these seven filters varies slightly. In general, the filtering performance of each of these fuzzy filters is quite consistent among images of similar characteristics. In practice, the edges and details of an image can be preserved when the window width is small (for $N = 3$). As the window width increases (for $N = 5$ or 7), filtered images become blur, but under a high-level noise, the filtering capability of the majority of these fuzzy filters increases. As a general guideline, a small window width appears to be appropriate for a low level of noise, and a larger window width may be considered for a higher level of noise.

Filters	N	Density of Impulse Noise		
		Low - 0.03	Medium - 0.15	High - 0.3
Noisy Image		691.92	3476.10	6994.90
MED	3	35.69	121.74	515.34
	5	47.27	147.12	352.10
	7	63.34	182.22	485.63
GMED	3	37.16	114.27	457.10
	5	51.05	120.26	257.08
	7	64.91	129.56	294.55
TMED	3	39.28	131.50	530.02
	5	51.99	148.94	360.47
	7	75.27	175.78	450.71
ATMED	3	37.29	117.75	468.14
	5	53.77	136.21	305.23
	7	80.11	166.06	401.81
MAV	3	251.49	703.64	1534.50
	5	322.66	605.71	1220.60
	7	423.40	664.63	1229.40
GMAV	3	71.87	188.01	663.33
	5	117.18	161.01	300.92
	7	160.93	197.18	291.92
TMAV	3	38.95	128.24	493.12
	5	69.23	143.21	295.16
	7	105.05	159.00	328.83
ATMAV	3	219.87	572.49	881.27
	5	305.30	531.55	756.24
	7	395.24	596.24	831.89
DWMAV1	3	184.01	657.97	1516.90
	5	195.75	483.83	1114.60
	7	236.31	471.80	1037.90
DWMAV2	3	189.57	645.05	1483.60
	5	206.98	484.30	1100.20
	7	252.79	480.73	1038.30
DWMAV3	3	192.05	644.24	1478.30
	5	212.15	485.46	1096.40
	7	261.44	487.07	1042.00

Table 1. MSE of original and filtered noisy Slope images contaminated with impulse noise

Filters	N	Variance of Random Noise		
		Low - 0.0052	Medium - 0.021	High - 0.106
Noisy Image		296.11	1076.80	4352.00
MED	3	92.29	256.66	1056.70
	5	80.59	173.63	587.36
	7	92.45	174.77	514.35
GMED	3	87.27	233.15	919.25
	5	77.11	151.03	476.51
	7	85.25	145.30	389.20
TMED	3	92.65	253.90	1041.80
	5	72.78	143.92	501.98
	7	79.48	133.65	414.92
ATMED	3	92.32	253.20	1082.90
	5	99.71	246.46	1045.60
	7	123.22	292.44	1192.80
MAV	3	211.93	336.88	989.13
	5	311.40	390.77	842.87
	7	419.47	487.99	897.64
GMAV	3	107.87	232.45	822.83
	5	136.23	200.80	504.89
	7	174.98	225.80	466.57
TMAV	3	84.20	218.79	856.90
	5	92.13	167.99	519.96
	7	116.59	187.50	504.52
ATMAV	3	154.63	344.13	1461.00
	5	206.62	387.11	1598.00
	7	268.41	470.23	1797.40
DWMAV1	3	138.57	269.59	943.32
	5	178.97	257.84	718.30
	7	227.82	292.10	695.74
DWMAV2	3	146.45	272.27	926.17
	5	191.99	267.50	714.76
	7	245.30	307.71	703.91
DWMAV3	3	149.59	274.18	923.77
	5	197.89	272.65	716.16
	7	254.16	315.84	709.85

Table 2. MSE of original and filtered noisy Slope image contaminated with random noise

Filters	N	Density of Impulse Noise		
		Low - 0.03	Medium - 0.15	High - 0.3
Noisy Image		625.45	3063.50	6065.40
MED	3	55.20	108.78	422.51
	5	113.29	154.79	260.17
	7	189.41	244.77	358.56
GMED	3	57.19	114.28	407.52
	5	115.11	148.26	244.76
	7	186.07	218.98	296.57
TMED	3	56.78	115.11	428.63
	5	118.65	157.51	259.68
	7	196.17	229.78	307.68
ATMED	3	69.12	118.12	355.70
	5	152.54	180.88	211.20
	7	244.33	269.01	290.25
MAV	3	190.52	539.94	1106.20
	5	265.89	459.92	831.00
	7	377.37	532.13	844.24
GMAV	3	79.83	183.63	575.72
	5	157.15	191.04	319.30
	7	243.12	272.80	346.06
TMAV	3	60.55	119.99	420.31
	5	139.40	183.79	275.29
	7	234.35	271.89	323.98
ATMAV	3	117.76	172.22	192.56
	5	254.99	292.09	309.84
	7	378.16	413.66	440.42
DWMAV1	3	170.95	547.87	1146.70
	5	204.86	416.80	816.08
	7	268.21	435.20	773.54
DWMAV2	3	176.81	537.51	1117.60
	5	220.26	422.39	808.44
	7	291.66	452.82	781.40
DWMAV3	3	179.67	537.21	1113.60
	5	227.39	426.98	809.28
	7	303.66	462.67	787.79

Table 3. MSE of original and filtered noisy Peppers images contaminated with impulse noise

Filters	N	Variance of Random Noise		
		Low - 0.0052	Medium - 0.021	High - 0.106
Noisy Image		326.95	1227.10	4750.20
MED	3	120.27	305.96	1192.10
	5	148.39	250.73	670.68
	7	219.72	310.60	613.87
GMED	3	113.79	287.28	1085.20
	5	140.92	228.57	590.20
	7	205.74	275.00	503.70
TMED	3	120.60	318.85	1257.50
	5	139.88	231.44	617.66
	7	208.71	279.19	521.38
ATMED	3	116.02	270.99	999.48
	5	174.04	283.28	840.86
	7	264.01	384.02	960.57
MAV	3	150.86	263.43	803.35
	5	247.05	297.70	595.62
	7	364.02	399.72	632.76
GMAV	3	117.84	253.87	863.83
	5	174.92	239.27	512.47
	7	257.05	306.24	479.97
TMAV	3	108.83	263.94	956.11
	5	161.23	236.96	526.63
	7	247.42	307.14	497.99
ATMAV	3	138.15	283.56	1034.90
	5	228.31	328.97	1015.30
	7	336.84	443.01	1164.00
DWMAV1	3	127.60	251.97	828.62
	5	182.98	241.65	570.75
	7	252.81	293.64	549.32
DWMAV2	3	135.41	253.50	807.06
	5	199.84	254.26	567.97
	7	277.38	315.86	561.05
DWMAV3	3	138.63	255.40	804.47
	5	207.48	260.64	569.79
	7	289.71	327.38	568.85

Table 4. MSE of original and filtered noisy Peppers images contaminated with random noise

Filters	N	Density of Impulse Noise		
		Low - 0.03	Medium - 0.15	High - 0.3
Noisy Image		578.40	2894.60	5841.80
MED	3	55.68	102.92	387.44
	5	122.62	159.92	254.00
	7	190.35	235.65	337.69
GMED	3	53.69	104.50	367.21
	5	118.85	150.75	230.05
	7	184.42	214.00	277.41
TMED	3	55.60	105.02	386.14
	5	124.21	158.22	245.93
	7	193.37	220.91	284.32
ATMED	3	69.17	109.12	319.19
	5	150.48	161.97	190.27
	7	226.52	228.68	253.53
MAV	3	167.54	486.44	1025.60
	5	242.14	406.55	745.69
	7	339.47	461.70	742.63
GMAV	3	64.46	157.20	520.40
	5	138.85	170.63	280.98
	7	214.73	240.04	306.87
TMAV	3	57.57	108.73	382.35
	5	135.52	169.61	251.08
	7	216.21	241.85	289.91
ATMAV	3	96.58	128.39	142.49
	5	202.80	208.50	214.30
	7	293.10	285.62	307.75
DWMAV1	3	138.88	484.09	1054.00
	5	169.27	354.52	720.39
	7	225.45	365.24	671.28
DWMAV2	3	141.93	472.10	1023.10
	5	183.01	358.48	711.53
	7	246.27	379.78	676.75
DWMAV3	3	143.71	471.14	1018.60
	5	189.37	362.20	711.34
	7	256.78	388.08	681.97

Table 5. MSE of original and filtered noisy Lena images contaminated with impulse noise

Filters	N	Variance of Random Noise		
		Low - 0.0052	Medium - 0.021	High - 0.106
Noisy Image		324.97	1248.80	4782.40
MED	3	120.90	312.97	1174.50
	5	157.38	258.12	673.65
	7	223.36	267.79	613.52
GMED	3	111.29	284.18	1052.70
	5	144.15	225.47	575.41
	7	203.91	259.65	492.34
TMED	3	121.61	321.53	1222.60
	5	146.04	227.68	591.19
	7	205.87	258.15	489.32
ATMED	3	111.81	263.74	958.55
	5	166.04	261.22	790.65
	7	238.10	334.26	884.61
MAV	3	133.05	244.98	751.83
	5	225.00	271.95	546.10
	7	327.27	267.79	567.96
GMAV	3	104.14	237.63	820.96
	5	157.04	217.54	477.50
	7	228.25	270.04	440.88
TMAV	3	106.38	257.58	916.52
	5	154.54	218.87	490.08
	7	226.01	270.10	450.10
ATMAV	3	116.40	253.71	956.70
	5	189.38	280.84	919.77
	7	273.14	367.86	1039.40
DWMAV1	3	102.06	223.87	770.79
	5	149.95	204.16	509.79
	7	211.37	245.91	483.01
DWMAV2	3	106.63	222.31	746.01
	5	164.79	214.70	506.21
	7	232.99	264.87	492.57
DWMAV3	3	108.62	223.20	742.33
	5	171.55	220.23	507.85
	7	243.76	274.69	499.34

Table 6. MSE of original and filtered noisy Lena images contaminated with random noise

Filters	Low	Medium	High
MED	1	3	4
GMED	2	1	1
TMED	5	5	5
ATMED	3	2	2
MAV	11	11	11
GMAV	6	6	6
TMAV	4	4	3
ATMAV	10	7	7
DWMAV1	7	10	10
DWMAV2	8	9	9
DWMAV3	9	8	8

Table 7. MSE ranking of filtered Slope images contaminated with Low-Medium-High level of impulse noise

Filters	Low	Medium	High
MED	3	6	9
GMED	2	3	3
TMED	5	5	8
ATMED	4	4	10
MAV	11	10	7
GMAV	6	2	1
TMAV	1	1	2
ATMAV	10	11	11
DWMAV1	7	7	6
DWMAV2	8	8	5
DWMAV3	9	9	4

Table 8. MSE ranking of filtered Slope images contaminated with Low-Medium-High level of random noise

References

1. I. Pitas and A. N. Venetsanopoulos, *Nonlinear digital filters*, Kluwer Academic Publishers, 1990.
2. S. Agaian, J. Astola, and K. Egiazarian, *Binary polynomial transformations and nonlinear digital filters*, Marcel Dekker, Inc., 1995.
3. S. K. Mitra and G. Sicuranza, Eds., *Nonlinear Image Processing*, Academic Press, 2000.
4. E. E. Kerre and M. Nachtegael, Eds., *Fuzzy techniques in image processing*, Series on Studies in Fuzziness and Soft Computing, Vol. 52, Springer-Verlag, 2000.
5. F. Russo, *Recent advances in fuzzy techniques for image enhancement*, IEEE Transactions on Instrumentation and Measurement, vol. 47, no. 6, pp. 1428-1434, Dec. 1998.

Fig.1. Original Slope image

Fig.2. Original Peppers image

Fig.3. Original Lena image

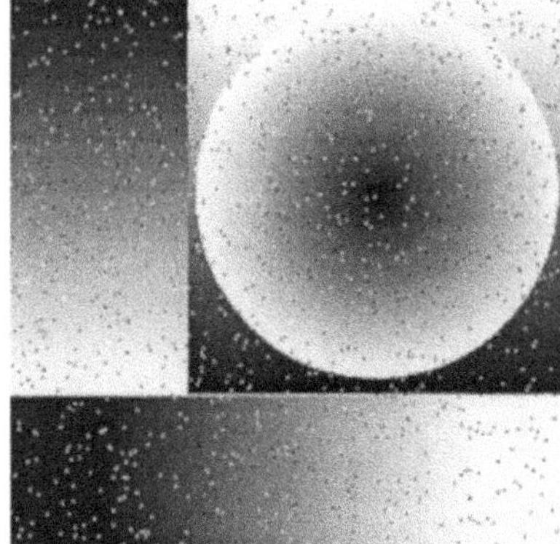

Fig.4. Slope with low impulse noise

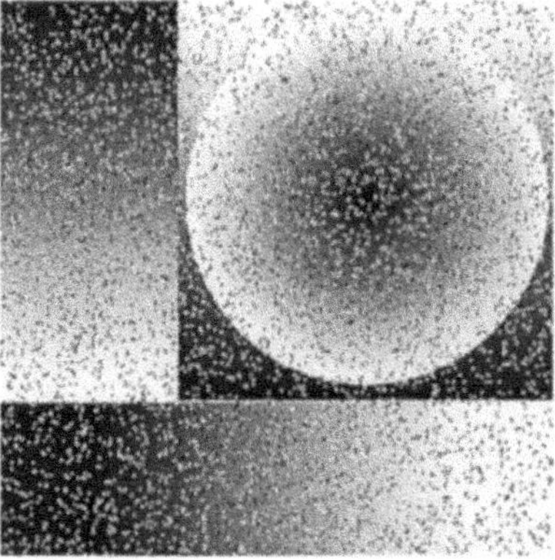

Fig.5. Slope with medium impulse noise

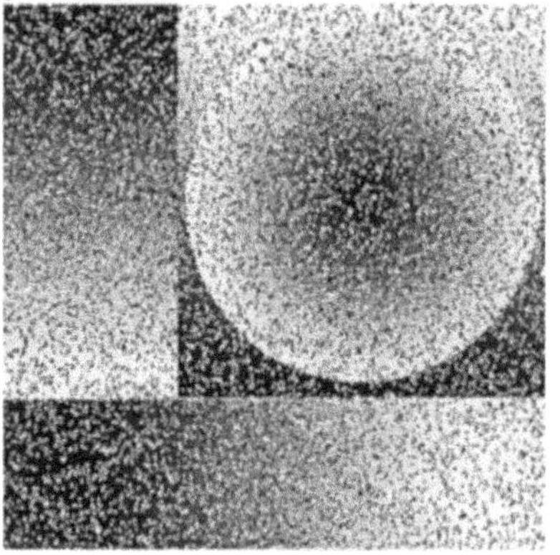

Fig.6. Slope with high impulse noise

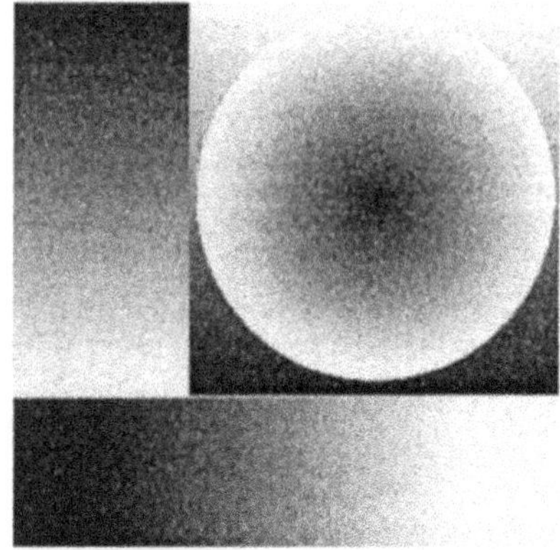

Fig.7. Slope with low random noise

Fig.8. Slope with medium random noise

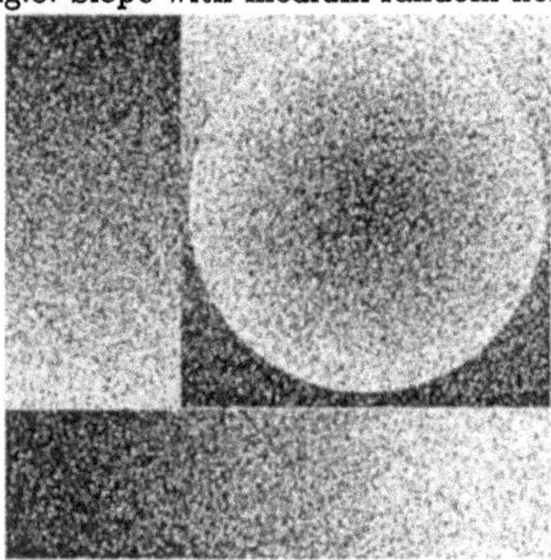

Fig.9. Slope with high random noise

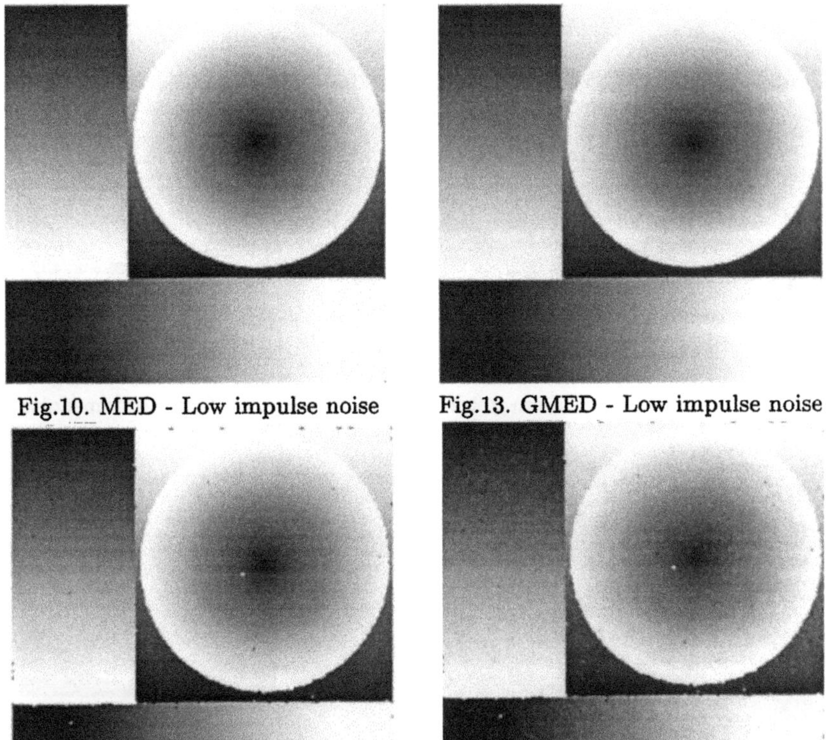

Fig.10. MED - Low impulse noise

Fig.13. GMED - Low impulse noise

Fig.11. MED - Medium impulse noise

Fig.14. ATMED - Medium impulse noise

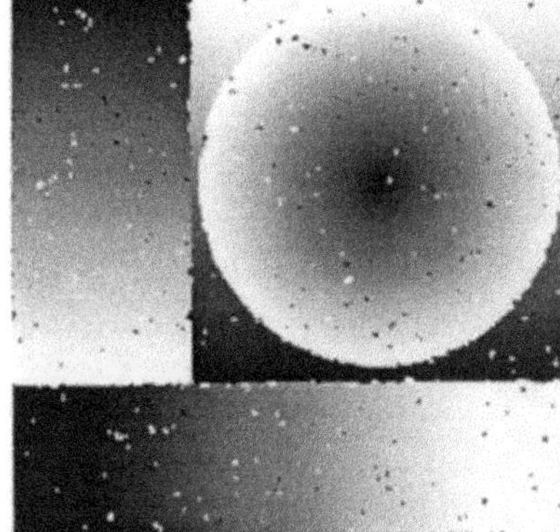

Fig.12. MED - High impulse noise

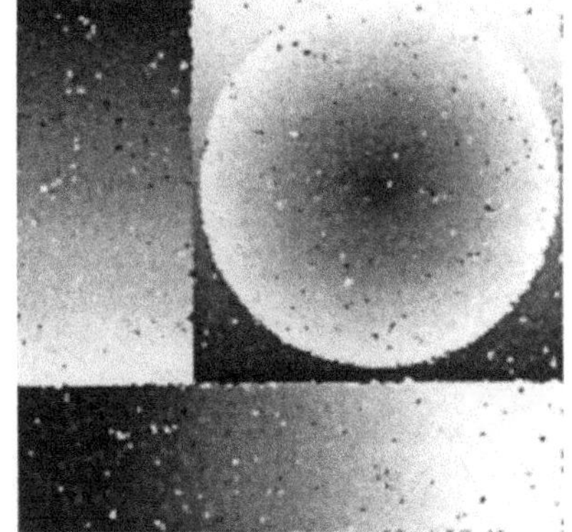

Fig.15. ATMED - High impulse noise

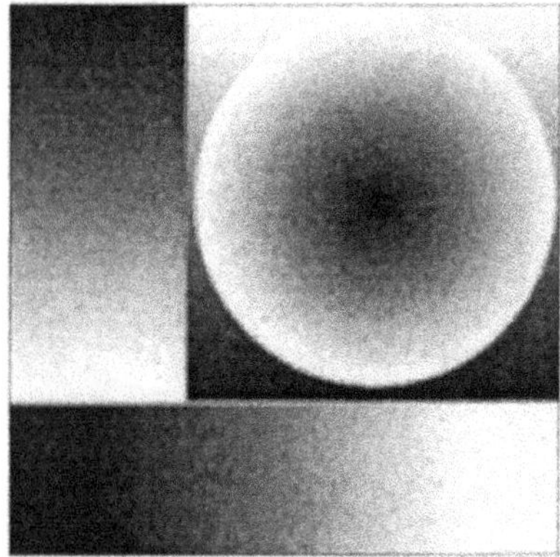

Fig.16. MAV - Low random noise

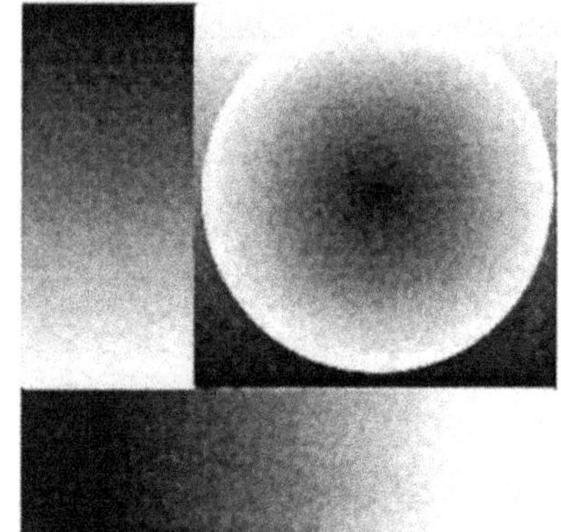

Fig.19. TMAV - Low random noise

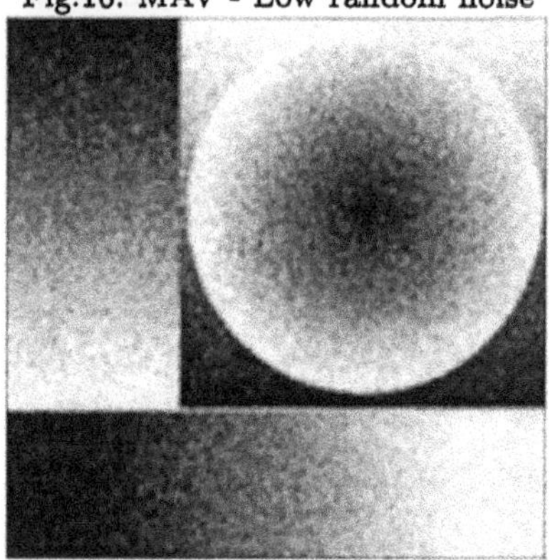

Fig.17. MAV - Medium random noise

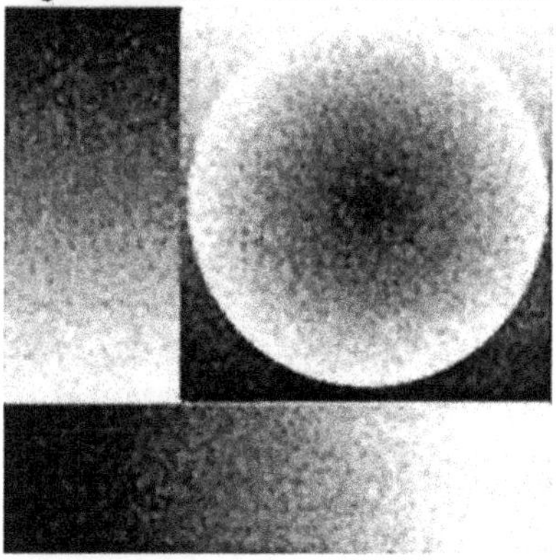

Fig.20. TMAV - Medium random noise

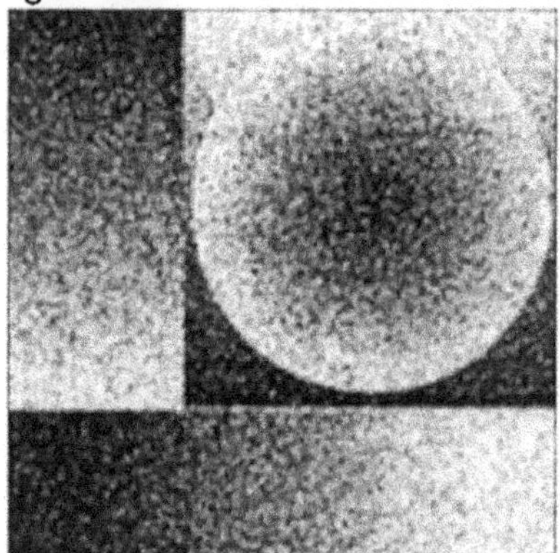

Fig.18. MAV - High random noise

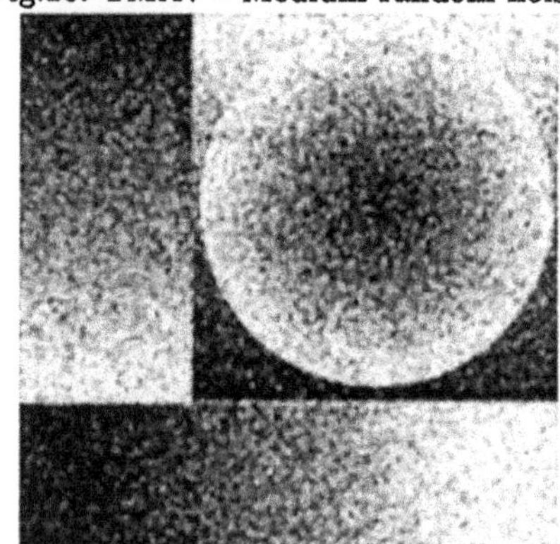

Fig.21. GMAV - High random noise

Fig.22. Peppers with low impulse noise

Fig.25. Peppers with low random noise

Fig.23. Peppers with medium impulse noise

Fig.26. Peppers with medium random noise

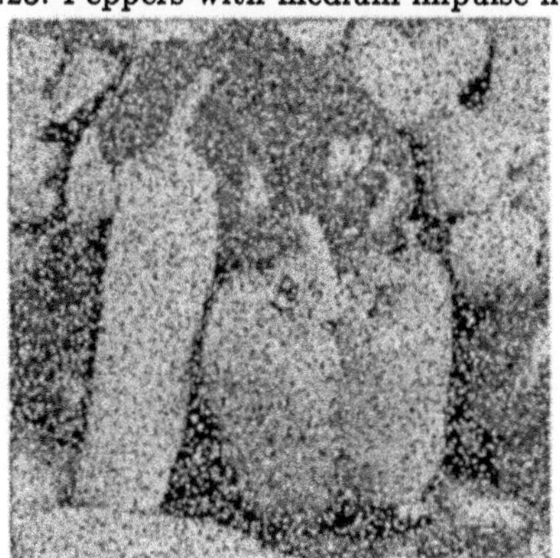

Fig.24. Peppers with high impulse noise

Fig.27. Peppers with high random noise

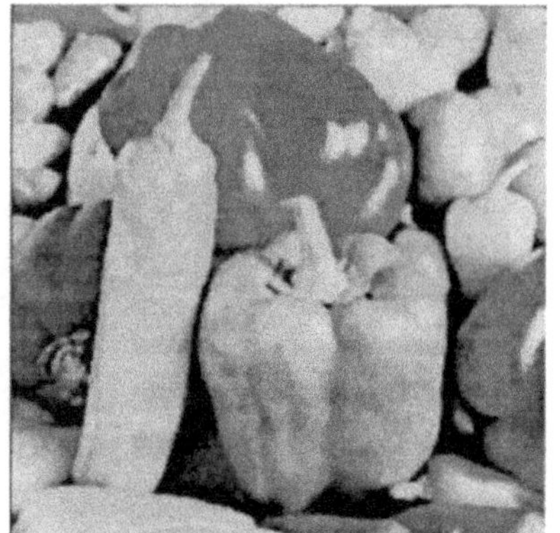

Fig.28. MED - Low impulse noise

Fig.31. GMED - Low impulse noise

Fig.29. MED - Medium impulse noise

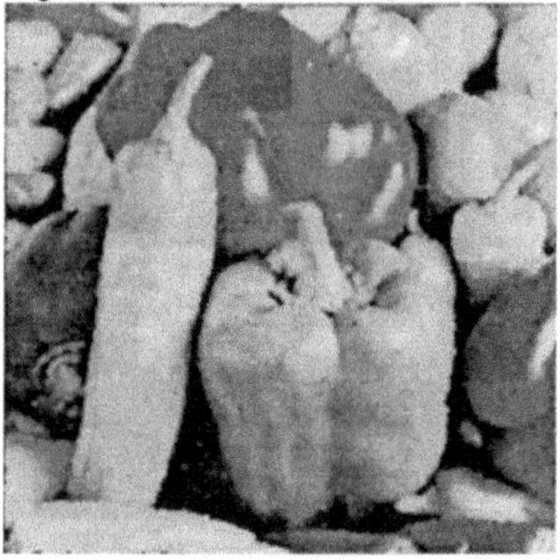

Fig.32. GMED - Medium impulse noise

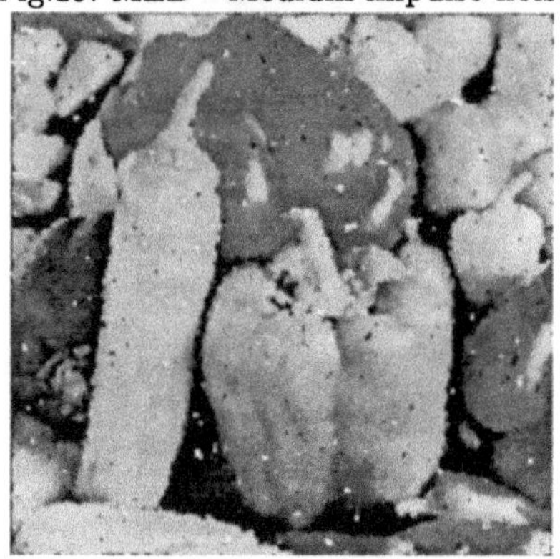

Fig.30. MED - High impulse noise

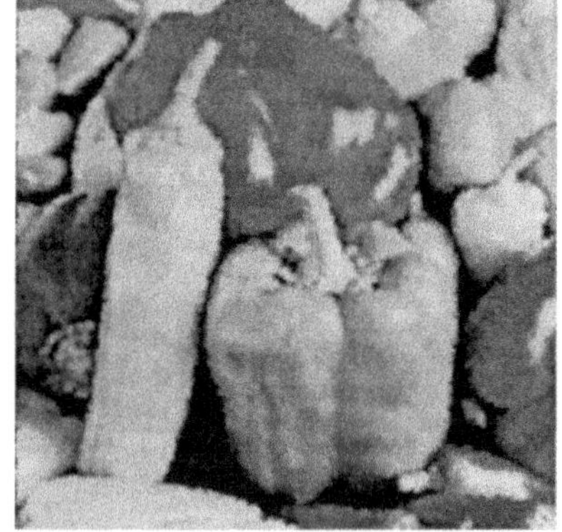

Fig.33. ATMAV - High impulse noise

Fig.34. MAV - Low random noise

Fig.37. GMAV - Low random noise

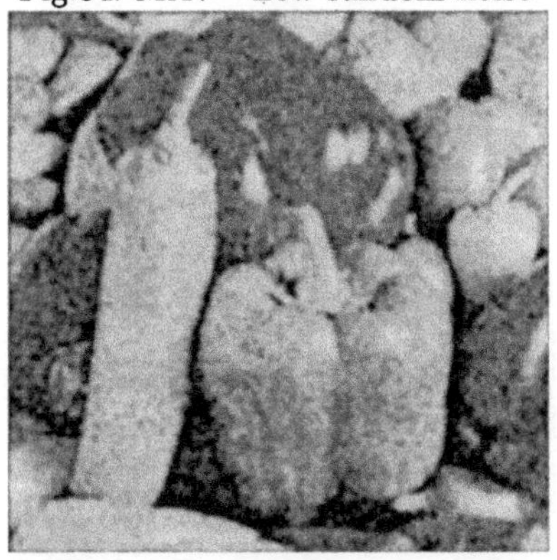

Fig.35. MAV - Medium random noise

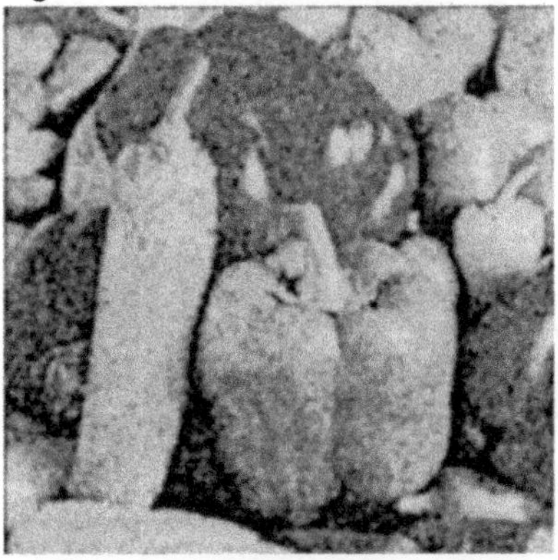

Fig.38. DWMAV1 - Medium random noise

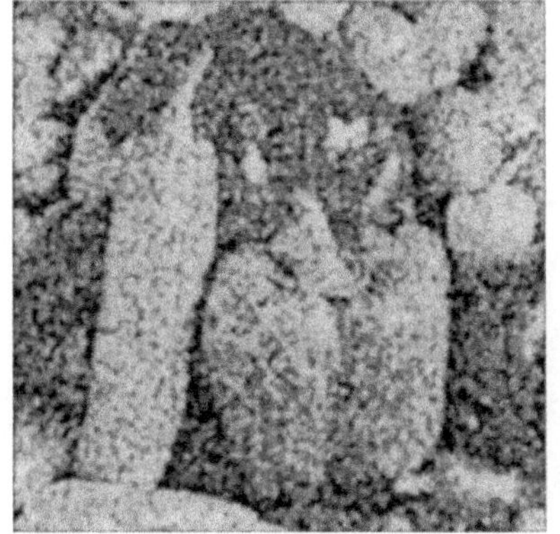

Fig.36. MAV - High random noise

Fig.39. DWMAV3 - High random noise

Fig.40. Lena with low impulse noise

Fig.41. Lena with medium impulse noise

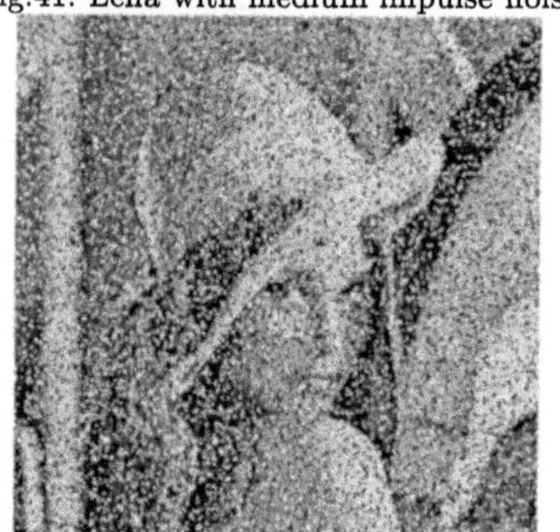

Fig.42. Lena with high impulse noise

Fig.43. Lena with low random noise

Fig.44. Lena with medium random noise

Fig.45. Lena with high random noise

Fig.46. MED - Low impulse noise

Fig.49. GMED - Low impulse noise

Fig.47. MED - Medium impulse noise

Fig.50. TMED - Medium impulse noise

Fig.48. MED - High impulse noise

Fig.51. ATMAV - High impulse noise

Fig.52. MAV - Low random noise

Fig.55. DWMAV1 - Low random noise

Fig.53. MAV - Medium random noise

Fig.56. DWMAV2 - Medium random noise

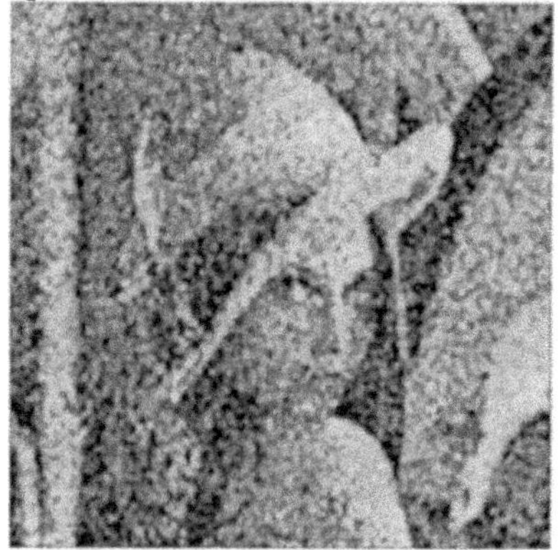

Fig.54. MAV - High random noise

Fig.57. DWMAV3 - High random noise

Filters	Low	Medium	High
MED	1	1	5
GMED	3	2	3
TMED	2	3	6
ATMED	5	4	2
MAV	11	10	8
GMAV	6	7	7
TMAV	4	5	4
ATMAV	7	6	1
DWMAV1	8	11	11
DWMAV2	9	9	10
DWMAV3	10	8	9

Table 9. MSE ranking of filtered Peppers images contaminated with Low-Medium-High level of impulse noise

Filters	Low	Medium	High
MED	5	10	10
GMED	2	9	9
TMED	6	11	11
ATMED	3	7	7
MAV	11	5	1
GMAV	4	3	5
TMAV	1	6	6
ATMAV	9	8	8
DWMAV1	7	1	4
DWMAV2	8	2	3
DWMAV3	10	4	2

Table 10. MSE ranking of filtered Peppers images contaminated with Low-Medium-High level of random noise

6. M. Nachtegael, D. Van der Weken, A. Van De Ville, E. Kerre, W. Philips, I. Lemahieu, *An overview of classical and fuzzy-classical filters*, Proceedings of IEEE International Conference on Fuzzy Systems, pp. 3-6, 2001.
7. M. Nachtegael, D. Van der Weken, A. Van De Ville, E. Kerre, W. Philips, I. Lemahieu, *An overview of fuzzy filters for noise reduction*, Proceedings of IEEE International Conference on Fuzzy Systems, pp. 7-10, 2001.
8. M. Nachtegael, D. Van der Weken, A. Van De Ville, E. Kerre, W. Philips, I. Lemahieu, *A comparative study of classical and fuzzy filters for noise reduction*, Proceedings of IEEE International Conference on Fuzzy Systems, pp. 11-14, 2001.
9. J. S. Lim, *Two-dimensional signal and image processing*, Prentice-Hall, pp. 536-540, 1990.
10. K. Arajawa, *Median filter based on fuzzy rules and its application to image restoration*, Fuzzy Sets and Systems, Vol. 77, pp. 3-13, 1996.

Filters	Low	Medium	High
MED	3	1	6
GMED	1	2	3
TMED	2	3	5
ATMED	6	5	2
MAV	11	11	10
GMAV	5	7	7
TMAV	4	4	4
ATMAV	7	6	1
DWMAV1	8	10	11
DWMAV2	9	9	9
DWMAV3	10	8	8

Table 11. MSE ranking of filtered Lena images contaminated with Low-Medium-High level of impulse noise

Filters	Low	Medium	High
MED	9	10	10
GMED	6	9	9
TMED	10	11	11
ATMED	7	8	8
MAV	11	5	3
GMAV	2	4	5
TMAV	3	7	6
ATMAV	8	6	7
DWMAV1	1	3	4
DWMAV2	4	1	2
DWMAV3	5	2	1

Table 12. MSE ranking of filtered Lena images contaminated with Low-Medium-High level of random noise

Image	Ranking	Low	Medium	High
Slope	1	MED	GMED	GMED
	2	GMED	ATMED	ATMED
	3	ATMED	MED	TMAV
Peppers	1	MED	MED	ATMAV
	2	TMED	GMED	ATMED
	3	GMED	TMED	GMED
Lena	1	GMED	MED	ATMAV
	2	TMED	GMED	ATMED
	3	MED	TMED	GMED

Table 13. Top 3 filters for Low-Medium-High level of impulse noise

Image	Ranking	Low	Medium	High
Slope	1	TMAV	TMAV	GMAV
	2	GMED	GMAV	TMAV
	3	MED	GMED	GMED
Peppers	1	TMAV	DWMAV1	MAV
	2	GMED	DWMAV2	DWMAV3
	3	ATMED	GMAV	DWAMV2
Lena	1	DWMAV1	DWMAV2	DWMAV3
	2	GMAV	DWMAV3	DWMAV2
	3	TMAV	DWMAV1	MAV

Table 14. Top 3 filters for Low-Medium-High level of random noise

11. K. Arajawa, *Fuzzy ruled-based image processing with optimization*, in Fuzzy Techniques in Image Processing, Edited by E. E. Kerre and M. Nachtegael, Springer-Verlag, pp. 222-247, 2000.
12. C.-S. Lee, Y.-H. Kuo, and P.-T. Yu, *Weighted fuzzy mean filters for image processing*, Fuzzy Sets and Systems, Vol. 89, pp. 157-180, 1997.
13. C.-S. Lee, Y.-H. Kuo, *Adaptive fuzzy filter and its application to image processing*, in Fuzzy Techniques in Image Processing, Edited by E. E. Kerre and M. Nachtegael, Springer-Verlag, pp. 172-193, 2000.
14. F. Russo and G. Ramponi, *A fuzzy filter for images corrupted by impulse noise*, IEEE Signal Processing Letters, Vol. 3, No. 6, pp. 168-170, June 1996.
15. F. Russo and G. Ramponi, *Removal of impulse noise using a fire filter*, Proceedings of IEEE International Conference in Image Processing, pp. 975-978, 1996.
16. F. Russo, *FIRE operators for image processing*, Fuzzy Sets and Systems, Vol. 103, pp. 265-275, 1999.
17. F. Russo, *Noise cancellation using nonlinear fuzzy filters*, Proceedings of IEEE Instrumentation and Measurement Technology Conference, Ottawa, Canada, pp. 772-777, May 1997.
18. F. Farbiz and M. B. Menhaj, *A fuzzy logic control based approach for image filtering*, in Fuzzy Techniques in Image Processing, edited by E. E. Kerre and M. Nachtegael, Springer-Verlag, pp. 194-221, 2000.
19. D. Van De Vile, M. Nachtegael, D. Van der Weken, W. Philips, I. Lemahieu, E. E. Kerre, *A new fuzzy filter for Gaussian noise reduction*, Proceedings of International SPIE Conference on Electronic Imaging, pp. 1-9, 2001.
20. A. Taguchi, H. Takashima, and Y. Murata, *Fuzzy filters for image smoothing*, in Proceedings of SPIE Conference on Nonlinear Image Processing V, San Jose, CA, pp. 332-339, Feb. 1994.
21. A. Taguchi, H. Takashima, and F. Russo, *Data dependent filtering using the fuzzy inference*, in Proceedings of IEEE Instrumentation Measurement Technology Conference, Waltham, MA, pp. 752-756, April 1995.
22. H. K. Kwan and Y. Cai, *Median filtering using fuzzy concept*, Proceedings of 36th Midwest Symposium on Circuits and Systems, Detroit, Michigan, USA, vol. 2, August 15-18, 1993, pp. 824-827.
23. H. K. Kwan and Y. Cai, *Fuzzy filters for image filtering*, Proceedings of 45th Midwest Symposium on Circuits and Systems, Oklahoma, August 25-28, 2002.

Chapter 3

Real-time Image Noise Cancellation Based on Fuzzy Similarity

Ivan Kalaykov and Gustav Tolt

Örebro University
Center for Applied Autonomous Sensor Systems
Department of Technology
SE-701 82, Örebro, Sweden
www.aass.oru.se

Summary. We propose a new filter structure for the reduction of mixed noise in images. It is based on the evaluation of fuzzy similarities between pixels in a local processing window and is suitable for high-speed hardware implementation. The filter involves two tunable parameters and is fairly robust against changes in noise distribution. Furthermore, we outline a modular hardware architecture for general high-speed image processing tasks.

1 Introduction

Real images are often corrupted by noise. For example, they may contain noise introduced by image sensors, noise caused by loss of information due to noisy transmission channels or a combination of both. Hence, the amount and type of noise may be very different from one application to another and robustness against changes in the noise distribution is a desirable property of a noise reduction filter.

In this paper, we present a filter structure based on the evaluation of fuzzy similarities between pixels within a local processing window. The filter output is a combination of outputs of several subfilters, the weights of which are related to the degrees of similarity extracted from the corresponding pixels in the window. The filter is designed to reduce mixed noise (Gaussian and impulse) and able to operate in real-time at frame rates considerably higher than normal video frame rates. This chapter is organized as follows. First, a brief overview of existing filtering techniques and implementations is given in Sects. 1.1 and 1.2. Then different noise models are discussed in Sect. 1.3. In Sect. 1.4, we introduce the fuzzy similarity concept along with some mathematic formulations. In Sect. 2, the fuzzy-similarity-based filter structure and in Sect. 3, some test examples are presented. Section 4 concerns the implementation of the filter on an FPGA device. Finally, concluding remarks are given in Sect. 5.

1.1 Filters based on fuzzy logic

In the recent years, filters based on fuzzy logic have shown to be able to provide efficient image filtering. In many of these filters, the "fuzziness" enter as a fuzzification of "classical" filters or a fuzzy weighted combination of the outputs of several subfilters, e.g. mean, median, "identity" and midpoint filters. The fuzzy weighted mean filter [2], the fuzzy median filter [1] and the (adaptive) weighted fuzzy mean (WFM, AWFM) [11,10] filter all belong to this type of filters, as do also the filters proposed by Taguchi/Meguro [16] and Choi/Krishnapuram [4]. Another example is the fuzzy cluster filter [6] that can be seen as a fuzzy weighted mean filter applied iteratively in each processing window. The weights are updated depending of the difference between the cluster center and the intensity of each pixel in the window. Muneyasu *et al.* proposed another filter [13], based on the weighted mean filter.

There are also fuzzy filters based directly on a set of rules, not involving any classical (sub)filter in the computation of the output. The filter proposed by Russo [14] is an example of a filter from this class.

Other filters, like the one presented in [15], can be seen as combinations of two filters, one from each category mentioned above.

1.2 Hardware implementation of non-linear filters

Real-time applications, e.g. visual servoing tasks, may require image processing at speeds higher than normal video frame rates of 25-30 Hz. The implementation of filters in hardware devices, such as Field-Programmable Gate Arrays (FPGAs) or ASICs is a way to provide high-speed image processing tools.

The WFM filter, designed to remove heavy impulse noise, was reported to operate at 90 256×256 frames per second (≈5.9 Mpixels per second) [11]. The hardware implementation of the AFWM filter allowed for a throughput of 6.6 Mpixels per second [10]. An ASIC capable of performing various rank order filterings and different types of standard and fuzzy morphological operations presented in [7] was reported to execute 3.5×10^6 non-linear filter operations per second. Khriji *et al.* [9] proposed an FPGA-based filtering structure being able to operate at frame rates of 50 720×576 frames per second (≈20.7 Mpixels per second). Delva *et al.* [5] presented an FPGA implementation of a filter producing a fuzzy weighted mean of 4 median values (one for each of 4 subgroups of pixels in a 5×5 neighborhood). This filter was claimed to operate at 60 1024×1024 frames per second (60 Mpixels per second), when implemented of an FPGA.

1.3 Noise models

The general goal for noise cancellation [1] is to provide an estimate of an uncorrupted signal $s(n)$ based on noisy observations $x(n)$, using the model

$$x(n) = s(n) + w(n), \tag{1}$$

where $w(n)$ is the noise signal, popularly modelled as either zero mean additive Gaussian noise or impulse noise, or a combination of both.

The *saturated* impulse noise, known as salt-and-pepper noise, is modelled as

$$x(n) = \begin{cases} x(n) \ , & \text{with prob. 1-P} \\ x_{max} \ , & \text{with prob. P/2} \\ x_{min} \ , & \text{with prob. P/2} \end{cases} \tag{2}$$

where P is the probability that a pixel is corrupted.

There are also other types of noise models. For example, multiplicative noise is modelled as

$$x(n) = s(n)(1 + v(n)). \tag{3}$$

Television raster degradation can be described by this model.

In this chapter, we will consider a noise distribution of the following type:

$$x(n) = \begin{cases} s(n) + v(n) \ , & \text{with prob. 1-P} \\ p(n) \ , & \text{with prob. P} \end{cases} \tag{4}$$

where $v(n)$ is Gaussian noise $N(0, \sigma^2)$ and $p(n)$ is impulse noise. Two impulse noise types are considered:
a) noise pulses whose amplitudes are uniformly distributed on $[x_{min}, x_{max}]$
b) salt-and-pepper noise.

1.4 Fuzzy similarity as a basic approach in image processing

In fuzzy-similarity-based (FSB) image processing, pixels in a local processing window are treated differently according to the degree of similarity between them. The term "fuzzy similarity" reflects that the similarity between two objects (e.g., pixels) is allowed to take values between 0 and 1. By introducing a set of templates and computing the similarity between the respective pixels, we can analyze the intensity distribution within the window.

Let X be a grayscale image and let x_n denote a pixel in X. Throughout this chapter, x_n will be used also to denote the *intensity* of this pixel, since the chance of confusion should be small. Furthermore, x_0 denotes the central pixel of the moving processing window.

[1] The widely used term *cancellation* may be somewhat misleading, in the sense that a total cancellation of the noise is generally impossible, due to the random nature of the noise. Noise *reduction* might be a better term, but it should be clear that it is not implied that *all* noise is removed.

Definition 1 *A is similar to B* $\Leftrightarrow A \simeq B$

The fuzzy similarity is defined by a fuzzy relation $R(\cdot,\cdot) : \mathbf{R}^2 \to [0,1]$. Furthermore, we define the similarity between a pixel x_i and a *set* of pixels $S = \{x_1, \ldots, x_m\}$ as an aggregation of similarities, each representing the similarity between x_i and a pixel in S:

$$R_{x_i \simeq S} = R(x_i, x_1) \cap R(x_i, x_2) \cap \ldots \cap R(x_i, x_m). \tag{5}$$

Theoretically, R can be represented by any 2-D relation, e.g. a triangular membership function whose width varies across the surface. We will focus on the case when the fuzzy relation is represented by a triangular membership function (Fig. 1), as this leads to a simple implementation:

$$\mu_{x_i \simeq x_j} = \begin{cases} 1 - |x_i - x_j|/\alpha\,, & \text{if } |x_i - x_j| < \alpha \\ 0\,, & \text{elsewhere} \end{cases} \tag{6}$$

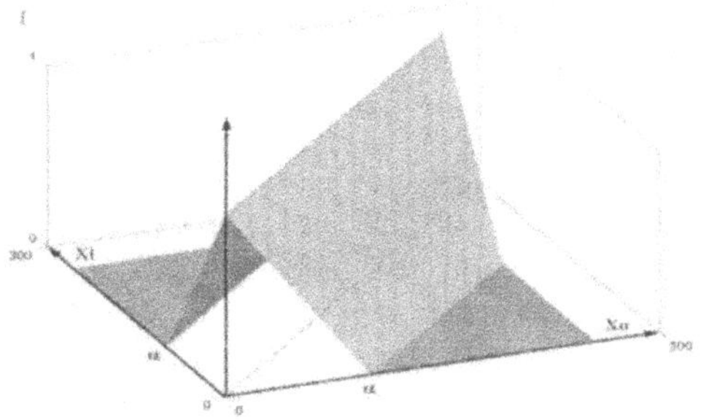

Fig. 1. The fuzzy similarity relation $\mu_{x_0 \simeq x_i}$ described in Eq. 6.

2 Fuzzy-similarity-based noise cancellation

In this section, the FSB concept for noise cancellation will be described.

2.1 Basic concept

The basic idea in the FSB filter approach is to check to which degree the hypothesis $\mathbf{H_0}$ is true:

$\mathbf{H_0}$: The pixels within a local neighborhood have similar intensities where local neighborhood refers to a number of equidistant pixels to any randomly taken pixel, and similarity corresponds to a fuzzy relation "A is similar to B", as described in Sect. 1.4. $\mathbf{H_0}$ expresses the local properties of the image; hence $\mathbf{H_0}$ appears to be more true when the neighborhood size is small and less true as the size increases. We assume that in the noise-free case, the interesting image details are locally smooth, so that $\mathbf{H_0}$ is more true. When the image is corrupted by noise, many pixels change their values and $\mathbf{H_0}$ is disregarded to a degree depending on the noise intensity. Hence, the noise cancellation task is to analyze the pixel similarities in all local neighborhoods by scanning the entire image. Each local neighborhood can be represented by a $w \times w$ pixels window, where w is small enough to limit the window to the local image properties and odd in order to allocate a central point. As the true pixel intensities are unknown a priori, their distribution within the window is unknown too. Take the example of a window containing an edge (Fig. 3). This edge causes the hypothesis H_0 to be "less true". On the other hand, some pixels in the window may still be similar. Hence, it is straightforward to analyze how the intensity is distributed within a number of templates. A template $\chi = \{x_k, x_l, x_m, \ldots, x_n\}$ contains a number of window pixels x_i. The variety of the templates corresponds to the variety of the image local properties. Therefore, we must consider several different template configurations when analyzing the need to reduce the noise by correcting the intensity of the central pixel x_0.

It is advantageous to have templates of equal size, so that a possible hardware implementation becomes regular. The number of pixels in the templates can be chosen arbitrarily. However, in order to be able to portray the local image properties, this number cannot be too small, nor too big. The FSB filter approach presented here employs templates that contain 4 pixels. Some examples of 4-pixel templates in a 3×3 pixels window are shown in Fig. 2.

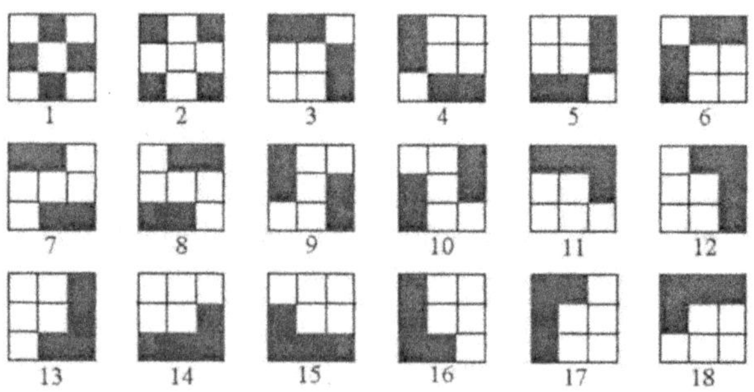

Fig. 2. Some examples of 4-pixel templates.

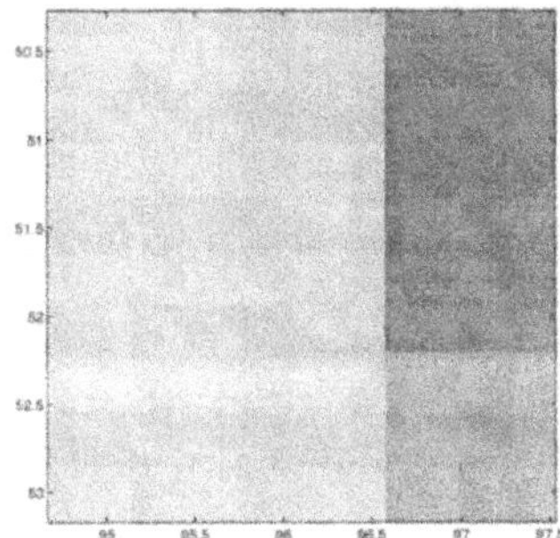

Fig. 3. An edge passing through a processing window.

2.2 Rules for noise cancellation

The basic, conceptual rules of the FSB filter are the following:

$\mathbf{R_1}$: IF the central window pixel and all templates have similar intensities;
THEN the window has nearly uniform intensity distribution
AND do not change the central pixel;

$\mathbf{R_2}$: IF the window has not uniform intensity distribution;
THEN the max similarity templates best portray the local image properties
AND change the central pixel according to the intensities of the max similarity templates;

The rule R_1 is only taken into account implicitely in the defuzzification process. However, the rule R_2 implies the computation of the degree of similarity between the central pixel and the pixels in each template.

$\mathbf{R_{2,j}}$: IF sim_j is high;
THEN $\Delta x_{2,j} = g(\chi_j)$;

In the experiments, the function $g(\chi_j)$ was chosen to be the median operator. The rules R_1, R_2 take into account only the similarities between the central pixel and pixels in the templates. However, the pixels that are most similar to a noisy central pixel can be noisy themselves. This is particularly noticable at higher noise levels. This means that even if there is a majority of pixels within the window that have similar intensities, this fact is not explicitly taken into account by these rules. Therefore, we define an additional set of rules, based on the idea of similarity between pixels *within* the templates, or template *homogeneity*:

$\mathbf{R_{3,j}}$: IF template j is not homogeneous
AND the central pixel is not similar to template j;
THEN template j does not contribute to the noise cancellation;

R$_{4,j}$: IF template j is homogeneous
AND the central window pixel is not similar to template j;
THEN template j portrays the local image characteristics
AND change the central pixel according to a function h of the pixels in template j;

R$_{5,j}$: IF template j is homogeneous
AND the central window pixel is similar to template j;
THEN central pixel and template j both portray the local image characteristics
AND change the central pixel according to a function h of the pixels in template j and itself;

Generally, the homogeneity of the template j can be computed as an aggregation of fuzzy similarity relations between pixels within the template. This allows for an analysis of the intensity distribution in the template. In this work, the homogeneity is computed as the similarity between the template pixels with maximum and minimum intensity, respectively, using the fuzzy relation $R(\cdot,\cdot)$ (Eq. 5):

$$R_{hom_j} = R(\max_{x_i \in \chi_j}(x_i), \min_{x_i \in \chi_j}(x_i)). \tag{7}$$

Hence, this way of computing homogeneity assigns the same values to the template $\{100, 100, 100, 120\}$ as to $\{100, 105, 110, 120\}$.

Let hom_j and sim_j denote the homogeneity of template j and similarity between the central pixel and template j, respectively. Then the rules can be written more compactly as:

R$_{3,j}$: IF hom_j is low
AND sim_j is low;
THEN template j does not contribute to the noise cancellation;

R$_{4,j}$: IF hom_j is high
AND sim_j is low;
THEN $\Delta x_{4,j} = h(\chi_j)$;

R$_{5,j}$: IF hom_j is high
AND sim_j is high;
THEN $\Delta x_{5,j} = h(x_0, \chi_j)$;

The function $h(\cdot)$ used in this work is the mean function. The firing strength of rule $R_{3,j}$ is not calculated explicitly, and is not used in the filtering process, simply because there is no logical consequent part to be associated with this rule. Nevertheless, it serves here as a way to provide intuitive understanding of the operation of the filter. We also define an additional rule, associated with the *dissimilarity* within the window, i.e. when the degrees of activation of the rules $R_{2,\cdot}$, $R_{4,\cdot}$ and $R_{5,\cdot}$ rules are low.

R_{dis}: IF no other rule is activated;
THEN change the central pixel according to a default function (or method), f_{dis}, of the pixels in the templates and itself.

A high degree of activation of rule R_{dis} indicates the presence of long-tailed noise within the processing window. Hence, f_{dis} can generally be any function or method suitable for the removal of long-tailed nosie (e.g., the median operator). Analogously, any function or method suitable for cancellation of short-tailed noise can be used as function h (e.g., the mean operator). The firing strength of this rule is computed as the degree to which the other rules are *not* activated:

$$\mu_{dis} = 1 - \max_{i,j} (\mu_{i,j}). \tag{8}$$

In order words, μ_{dis} is the degree of dissimilarity in the window.

2.3 Recursive implementation

If the filter is "recursively" implemented, i.e. if the intensity value of a filtered pixel is directly used for further filtering, the template consisting of the pixels $\{x_1, x_2, x_3, x_4\}$ (template 18 in Fig. 2) deserves some extra attention. Let χ_p denote this template. Then this template contains only *already processed* pixels, which are more likely to be similar to each other, than pixels in other templates. Hence, rule $R_{4,p}$ will be more strongly activated than the others, which favors a pixel update that does not depend on the new data $\{x_5, x_6, x_7, x_8\}$ (template 14 in Fig. 2). In other words, the update of the central pixel intensity is driven towards a filtering deterioration based completely on filtered data. This may lead to a destruction of image details at high noise levels as the level of similarity between the central pixel and the templates and, subsequently, the degrees of activation of the other rules, are generally low.

The above implies that this template should to be treated differently when the noise intensity is high. By adding an additional antecedent in the rules $R_{4,p}$ and $R_{5,p}$, we suggest a reduction of the dependence on already filtered data:

$R_{4,p}$: IF template p is homogeneous
AND template p is similar to at least P pixels in template r
AND the central window pixel is not similar to template j;
THEN template j portrays the local image characteristics
AND change the central pixel according to a function h of the pixels in template j;

$R_{5,p}$: IF template j is homogeneous
AND template p is similar to at least P pixels in template r
AND the central window pixel is similar to template j;

THEN the central pixel and template j both portray the local image characteristics
AND change the central pixel according to a function h of the pixels in template j and itself;

In other words, *if* the "new" data supports the central pixel update implied by the "old" data, *then* this value is considered reliable. However, for the noise levels in our experiments, this modification of the rules did not improve the results. Hence, it was not used in the tests. Potentially, the pixels in template p can also provide useful information about intensity changes in the window. Low similarity between these pixels indicates a non-homogenous window, e.g. the existence of an edge structure (true or not) in the window (Fig. 4).

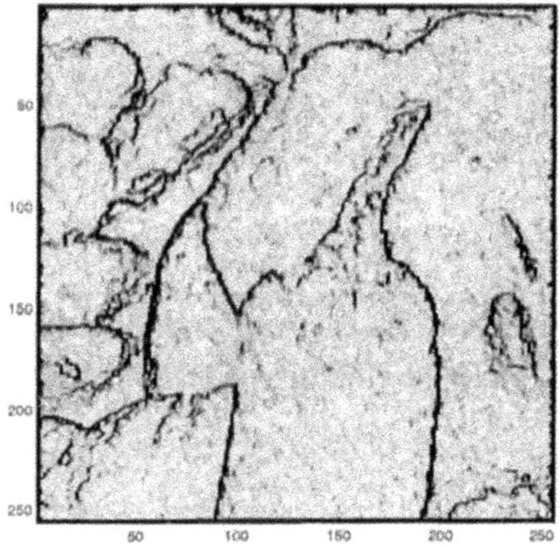

Fig. 4. Upper left part of the 'Peppers' image. The dark areas correspond to low similarity between the already processed pixels in the window.

2.4 Choosing filter parameters

The parameters steering the operation of the FSB filter are related to the width of the fuzzy similarity relation function (Eq. 6). The parameter α is associated with the similarity between the central pixel and a template pixel whereas β is associated with the similarity between two pixels within the template. Hence, β determines the degree of template homogeneity; the smaller the β, the lower the homogeneity of the template. The parameters α and β can be chosen by analyzing MSE surfaces (Fig. 5) and choosing the parameters that give the lowest errors. As expected, there is no single optimal set of parameters for all noise types, but depending on the nature and amount of noise in the image, the MSE surfaces have different structures.

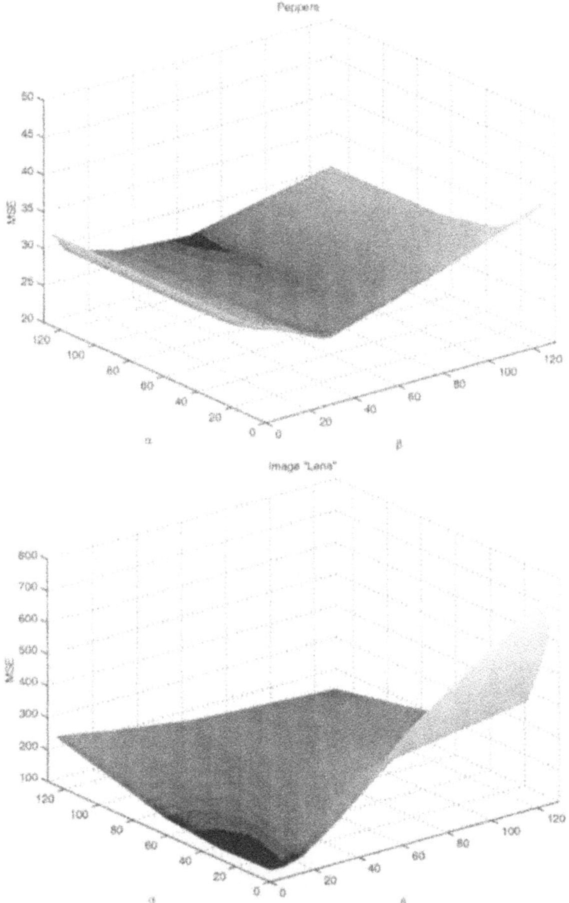

Fig. 5. Top: MSE surface for the 'Peppers' image with Gaussian noise (σ^2=100). Bottom: MSE surface for the 'Lena' image with Gaussian noise (σ^2=300) and 10% impulse noise with random amplitude.

As α and β participate in division arithmetic operations, their values should be selected as $\alpha = 2^k$ and $\beta = 2^l$, ($k, l \in \{0, 1, \dots, 8\}$) in order to provide high-speed hardware implementation of the filter.

2.5 Pixel correction

In the FSB filter approach, the correction of the intensity of the central pixel, Δx, is generally obtained through centroid defuzzification:

$$\Delta x = \frac{\sum_{i=2,4,5} \sum_{j=1}^{n} \mu_{i,j} \Delta x_{i,j} + \mu_{dis} \Delta x_{dis}}{\sum_{i=2,4,5} \sum_{j=1}^{n} \mu_{i,j} + \mu_{dis}}, \tag{9}$$

where $\mu_{i,j}$ is the firing strength of rule $\mathbf{R_{i,j}}$. For implementation reasons, this expression can be approximated by only considering the rule that contributes the most to the filtering. This gives a simplified defuzzification, that does not incorporate any division operations:

$$\Delta\hat{x} = \mu_{max}\Delta x_{max} + \mu_{dis}\Delta x_{dis}, \tag{10}$$

where μ_{max} is the firing strength of the rule with the highest degree of activation and Δx_{max} is the corresponding change implied by this rule. Experiments show that the MSE is typically somewhat higher when using the simplified defuzzification. Though, in some cases (cf. Table 3), this defuzzification yields better results, since it involves less averaging.

3 Experiments

In order to investigate the performance of the filter, a series of filtering simulations were carried out. The well-known test images 'Lena', 'Peppers' and 'Cameraman' were contaminated by adding noise according to the noise model in Eq. (4) in Sect. 1.3. The mean square error (MSE) was used as the objective performance indicator.

The filters used for comparison were Russo97 [14], Russo00 [15], Delva [5] and the well-known, non-fuzzy DW-MTM filter [12]. The window sizes of the DW-MTM filter were 3×3 and 5×5 pixels, respectively. For all of these filters, except the DW-MTM filter, recursive implementation yielded better results and was hence used in the experiments.

For each image, the parameters of the Russo97, Russo00, Delva and DW-MTM filters were chosen so as to minimize the MSE for a 256×256 pixels part of the respective image contaminated with Gaussian noise (σ^2=100) and salt-and-pepper noise (P=0.1). By optimizing the parameters for each noise type and level, we would achieve lower MSE values, but we are interested in analyzing the performance of the filters with constant parameters as the noise distribution changes.

Three different FSB filters were used in the examples. The numbers of the templates used in the experiments are given according to the numbering in Fig. 2. The parameters α and β were chosen to provide efficient overall filtering. The minimum operator was used for the aggregation operations.

FSB 1: Rules: R_2, R_4 and R_5. Templates: {1,2,12,14,16,18}.
Centroid defuzzification (Eq. 9).
α=64, β=64. Recursive implementation.

FSB 2: Rules: R_2. Templates: {1,2,14,18}.
Simplified defuzzification (Eq. 10).
α=64, β=64. Recursive implementation.

FSB 3: Rules: R_2. Templates: {1,2,12,14,16,18}.
Centroid defuzzification (Eq. 9).
α=128, β=128. Non-recursive implementation.

The MSE values for different noise intensities are given in Tables 1, 2 and 3. Some images are shown in Figs. 6 and 7.

Table 1. MSE results for the 'Peppers' image with different noise distributions.

Filter	σ^2=300 P=0	σ^2=100 P=0.1, s&p	σ^2=100 P=0.1, unif	σ^2=0 P=0.3, unif	σ^2=500 P=0.2, unif
No filtering	292.1	1995.9	503.0	1215.4	1263.5
FSB 1	48.4	36.6	31.8	28.9	81.0
FSB 2	56.3	47.4	34.0	29.3	102.7
FSB 3	55.1	64.3	38.9	48.5	109.3
Russo97	97.1	43.0	35.6	45.4	216.1
Russo00	79.1	29.0	39.8	54.3	204.9
Delva	72.2	44.6	42.6	41.5	131.5
DW-MTM	49.7	35.0	30.2	32.1	100.1

Table 2. MSE results for the 'Lena' image with different noise distributions.

Filter	σ^2=300 P=0	σ^2=100 P=0.1, s&p	σ^2=100 P=0.1, unif	σ^2=0 P=0.3, unif	σ^2=500 P=0.2, unif
No filtering	292.0	2046.8	547.5	1353.3	1330.0
FSB 1	73.3	65.7	57.0	60.3	112.6
FSB 2	74.3	64.8	51.4	54.0	129.3
FSB 3	72.1	81.9	57.2	73.8	129.0
Russo97	110.0	58.0	48.2	64.8	229.7
Russo00	105.0	43.5	54.0	74.1	240.0
Delva	108.3	77.3	73.5	74.1	177.4
DW-MTM	81.3	59.2	53.0	55.6	144.1

3.1 Comparison

It must be pointed out that all filters in the comparison were originally designed and tested on some particular noise types and levels. Hence, they may perform better in some cases and worse in others. The filters Russo97 and Russo00 were designed to deal with a mixture of Gaussian noise and impulse noise, but only tested on mixed Gaussian and salt-and-pepper noise. The Delva filter was tested on a mixture of Gaussian and salt-and-pepper noise.

We note that the Russo97 and Russo00 filters perform well for some noise distributions, but in order to achieve lower MSE values for others, retuning of the parameters is necessary. By optimizing/tuning the parameters for

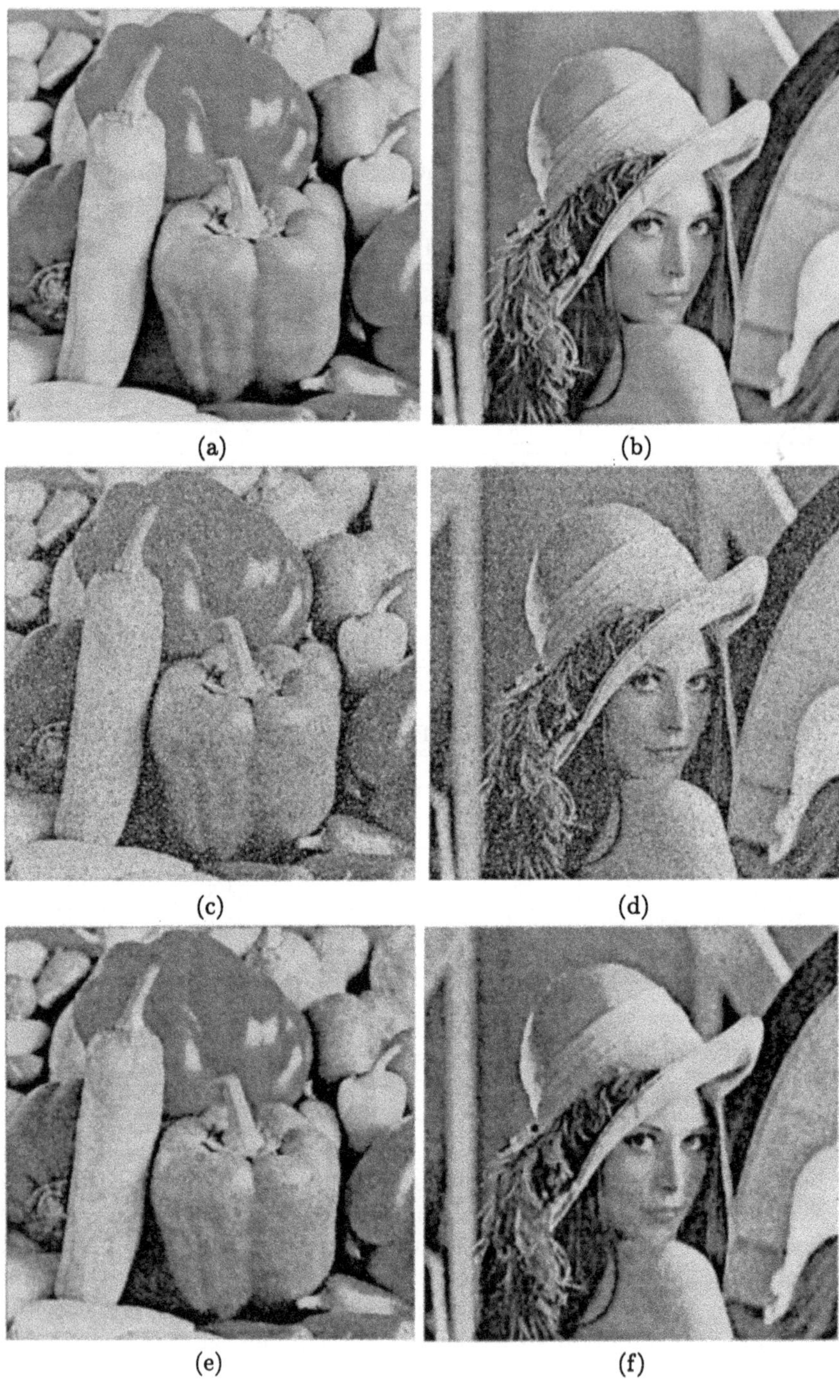

Fig. 6. (a),(b) Original images 'Peppers' and 'Lena'. (c),(d) Images contaminated with Gaussian (σ^2=500) and random impulse (P=0.2) noise. (e),(f) Filtered images.

Fig. 7. (a),(b) Original images 'Cameraman' and 'Lena'. (c) Image 'Cameraman' contaminated with Gaussian noise (σ^2=300). (d) Image 'Lena' contaminated with Gaussian (σ^2=100) and salt-and-pepper (P=0.1) noise. (e),(f) Filtered images.

Table 3. MSE results for the 'Cameraman' image with different noise distributions.

Filter	σ^2=300 P=0	σ^2=100 P=0.1, s&p	σ^2=100 P=0.1, unif	σ^2=0 P=0.3, unif	σ^2=500 P=0.2, unif
No filtering	282.7	2104.3	572.1	1434.8	1338.5
FSB 1	171.6	186.8	167.5	187.0	230.2
FSB 2	121.0	137.4	108.6	137.3	195.5
FSB 3	147.1	176.1	141.1	175.5	221.8
Russo97	152.3	117.8	97.3	160.5	351.0
Russo00	180.8	148.0	137.3	188.4	348.2
Delva	184.0	200.7	180.0	221.9	283.0
DW-MTM	179.0	182.2	158.1	183.9	256.0

each noise type and level we would achieve lower MSE values, but in order to study the robustness of the filter performances to noise changes, the parameters were held constant as the noise distribution was changed. The FSB filters display a more robust behaviour for the noise distributions used in the examples. The simpler FSB 2 filter performs better than the FSB 1 filter when applied to the image 'Cameraman'. This is mainly because the FSB 1 filter, involving a greater amount of averaging of subfilter outputs, smears the lower part of this image (cf. Fig 7). Generally, recursive implementation of the FSB filters resulted in lower MSE values.

4 Hardware implementation of FSB filters

As visual servoing systems require real-time image processing and pattern recognition, we defined a goal to achieve rates beyond the boundary of 100 frames per second with frame size at least 512×512 pixels. This defined the need to acquire a dedicated vision sensor able to perform a high frame rate and good resolution. High frame rate is possible if the sensor has a digital output on which the data are provided by parallel functioning ports typically column wise or line wise. However, such kind of parallelism puts a lot of constraints and requirements for the next layers of the system. To perform the scanning of the entire frame by moving a window, it is necessary to include line buffers for memorizing temporarily the current lines on which the moving window is allocated, as the data are read column wise line by line. The number of line buffers depends on the size of the window. In our case it is 3, in rare cases more than 5. The vision sensor and the fuzzy processing layer have different working speeds and data formats, therefore the best solution is to use FIFO memories with bus transformation property (i.e. input and output data can have different word length) as line buffers. The FIFO volume must be at least equal to the number of pixels in a line. Additionally, in order to provide the pixels in the window to the input of the fuzzy processing layer in parallel, there is a need to include delays by separate shift registers. The number of delay stages depends on the size of the moving window.

Having all these issues we propose a general structure of a *fuzzy-image-processing-structure* (FIPS) shown in Fig. 8. Actually, the block marked by

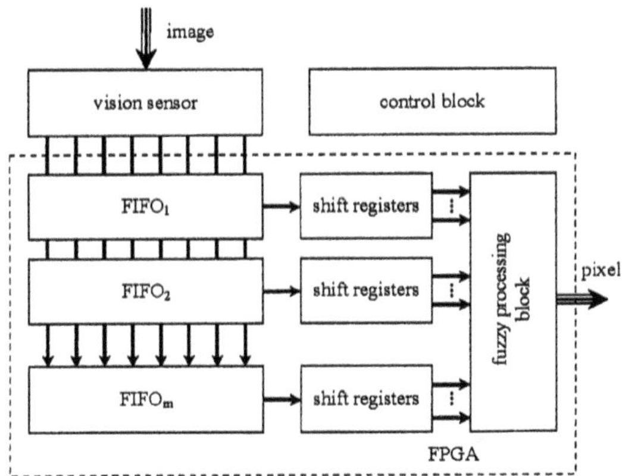

Fig. 8. Fuzzy image processing structure.

the dashed line in Fig. 8 can implement various type of low-level image processing algorithms depending on the application task. This variety requires that the structure has to be flexible and opened for arranging and re-arranging. Therefore, we suggest a family of substructures that perform separate image processing tasks and can be connected in a chain. It is very convenient to implement it on an FPGA that gives also the possibility to build bigger systems similar to LEGO® cubes put together.

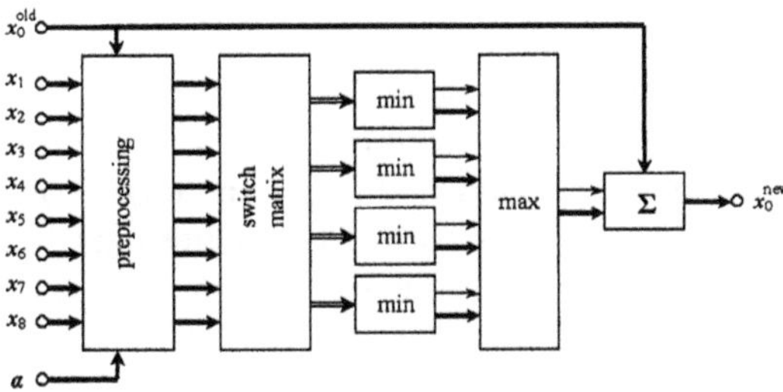

Fig. 9. Simple fuzzy-similarity-based image noise cancellation filter.

The substructure in Fig. 9 implements a simple FSB noise filtering algorithms. The pre-processing block evaluates the two-dimensional FSB relations

(Eq. 6). The switch matrix configures the predefined templates connecting the respective similarity measures; in this case four templates are used. Then the "min" blocks evaluate the contribution of each template and the "max" block makes the final decision about the correction term added to the old value of the central pixel.

To estimate the performance of the FSB noise cancellation filter we need to estimate the execution time of the circuit in Fig. 8. Due to the parallelism and pipelining involved [8], one pixel is filtered for about 10-12 nsec. The additional time for communication with the FIFO memories and shift registers is estimated to be not more than 50% overheads giving 15-18 nsec total time for one pixel. In Table 4, the performance with these values is shown. The figures indicate that higher frame rate for the full resolution of 1 Mpixel can be achieved if providing further parallelism in the processing block with several FSB filters. A laboratory prototype of this filter is implemented on a Virtex FPGA of Xilinx (www.xilinx.com) and further development is planned for the advanced Virtex-II.

Table 4. Performance of an FSB noise cancellation filter implemented on a Xilinx Virtex FPGA.

Frame size	Time per frame	Frame rate
1024 × 1024	16-19 ms	52-62 fps
768 × 560	6.5-8 ms	125-150 fps
512 × 512	4-5 ms	200-250 fps
320 × 256	1.3-1.5 ms	650-750 fps

5 Conclusions

We find the FSB filter approach useful for reduction of noise in images with a number of different noise types and levels. The simple structure of the FSB filter makes it fairly robust against changes in noise distribution. On the other hand, this also prevents the filter from being particularly optimizable for a known noise type. The experiments indicate that the filtering quality is comparable to or slightly better than other existing methods. In addition, the structure of the FSB filter is suitable for hardware implementation on an FPGA device, implying that the filter can be used in real-time, at frame rates higher than normal video rate. Simulations and experiments on a laboratory prototype imply that a simple FSB filter is able to operate at 60 frames per second at the resolution of 1024×1024 pixels. However, the performance of the FSB filters is less efficient for high levels of salt-and-pepper noise. This is due to the fact that a template consisting of noisy, bright pixels is still homogeneous. Subsequently, the intensity change implied by the corresponding rule will be assigned a large weight in the defuzzification, implying an inaccurate central pixel update.

References

1. K. Arakawa, *Median filter based on fuzzy rules and its applications to image restoration*, Fuzzy Sets and Systems 77, pp. 3-13, 1996
2. K. Arakawa, *Fuzzy rule-based image processing with optimization*, Fuzzy Techniques in Image Processing, Springer Verlag, pp. 222-247, 2000
3. C-L Chen, C-S Lee and Y-H Kuo, *Design of high speed weighted fuzzy mean filters with generic LR fuzzy cells*, Proc. IEEE International Conference on Image Processing, New York, NY, USA, Vol. 2, pp. 1027-1030, 1996
4. Y. Choi and R. Krishnapuram, *A robust approach to image enhancement based on fuzzy logic*, IEEE Trans. Image Processing, Vol. 6, No. 6, pp. 808-825, 1997
5. J. G. R. Delva, A. M. Reza and R. D. Turney, *FPGA Implementation of a Nonlinear Two Dimensional Fuzzy Filter*, Proc. IEEE Conference on Acoustics, Speech and Signal Processing ICASSP'99, Piscataway, NJ, USA, Vol. 4 pp. 2143-2146, 1999
6. M. Doroodchi and A. M. Reza, *Fuzzy cluster filter*, Proc. IEEE Conference on Image Processing ICIP'96, Lausanne, Switzerland, pp. 939-942, 1996
7. A. Gasteratos, I. Andreadis, *Non-linear image processing in hardware*, Pattern Recognition 33, pp. 1013-1021, 2000
8. I. Kalaykov, *Parallelism for Very Fast Fuzzy Hardware*, Proc. IASTED Conference on Artificial Intelligence and Soft Computing, ASC'2001, Cancun, Mexico, pp 123-127, 2001
9. L. Khriji, G. Bernacchia, M. Gabbouj and G. Sicuranza, *A dedicated hardware system for a class of nonlinear order statistics rational hybrid filters with applications to image processing*, Proc. IEEE International Conference on Image Processing, Piscataway, NJ, USA, Vol. 2, pp. 419-423, 1999
10. Y-H Kuo, C-S Lee and C-L Chen, *High-stability AWFM filter for signal restoration and its hardware design*, Fuzzy Sets and Systems 114(2), pp. 185-202, 2000
11. C-S Lee, Y-H Kuo and P-T Yu, *Weighted fuzzy mean filters for image processing*, Fuzzy Sets and Systems 89(2), pp. 157-180, 1997
12. Y-H Lee and S. A. Kassam, *Generalized Median Filtering and Related Nonlinear Filtering Techniques*, IEEE Trans. Acoustics, Speech and Signal Processing, Vol. ASSP-3, No. 3 pp. 672-683, 1985
13. M. Muneyasu, Y. Wada and T. Hinamoto, *Edge-preserving smoothing by adaptive nonlinear filters based on fuzzy control laws*, Proc. IEEE International Conference on Image Processing ICIP'96, Lausanne, Switzerland, pp. 785-788, 1996
14. F. Russo, *Noise cancellation using nonlinear fuzzy filters*, Proc. IEEE Instrumentation and Measurement Technology Conference IMTC'97, Ottawa, Canada, Vol. 2 pp. 772-777, 1997
15. F. Russo, *A technique for image restoration based on recursive processing and error correction*, Proc. IEEE Instrumentation and Measurement Technology Conference IMTC'00, Piscataway, NJ, USA, Vol. 3, pp 1232-1236, 2000
16. A. Taguchi and M. Meguro, *Adaptive L-filters based on fuzzy rules*, Proc. IEEE Symposium on Circuits and Systems ISCAS'95, Seattle, USA, pp. 961-964, 1995

Chapter 4

Fuzzy Rule-Based Color Filtering Using Statistical Indices

Spiros Fotopoulos, Antony Fotinos and Socrates Makrogiannis

University of Patras
Department of Physics
Electronics Laboratory
GR-26500, Patras
Greece
email: spiros@physics.upatras.gr

Summary. The following chapter describes the design and evaluation of three fuzzy color filters that are based on a specific local image model. It is assumed that an ideal image is a set of flat regions, edges and ramps. However in natural images these regions are not strictly defined so a sense of ambiguity appears which could be handled by a fuzzy rule-based system. In addition, the existence of the noise consolidates the fuzziness. We introduce three different statistical indices as fuzzy variables which are used to detect flat regions, edges and ramps. Our purpose is to remove the noise while preserving the details of the color images. The idea is that these regions from the filtering point of view should be handled in different ways. As a result three fuzzy color filters are introduced and their fuzzy variables are extracted from the estimation of the local distribution that is carried out using the Parzen estimators-Potential functions. The summation of the Potential functions, the maximum/minimum value of the local distribution and the Relative entropy are the indices that used to determine the type of the local region. Finally the fuzzy filters are evaluated with qualitative and quantitative criteria using natural color images corrupted with different types of noise.

1 Introduction

Fuzzy logic [37] has been extensively employed in computer science and telecommunications applications. In image processing particularly a plethora of works have been reported that employ fuzzy logic. In [14] and [15] the topic of fuzzy theory applications in several fields of image processing is reviewed and in [22] the interest is focused on segmentation and pattern recognition methods that make use of fuzzy logic. In [20], it is also pointed out that fuzzy logic represents a promising framework for image segmentation applications. Regarding image analysis, it has been widely used in low and medium level image processing approaches and more recently in higher level processing (e.g. search engines in data retrieval). In this chapter techniques for noise removal and image enhancement are reported.

The existence of noise increases the amount of ambiguity in a signal, therefore fuzzy logic is considered as the appropriate solution for these problems. Fuzzy filters are classified into the category of adaptive filters and their objective is to remove noise while preserving the important details of the image.

The first family of filters combine fuzzy inference with the basic principles of Adaptive L-filters [23], Adaptive Center Weighted Average (ACWA) and Modified Trimmed Mean (MIM). These are divided into two groups, that is Adaptive Fuzzy L-filters [30] and the Adaptive Fuzzy Weighted Median/Average filters [19]. The Adaptive Fuzzy L-filters are based on the corresponding L-type filters which are a generalized version of the median type filter. These are non-linear filters that include median value, mean value and alpha-trimmed filters. Furthermore, they may be combined to remove simultaneously several types of noise due to the fact that they are more effective for specific types of noise. The structure of Adaptive Fuzzy Weighted Median/Average filters is similar to that of Adaptive Fuzzy L-filters, while their main difference is that the former does not use specific filter structures as output but the weighted mean and median value.

Apart from that, the Weighted Fuzzy Mean Filter [4] was also proposed for the case of excessively high impulsive noise (>4%). According to this method, some additional information about the ideal signal is required, which is usually related to the histogram boundaries. This is usually of limited size (approximately 10 bytes per image) and may be easily included in the initial image and supplied to the filter system.

In the work of Y. Choi and R. Krishnapuram [5] an adaptive noise removal algorithm is described, where the local image features are controlled by an index that comes from Robust theory and the Least Squares method (LS). The Adaptive Multilevel Fuzzy Median Filters [35] are based on the simple Multilevel Median Filter (MLMF)) [1] and are effective for removing "line-like" noise.

Another family of fuzzy filters are the non-linear fuzzy operators [26,27], which apply to three levels of processing for scalar images: i) image restoration ii) image enhancement and iii) edge detection, also known as FIRE filters. According to this approach the correlation of a region with a specific signal type is examined using statistical (intensity) and spatial information as well.

Fuzzy theory has also been employed for multiple-channel signals. The proposed methods are usually related to color image segmentation [3,17] image enhancement [29] and noise removal [24,34].

The filters proposed in [24] are developed for multichannel signals and produce efficient results for several types of noise. This approach does not take into account the local signal and noise features. Its basic structure results from an adaptive process using transformations that are based on the Vector Median Filter [2] and the Vector Directional Filter [33].

The multichannel fuzzy filter proposed by Hung-Hsu Tsai and Pao-Ta Yu [34] utilizes fuzzy rules that were developed using empirical knowledge and it also makes use of the following multichannel filters: Vector Median [2], Vector Directional [33] and the Identity filter. This method is suitable for impulsive noise removal, whereas it is not very efficient for the case of normally distributed noise.

This chapter deals with noise removal techniques for color images that combine statistical features to model the local distribution of the image with fuzzy logic principles. The statistical indices are derived from the local distribution of pixels which is found by extensive use of a nonparametric density estimation method known as Parzen estimation or Potential Functions. These indices serve as the features to classify the region in one of three model classes and select the appropriate vector filter accordingly. Three filtering methods are proposed and studied.

2 Principles of the filtering method

2.1 General structure of a fuzzy filter

Fuzzy filters presented in the literature employ the typical structure of a fuzzy rule-based system. Figure 1 displays the general structure of the techniques presented in this chapter.

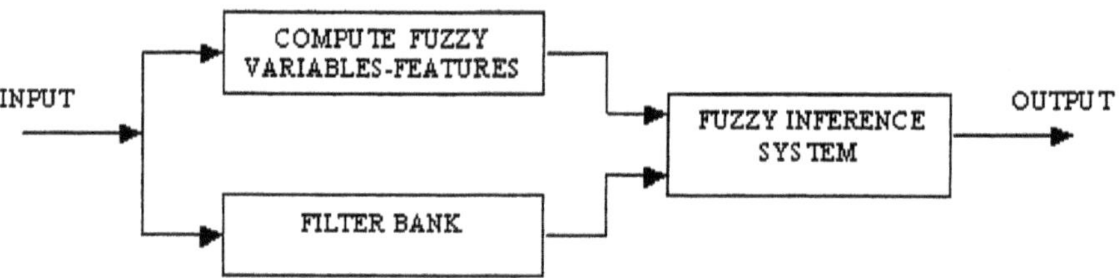

Fig. 1. General structure of a fuzzy filter.

A fuzzy filter consists of three basic stages, that is the computation of fuzzy variables, the filter bank and the fuzzy inference system. In the first two stages the calculation of fuzzy variables and the various filter outputs are performed in parallel. The fuzzy inference system determines the overall filter output by assigning weights to each filter of the filter bank. Alternative structures where the inference system operates on the image pixels prior to the filtering output, have also been reported [19].

2.2 Modelling of a color image

A key idea used in the presented fuzzy filtering methods is the adoption of a local image model which consists of three discrete cases defined in the following.

From the deterministic point of view the local image model may be classified into constant regions, edges and ramps, an example is given in Figure 2. This model which is rarely met in real images, in fact describes three extreme situations which could be considered as independent coordinates in a 3D space and where all local image shapes are represented as a combination of the above three model-cases. Variations of the pixel values due to the noise or other degradation effects are described the same way.

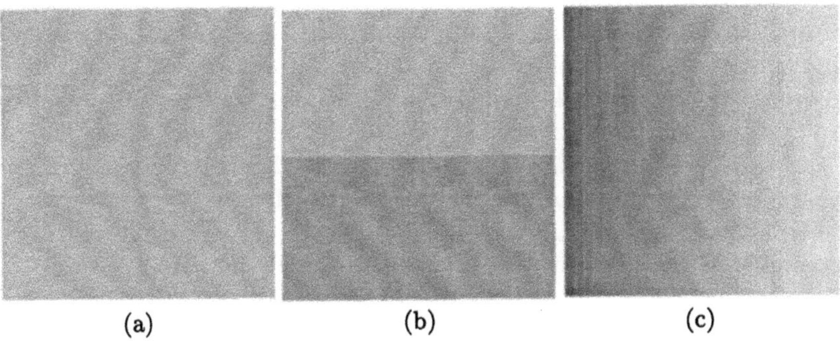

Fig. 2. Example of (a) constant region (b) well defined edge and (c) ramp.

In the above assumption the window size is very important and without any loss of generality the 3×3 window size is adopted here. In the case of a larger window the model should include more ideal cases. The defined 3-case model for the image regions corresponds to 3-types of pixel distributions. Inversely this information is also utilized to detect the region type from the local pixel distribution.

As constant region is considered the local neighborhood of a pixel where all the samples have the same intensity and color values. In this ideal case the corresponding histogram consists of a single peak (Figure 3a) and its height is equal to the amount of samples. A well defined edge may be modelled by a step function and its histogram consists of two peaks (Figure 3b). The height of each peak is determined by the number of samples located on each side of the edge. In the ramp case, the samples have several values and in most of the cases they form a plane of constant slope (Figure 3c).

Coming closer to a more realistic case, a constant region consists of pixels distributed around a mean value with a small variance. Similar conclusions are made for the edge and ramp cases. The histograms corresponding to this

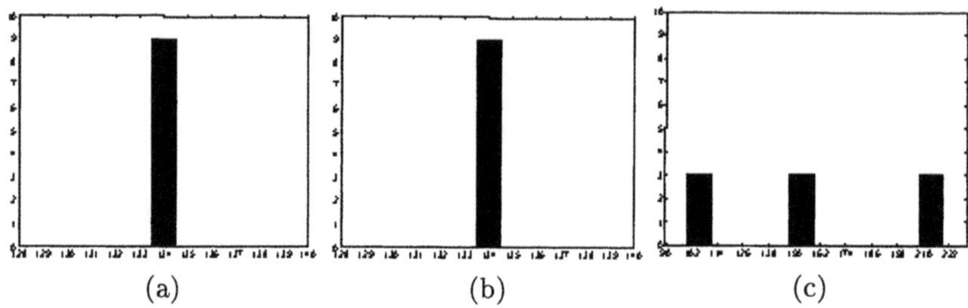

Fig. 3. Histograms of the ideal cases of the local image model: (a) constant region (b) well defined edge and (c) ramp.

improved model where IID noise is overlaid on the image are modified accordingly. In the case of normally distributed additive noise the corresponding histograms are displayed in Figure 4, i.e. for a constant region the distribution is unimodal (Figure 4a), for the well defined edge it is bimodal (Figure 4b) and for the ramp multimodal (Figure 4c). It is worth noting that the presence of an outlier brings about variations in the distribution such as the appearance of an additional peak with small height.

The above description of the image in terms of model cases and local histograms is utilized to provide rules for the fuzzy filtering process.

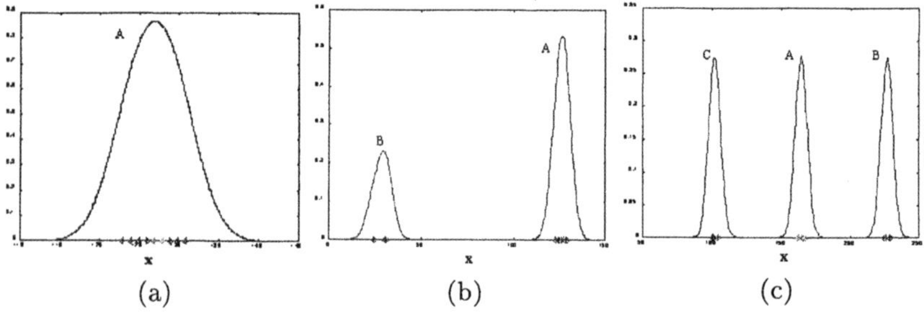

Fig. 4. Examples of histogram shapes for non ideal local neighborhoods of (a) constant region (b) well defined edge (c) ramp.

2.3 Nonparametric density estimation

In the previous paragraph the three ideal image cases and the corresponding histograms were presented. The next step is to consider the real case, compute their histograms and associate memberships to the above three-model case. However, due the small number of pixel samples (N=9) of the local window, the nonparametric probability density estimation using Parzen kernels [21] is employed. This estimator is also very well suited for multidimensional signals and in this case the color images. Let N be the data cardinality, the density estimation for the point X is given by the relation:

$$\tilde{f}(X) = \frac{1}{Nh^p} \sum_{i=1}^{N} K\left(\frac{X - X_i}{h}\right), \tag{1}$$

where *K(.)* in (1) is the Parzen kernel function that may have various forms, a standard Gaussian function is utilized and h is a parameter that controls the resolution of the density estimation. The optimal h value may be defined by an error minimization technique. The optimization process is significant for the final results and several techniques have been previously reported that result in h values of the same magnitude order. An efficient formula to define the value of the h parameter, is that proposed in [12]:

$$h = \left[\frac{4}{2d+1}\right]^{\frac{1}{d+4}} N^{-\frac{1}{d+4}} \sigma. \tag{2}$$

d represents dimension of the data and σ denotes the standard deviation of the data additive noise.

2.4 The filter bank

Filtering of color images is usually accomplished by vector techniques [2]. Most of these techniques are based on a vector ordering according to the distance metric employed. One well known filtering method is the Vector Median Filter (VMF) [2] which is based on the aggregate distance of each pixel. In that method the pixel - vector with the lowest aggregate distance is selected as the filter output. In another approach a subset of the ordered pixel-vectors is selected and averaged. This is the so called alpha-trimmed filter [23]. Other similar techniques like Vector Directional Filters (VDF) and versions of the above have been successfully used for Color Image Filtering. The VMF suppresses better the impulsive noise while averagers suppress better Gaussian noise. Each of the above filters has a different impact on the noise suppression of the image. Filters like alpha-trimmed or adaptive filters make a compromise between the above two extreme types of noise i.e impulsive and Gaussian.

The other major issue of image filtering is the preservation of the image characteristics during the filtering process. The above filters present a different behavior to the smoothing of the image details with the average being the worse. Improvement of the filtering action is achieved by adaptively selecting at each point the most appropriate filter according to the information provided at each point. The usual approach is to compute all the above filter outputs at each point and let the fuzzy system have the control for the final output. This operation is implemented by a weighting average of the vector filter outputs with weights provided by the fuzzy system. In an alternative way an average scheme may be employed where the fuzzy system assign weights to each pixel. Both operations are used in this study.

3 The proposed fuzzy color filters

Three fuzzy color filters are proposed and studied here. They are denoted as MFF (Multichannel Fuzzy Filter) [7] and [9], HMF (Histogram-based Multichannel Filter) [11], and RELF (Relative Entropy-based Filter) [8]. The last one has two versions RELF1 and RELF2.

3.1 Multichannel fuzzy filter (MFF)

In this method extensive use of Parzen estimators (potential functions) is made. It is based on classifying a set of pixels into a constant region, edge or ramp using combination of two indices [13]. The presented method removes Gaussian, impulsive noise and their combinations as well. The two statistical measures used are the standard deviation or dispersion and the sum of the Parzen estimators over each pixel of the corresponding window. The standard deviation $v(X)$ is routinely calculated by the relation:

$$v(X) = \frac{1}{N} \sum_{n=1}^{N} (X_i - \overline{X})^2, \tag{3}$$

where $\overline{X}$ corresponds to the mean value vector. It is obvious that constant regions have small standard deviation values, while in edge regions the standard deviation is relatively large.

The summation $g(X)$ of the Parzen estimators is calculated as follows:

$$g(X) = \frac{1}{N} \sum_{i=1}^{N} P(X_i). \tag{4}$$

$P(X_i)$ is the probability density as it is given by the Parzen estimators (1).

It is worth noting that the $g(X)$ is a density estimation index and is related to the entropy of the samples. The normalization weights of the kernel functions may be omitted.

The parameter $g(X)$ takes large values in the case of well defined edges while $v(X)$ takes large values both to edges and ramps [13]. In the ideal cases these indices take specific values providing full knowledge for the type of the region while in real images or in the presence of noise these parameter values will vary in a specific range. This ambiguity in knowledge is handled by fuzzy logic formulation.

The combination of parameters $v(X)$ and $g(X)$ may describe the local distribution information according to the image model outlined in paragraph 2.2. In order to quantify the relations (3), (4) the following formulation for a window of N samples is introduced:

Ideal step edge:

$$X_i = \begin{cases} \boldsymbol{a}\,, & i = 1 : k \\ \boldsymbol{a} + \boldsymbol{d}\,, & i = k+1 : N \end{cases} \tag{5}$$

Ideal ramp:

$$X_i = \boldsymbol{a} + \left[\frac{(i-1)}{\sqrt{N}}\right] \boldsymbol{c} \qquad i = 1 : N \tag{6}$$

Ideal constant region:

$$X_i = \boldsymbol{a}\,, \qquad i = 1 : N. \tag{7}$$

$[x]$ represents the integer part of x. These cases are illustrated in Figure 5 in the RG plane for visualization purposes.

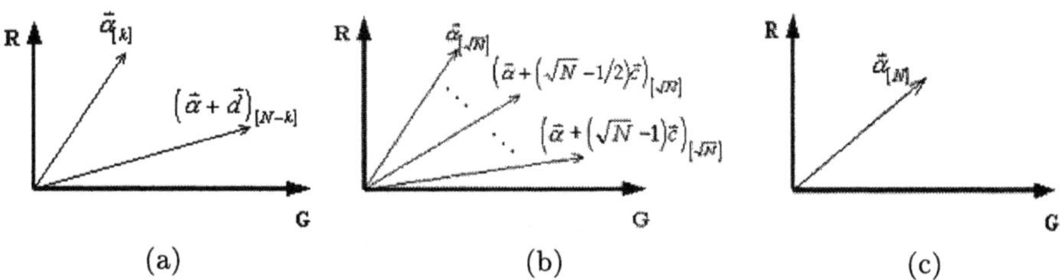

Fig. 5. *Graphical representation in the RG plane for the ideal case of (a) edge, (b) ramp and (c) constant region where $\vec{\alpha}_{[m]}$ is m times vector $\vec{\alpha}$.*

According to the previous discussion, the presented indices are reformulated as follows:

Ideal step edge:

$$v(X) = \frac{1}{N}\sum_{i=1}^{N} \left\|X_i - \overline{X}\right\|^2 = \frac{1}{2N^2}\sum_{j=1}^{N}\sum_{i=1}^{N} \left\|X_i - X_j\right\|^2$$

$$= \frac{1}{2N^2} \cdot \left\{2k\,(N-k)\ \|d\|^2\right\} \tag{8}$$

$$g(X) = \frac{1}{N}\sum_{i=1}^{N} P(X_i) = \frac{1}{N^2}\sum_{j=1}^{N}\sum_{i=1}^{N} \exp\left(^{-\|X_i - X_j\|^2}\!\big/_{2h^2}\right)$$

$$= \frac{1}{N^2}\ \left\{\ k^2 + \ 2\,k\,(N-k)\ \exp\left(-\,\|d\|^2\big/2h^2\right) + \ (N-k)^2\right\} \tag{9}$$

Ideal ramp:

$$v(X) = \frac{1}{2N^2}\sum_{j=1}^{\sqrt{N}}\sum_{i=1}^{\sqrt{N}} \|\, a + (i-1)\,c \ - \ (a + (j-1)\,c)\ \|^2 \ = \ \frac{(N-1)\ \|\,c\|^2}{12} \tag{10}$$

$$g(X) \ = \ \frac{1}{N}\sum_{j=1}^{\sqrt{N}}\sum_{i=1}^{\sqrt{N}} \exp\left(-\ ^{(i-j)^2\|c\|^2}\big/2h^2\right) \tag{11}$$

Ideal constant region:

$$v(X) \ = \frac{1}{2N^2}\sum_{j=1}^{N}\sum_{i=1}^{N} \|X_i - X_j\|^{\,2} = \ 0 \tag{12}$$

$$g(X) = \frac{1}{N^2}\sum_{j=1}^{N}\sum_{i=1}^{N} \exp\left(\frac{-\|X_i - X_j\|^2}{2h^2}\right) \ = \ 1. \tag{13}$$

It becomes apparent from the above relations that the employed indices have fixed values only for the constant region, whereas in the other two cases these values depend on other parameters such as the edge height or the samples' color distance when they form a ramp.

Assuming that

$$h^2 = v(X)\,\varrho, \tag{14}$$

where ρ is a constant, which will be defined later. The index *g(X)* receives specific values regardless of the edge height and color distance parameters. As a result, *g(X)* computed by (9) or (11) is given by:

$$gstep = \frac{1}{N^2}\ \left\{k^2 + \ 2\,k\,(N-k)\ \ \exp\left(-N^2/(2k\,(N-k)\,\rho)\,\right) + \ (N-k)^2\right\} \tag{15}$$

$$gramp = \ \frac{1}{N}\sum_{j=1}^{\sqrt{N}}\sum_{i=1}^{\sqrt{N}} \exp\left(-\ 6\ (i-j)^2\Big/(\rho\,(N-1))\right). \tag{16}$$

These parameters are considered to be scale invariant. In addition if $k{=}(N{+}1)/2$ i.e the edge is exactly positioned in the window center, then

$$gramp \quad < \quad gstep \quad < \quad 1. \tag{17}$$

It was observed that this relation is better emphasized when ρ takes values in the interval 0.1-0.8 [9].

Having defined the parameter ρ the values *gramp* and *gstep* are computed. These values provide the necessary information for the fuzzy classification of the region to the assumed model-classes.

At this stage we may define three classification rules in accordance with the value of the presented indices as displayed in Table 1.

$v(X)$	*g(X)*	**Class descrip-tor**	**Filter parameters**
large values	large values	Edge	$range_1 = \frac{\lVert d\rVert^2}{4} = \frac{N^2}{N^2-1}v(X)$

large values	small values	Ramp	$range_2 = \frac{\left(\sqrt{N}-1\right)^2 \|c\|^2}{4} = \frac{3\left(\sqrt{N}-1\right)}{\left(\sqrt{N}+1\right)} v(X)$
small values		Constant (flat) region	$range_3 = \max(c,\ v(X))$

Table 1: The classification rules and filter parameters.

The filter. For each one of these classes (i.e. ramp, edge, constant region) an appropriate vector filter of paragraph 2.4 could be selected. Most of them are based on a vector ordering approach. In this method it was found better to order the vectors by examining the distance between the samples and the maximum density vector. According to this distance a weight $m_k(X_i)$ is associated to each pixel of the window as follows:

$$m_k(X_i) = \quad 1 - \quad 1/(1 + \exp(-c_k(r_i - range_k))) \tag{18}$$

$$r_i = \|X_r - X_i\|^2$$

where k=1,2,3 represents each class, c_k is a constant and $range_k$ controls the amount of high weighted vectors.

The final output for each class is calculated as follows:

$$Y_k = \frac{\sum_{i=1}^{N} m_k(X_i) \cdot X_i}{\sum_{i=1}^{N} m_k(X_i)}. \tag{19}$$

Table 1 (4th column) contains the values of $range_k$ for these three classes. From this table it becomes obvious that $range_3 < range_1 < range_2$. This is a reasonable result since for large values of $range_k$ less vectors take part in the summation of equation 19.

It should be noticed that the above introduced filter suppress impulsive noise by associating small weights to impulses. Besides that, c is a constant used to exclude infinite weight values and it is set experimentally.

The fuzzy system. The observations of Table 1 are quantified by the fuzzy system which consists of two fuzzy variables, *v(X)* and *g(X)* each represented with two fuzzy sets SMALL and LARGE and complementary membership functions: $\mu_S = 1 - \mu_L$.

The employed fuzzy sets have trapezoidal shape, with boundaries defined by parameters a and b as displayed in Figure 6.

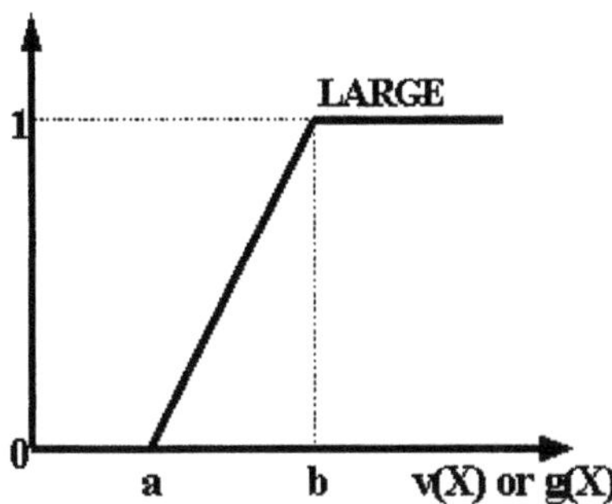

Fig. 6. Fuzzy set LARGE. It's complementary is defined as SMALL.

The fuzzy rule-base is displayed in Table 2. Let μ_S(v), μ_L(v), μ_S(g), μ_L(g) be the membership values of *v(X)* and *g(X)* to the fuzzy sets SMALL, LARGE respectively. The weights of the rules are subsequently given by:

$$\mu_1 = \min(\ \mu_L(v)\ ,\ \mu_L(g)\ )$$

$$\mu_2 = \min\left(\mu_L\left(v\right), \mu_S\left(g\right)\right) \tag{20}$$

$$\mu_3 = \mu_S(v)$$

and the overall filter output is calculated using the weighted mean inference method.

$$Y\left(X\right) = \frac{\sum\limits_{k=1}^{3} \mu_k\left(X\right) \cdot Y_k\left(X\right)}{\sum\limits_{k=1}^{3} \mu_k\left(X\right)}, \tag{21}$$

where k refers to the class and $Y_k(X)$ stands for the output of the corresponding filter of each class k.

Table 2. The rule base for MFF.

IF	$v(X)$ is LARGE	**AND**	$g(X)$ is LARGE	**THEN**	μ_1(class 1)
IF	$v(X)$ is LARGE	**AND**	$g(X)$ is SMALL	**THEN**	μ_2(class 2)
IF	$v(X)$ is SMALL			**THEN**	μ_3(class 3)

The presented method was tested on the color images PEPPERS and LENNA (24-bits/pixel) using additive uniformly distributed noise mixed with pulse noise. Results are given in the end of this study in tables 5 and 6. From these results it is concluded that MFF is efficient for removing normally distributed noise and impulsive noise with medium magnitude as well. Objective results are also given in corresponding color images of Figure 9.

3.2 Multichannel filter based on local histogram values (HMF)

The second color fuzzy filter presented in this study is named Histogram Multichannel Filter due to the histogram information which is used to discriminate between a ramp, an edge and a constant region.

Due to the fact that it is not computationally efficient to model the complete histogram information, some partial histogram features may be used such as its maximum and minimum values [11].

More precisely, the maximum value of the histogram may be directly used to determine which is the model of the image in the examined pixel neighborhood. Let N be the amount of samples, the maximum histogram value for a constant region is N, and for the case of a well defined edge is $(N\text{-}k)$, k being the amount of samples of the estimation point with lower cardinality i.e. $k<N$.

As it has been mentioned before, the ramp is a region that consists of more than two classes that have regularly the same height and its histogram values depend on the windows size. The maximum values for each case consequently are:

Constant region:

$$H_M(X) = N$$

Edge:

$$H_M(X) = (N - k) \tag{22}$$

Ramp:

$$H_M(X) < (N - k).$$

$H_M(X)$ is the maximum histogram value, which may be represented by a function in the p-dimensional space $h(X)$.

When a 3×3 window is employed the maximum histogram values are:

Constant region:

$$H_M(X) = 9$$

Edge:

$$H_M(X) = 6 \tag{23}$$

Ramp:

$$H_M(X) = 3.$$

In this discussion we did not take into consideration the occurrence of impulsive noise, which however does not practically change the above values.

At this point the problem that appears is the calculation of maximum probability. The histogram may be calculated using Parzen estimators. In this case the probability density is estimated and multiplied with the suitable coefficient, that is:

$$f(X_i) = \frac{1}{Nh^p} \sum_{j=1}^{N} K\left(-\frac{(X_j - X_i)^2}{2h^2}\right) \tag{24}$$

$$H(X_i) = Nh^p f(X_i). \tag{25}$$

The basic parameters of the Parzen density estimation are the type of function $K(.)$, the resolution parameter h and the amount of samples N. Even if the optimal values are used an error still occur between the estimation and the ideal distribution in the maximum (bias error) and the width of the distribution, which might affect the above result.

An additional problem that might influence the maximum probability density criterion H_M is the occurrence of noise pixels placed exactly on the edge that may mask the calculation of $H_M(X)$. This uncertainty is handled with fuzzy logic, by assuming that $H_M(X)$ is a fuzzy variable. Three fuzzy sets are consequently defined that correspond to the local model cases. In Figure 7(a) are depicted the membership functions of fuzzy sets RAMP, STEP and FLAT for a 3×3 window (N=9).

Apart from that the occurrence of impulsive noise has to be handled. This information is acquired by the probability density minimum H_m. Due to the previously reported problems of density estimation the $H_m(X)$ measure may be regarded as a fuzzy variable represented by the fuzzy sets SMALL and LARGE and with membership functions displayed in Figure 7b.

When the H_M has a large value and belongs to the FLAT set and H_m is small then the pixel neighborhood is constant corrupted with impulsive noise. When H_M is large and H_m is also large the outcome is a constant region with no impulsive noise.

When H_M is medium value and thus belongs to STEP and H_m is small, the examined region is considered to be a well defined edge with impulsive noise.

Finally when H_M is small and belongs to RAMP set and H_m is also small then the outcome is a ramp with impulsive noise, but if H_m is large the region is a ramp with impulsive noise.

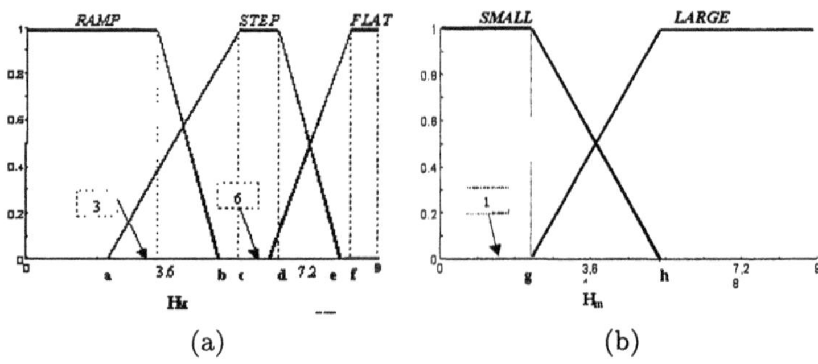

Fig. 7. Shape of the employed membership functions.

According to the previous discussion the following rule-base is developed (Table 3):

Table 3. The rule base for the HMF.

IF	H_M is FLAT	**AND**	H_m is SMALL	**THEN** Y is Y_1	8 samples
IF	H_M is FLAT	**AND**	H_m is LARGE	**THEN** Y is Y_2	9 samples
IF	H_M is STEP	**AND**	H_m is SMALL	**THEN** Y is Y_3	5 samples
IF	H_M is STEP	**AND**	H_m is LARGE	**THEN** Y is Y_4	6 samples
IF	H_M is RAMP	**AND**	H_m is SMALL	**THEN** Y is Y_5	2 samples
IF	H_M is RAMP	**AND**	H_m is LARGE	**THEN** Y is Y_6	3 samples

Y_m are the suitable (alpha-trimmed) filters for each case.

The parameter α translated in the number of samples selected from the ordered set of pixels for each case is given in the same Table 3 (last column). The basic idea is that when a constant region is inferred almost all the samples take part in the average, for the well defined edge almost half of the samples are used in the average, whereas for the ramp the pixels closest to the median sample are selected and averaged.

The final outcome from the fuzzy system is:

$$Y = \frac{\sum_{i=1}^{M} w_i Y_i}{\sum_{i=1}^{M} w_i}, \tag{26}$$

where M is the number of rules i.e. six rules and w_i the firing level of each rule defined by the relation:

$$w_i = min\{\mu_{im}\}, \tag{27}$$

where μ_{im} is the membership degree to the i-th fuzzy set in the m-th rule.

Objective and subjective results for the performance of this filter are given in Tables 5 and 6 and in Figure 9.

3.3 Fuzzy color filter using relative entropy (RELF)

This is the third type of fuzzy color filter based on indices derived from relative entropy. The idea is to classify the local region in one of the three model classes described previously (paragraph 2.2) using as distance-measure the relative entropy. This method produces two versions of fuzzy filters which for our convention are given the names RELF1 and RELF2.

The relative entropy or diverge was originally proposed by Kullback [16] and is also known in the literature as Kullback-Leiber measure. The relative entropy expresses the dissimilarity between two distribution. Let us assume two distributions for which the following probability ratio may be defined:

$$u_{ij} = \log \frac{p_i(X)}{p_j(X)}, \tag{28}$$

where $p_k(X)=p(X— \omega_k)$ is the probability that X belongs to class ω_k, and X is a random variable then the relative entropy measure is defined as the mean value of this ratio:

$$E_{rel}(i,j) = \int_X p_i(X) \log \frac{p_i(X)}{p_j(X)} dX. \tag{29}$$

For the discrete case the above relation is formulated as:

$$E_{rel}(i,j) = \sum_X p_i(X) \log \frac{p_i(X)}{p_j(X)}, \tag{30}$$

where

$$\int_X p_i(X)\,dX = 1 \Rightarrow \sum_X p_i(X) = 1, \quad \int_X p_j(X)\,dX = 1 \Rightarrow \sum_X p_j(X) = 1. \tag{31}$$

The relative entropy, which is considered as a distance measure, possesses the following interesting properties ([18]):

• It is always positive that is,

$$E_{rel}(i,j) \geq 0. \tag{32}$$

• It becomes zero if and only if the tested distributions $p_i(X)$ and $p_j(X)$ are exactly the same.

• It has only one minimum point for every distribution $p_j(X)$ that is $E_{rel}(i,j)$ is a useful measure for optimization methods. The relative entropy has been employed as a cost function to find the optimal resolution parameter h, in non parametric density estimation ([28]).

• It provides information for the complete distribution and not just for the first, second or higher order statistics to determine the examined data.

It should be noticed that the above properties of Relative Entropy is very well suited to the fuzzy logic formulation.

Relative entropy has been utilized in image processing for edge detection [10], image retrieval [6], image segmentation, data classification and pattern recognition [18], [32] and other fields. The use of relative entropy in producing the color (and multichannel) filter RELF is as follows:

Let a filter window (mask) W of length N that is applied to a color (multichannel) image. The main idea is to use the relative entropy measure E_{rel} to compare the local distribution $p(X)$ with the distribution associated to the previously presented local image models:

$$E_{rel}^F = \sum_X p(X) \log \frac{p(X)}{q_F(X)} \tag{33}$$

$$E_{rel}^S = \sum_X p(X) \log \frac{p(X)}{q_S(X)} \tag{34}$$

$$E_{rel}^{R} = \sum_{X} p(X) \log \frac{p(X)}{q_R(X)} \tag{35}$$

where E_{rel}^{F}, E_{rel}^{S}, E_{rel}^{R} denote the relative entropy values that result from the comparison of local distribution $p(X)$ with the model distributions of a constant (flat) region $q_F(X)$, (step) edge $q_S(X)$ and ramp $q_R(X)$ respectively.

The local distribution $p(X)$ is estimated using Parzen kernels and the model distributions are calculated as follows:

• For a constant region $q_F(X)$ is a (three-dimensional) normal distribution function:

$$q_F(X) = \frac{1}{(2\pi)^{0.5p} |\Sigma|^{1/2}} \exp\left\{-\frac{(X-\overline{X})^T \Sigma^{-1} (X-\overline{X})}{2}\right\} \tag{36}$$

or in a simplified form where the covariance matrix is represented by its diagonal elements,

$$q_F(X) = \frac{1}{(2\pi)^{1.5} \sigma} \exp\left\{-\frac{(X-\overline{X})^2}{2\sigma^2}\right\}. \tag{37}$$

• The ramp and edge case are similar in the sense that they are both represented by multimodal distributions with more than one peaks. In the general case this is

$$q(X) = \sum_{i=1}^{L} P(M_i)\, q(X/M_i), \tag{38}$$

$M_1, \ldots, M_L$ are the different classes (similarly to A, B or C in Figure 4) and $q(X/M_1), \ldots, q(X/M_L)$ stand for the conditional distributions represented by their mean and standard deviation values.

As a result, for the case of well defined edge $L=2$, while for a ramp $L>2$. The exact value of L depends for the ramp case on the window size N. Let us assume window of $N=9$ elements, then $L=3$, that is the ramp for the 3×3 may have three different values.

The question that arises is how the distribution parameters could be found in order to calculate the conditional distributions $q(X/M_i)$. The number of parameters depends on the number of classes. Lets assume a 3×3 window, where the relative entropy has to be calculated for $L=2$ and $L=3$, that is the ramp and edge case respectively. The parameters that have to be estimated are subsequently:

$$q_S(X) \to \begin{cases} \bar{X}_1^S, \sigma \\ \bar{X}_2^S, \sigma \end{cases} \tag{39}$$

$$q_R(X) \to \begin{cases} \bar{X}_1^R, \sigma \\ \bar{X}_2^R, \sigma \\ \bar{X}_3^R, \sigma \end{cases} \tag{40}$$

where $\bar{X}_1^S$ and $\bar{X}_2^S$ are the corresponding mean values of the classes that result from the edge local model, $\bar{X}_1^R$,$\bar{X}_2^R$ and $\bar{X}_3^R$ are the equivalent mean values for the ramp model and the standard deviation will remain the same for all the cases. The standard deviation may be regarded as the standard deviation of the assumed additive noise. The mean values have still to be calculated, though. This may be accomplished using the subtractive clustering method [35]. Therefore in order to calculate $\bar{X}_1^S$and$\bar{X}_2^S$ the following steps should be followed:

Step 1: The density values are estimated for each pixel of window W (N=9):

$$f(X_i) = \frac{1}{Nh^d} \sum_{j=1}^{N} K\left(\frac{X_j - X_i}{h}\right). \tag{41}$$

Step 2: The maximum density vector is selected:

$$X_{M1} = arg\{max(f(X))\}. \tag{42}$$

Step 3: The 5 closest to X_{M1} samples are selected.

Step 4: The mean value of the selected pixels $\bar{X}_1^S$ and the mean value of the rest pixels denoted by $\bar{X}_2^S$, are calculated.

Similar procedures are followed to calculate the $\bar{X}_1^R$,$\bar{X}_2^R$and $\bar{X}_3^R$.

The relative entropy may be subsequently calculated for each case as described previously. An additional factor that affects negatively the estimation of relative entropy is that the data included within the window boundaries do not sample the feature space uniformly and their number is small. This is overcome by adding new vectors X produced by combining the components of the existing vectors. This process is illustrated in Figure 8.

The new samples Y_k are produced using a combination of the maximum, mean and minimum coordinates of the original window samples. The resulting Y_k vectors for a 3×3 window are:

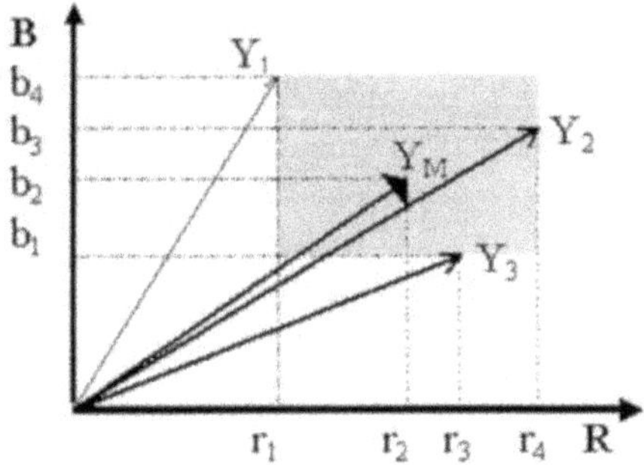

Fig. 8. Adaptive feature space sampling using the window data.

$$Y_1 = [\min(x_R), \min(x_G), \min(x_R)] \quad (43)$$

$$Y_2 = [\max(x_R), \max(x_G), \max(x_R)] \quad (44)$$

$$Y_3 = [\mathrm{mean}(x_R), \mathrm{mean}(x_G), \mathrm{mean}(x_R)]. \quad (45)$$

Although a pre-processing outlier removal step may be also used, it might negatively influence the calculation of relative entropy. The relative entropy is subsequently calculated for the three cases.

In the window width (N=9) we have assumed the constant region, edge and ramp cases might occur at three different situations. The following relative entropy values have to be calculated: $E_{rel}^{F}$for the constant region and E_{rel}^{E} for the edge. The ramp case may be inferred using the ELSE operator in the rule base.

Small values of E_{rel}^{F} correspond to a constant region. However, according to the calculation process discussion of the previous paragraph small values might occur for the other two cases as well. This attributed to the fact that for E_{rel}^{E} $q(X)$ will have two identical peaks, the mean value calculation $\bar{X}_1^S$and$\bar{X}_2^S$ will produce similar values. The same holds for the ramp case. If E_{rel}^{F} is large and $E_{rel}^{\mathbb{R}A}$ small the examined area corresponds to a well defined edge. The fuzzy rule-base subsequently is (Table 4) :

Based on this idea several rules may be added for an increasing window size. Two fuzzy sets are defined for each variable namely SMALL, LARGE.

Table 4. The rule-base for RELF.

IF	E^F_{rel} SMALL	**AND**	E^E_{rel} ,SMALL	**THEN**	FLAT region	9 samples
IF	E^F_{rel} LARGE	**AND**	E^E_{rel} ,SMALL	**THEN**	EDGE region	6 samples
ELSE					RAMP region	3 samples

Using a Sugeno type inference system [31] the weight of each rule may be determined by the relation:

$$\omega_m = min\left\{\mu_{mp}\right\}, \tag{46}$$

μ_{mp} is the membership degree to the m-th rule for the p-th fuzzy set and the final inference is calculated as:

$$Y = \frac{\sum\limits_{m=1}^{M} \omega_m Y_m}{\sum\limits_{m=1}^{M} \omega_m}. \tag{47}$$

M is the number of rules (M=3) and Y_m is the output of the employed filter type according to the deterministic model (see Table 4).

This method was implemented for two cases: in the first one the impulsive removal process was adopted to calculate the relative entropy (RELF1), whereas for the second it was not used (RELF2). These approaches were compared to other well known noise removal methods for color images such as the vector median (VMF), the mean (AF) and the fuzzy vector directional (FVDF) filter. Results are given in Tables 5 and 6 and Figure 9.

The method described in this paragraph makes use of the relative entropy measure to determine the local image model using distribution information. In contrast to methods in the previous paragraphs (3.1 and 3.2) where some local density parameters were adopted, in this method the complete distribution information is taken into account producing efficient results.

4 Experimental results

This paragraph contains qualitative and quantitative comparative results of the three presented fuzzy filters using two color test images (PEPPERS and LENNA) and different noise types. In addition, these methods are compared to well known color image noise removal techniques, such as the Vector Median Filter (VMF, [2]), the simple mean value filter (AF-average filter) and the Fuzzy Vector Directional Filter (FVDF, [24]). The employed color test images have size 256×256 pixels with 24-bit color depth.

The quantitative results are carried out by means of the Mean Square Error measure (MSE), calculated as:

$$MSE = \frac{1}{N \times M} \sum_{j=1}^{M} \sum_{i=1}^{N} \|Y(i,j) - X(i,j)\|^2. \tag{48}$$

X(i,j) are the samples-vectors of the original image (without any noise corruption), *Y(i, j)* are the samples-vectors of the outcome, M and N is the number of columns and rows of the image respectively.

The parameters of the fuzzy sets were tuned by means of the genetic algorithm optimization approach ([25]), using as training data a part of PEPPERS test image

Tables 5 and 6 contain the MSE values for several types of noise i.e. Gaussian, impulsive and mixed noise (Gaussian and impulse) of various intensities and the most effective results are indicated using bold letters. Furthermore, σ stands for the standard deviation value of Gaussian noise (mean value is always 0) and p denotes the probability of impulse noise.

Table 5. MSE values for the test image PEPPERS.

	σ=15-p=1%	σ=25-p=2%	σ=15	p=1%	σ=40	p=4%
AF	143.0	226.4	121.1	118.6	271.2	199.0
VMF	170.4	315.3	154.6	92.8	529.5	124.4
FVDF	109.4	186.5	104.4	66.8	324.5	78.6
MFF	105.6	164.3	100.7	66.0	266.8	80.6
HMF	98.4	**155.0**	91.7	**60.5**	**261.5**	72.5
RELF1	**95.0**	155.7	92.3	63.7	287.8	**71.5**
RELF2	98.8	168.0	**91.6**	69.8	263.9	88.8

Table 6. MSE values for the test image LENNA.

	σ=15-p=1%	σ=25-p=2%	σ=15	p=1%	σ=40	p=4%
AF	142.1	207.9	119.3	116.2	265.7	190.6
VMF	183.0	320.0	166.3	104.4	539.3	131.4
FVDF	116.2	188.2	112.3	80.3	322.3	90.5
MFF	114.3	168.9	108.0	79.0	262.8	91.9
HMF	110.0	163.6	107.7	**78.4**	262.7	88.8
RELF1	**108.6**	**163.5**	105.6	79.3	282.3	**87.1**
RELF2	112.2	179.1	**104.5**	87.4	**260.7**	105.2

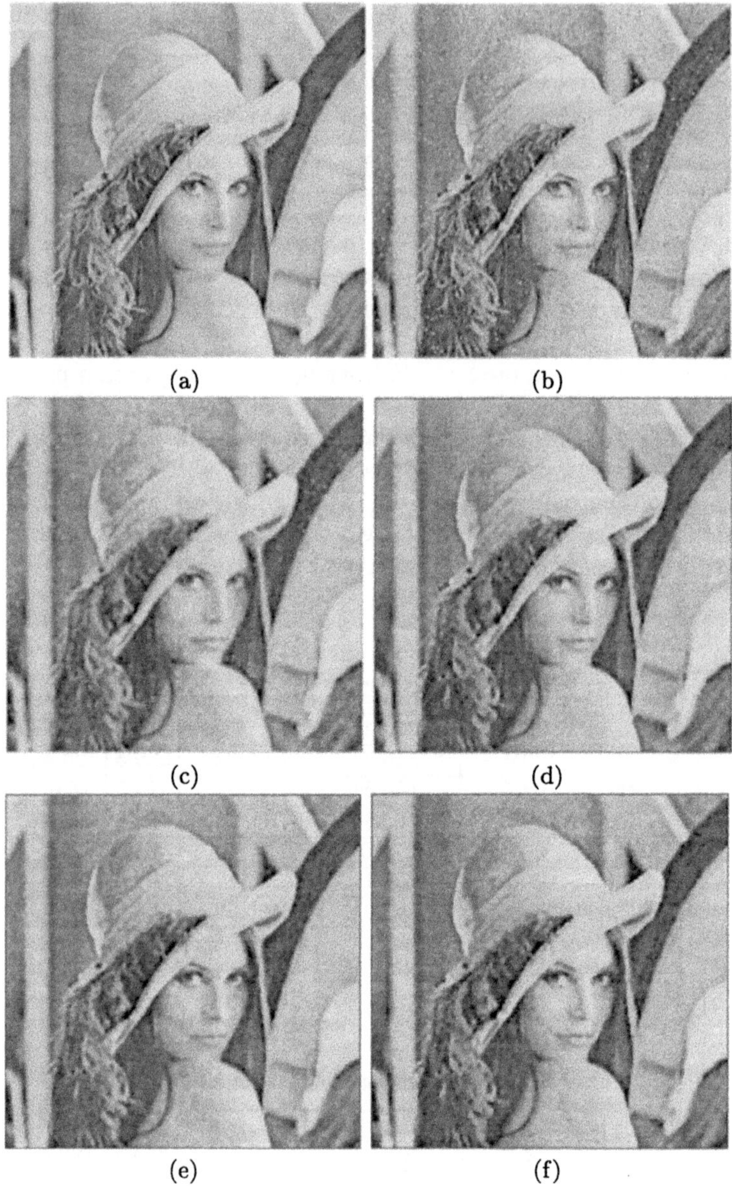

Fig. 9. (a) LENNA test image, 256x256 pixel, 24 bits/pixel, (b) corrupted with noise (N,0,225) and impulsive noise with probability 1% for each channel (c) AM-average filter (d)MFF, (e) HMF and (f)RELF outputs.

From these results it becomes obvious that the three presented filters produce more efficient results than the AM, VMF and FVDF. The RELF1 in particular, is the most effective and this can be attributed to the fact that the Relative Entropy measure is a more descriptive statistical measure than the features employed in the other two approaches The RELF2 version produces good results for Gaussian noise type, while its results are degraded a little by the presence of impulsive noise.

The MFF is less efficient than HMF and RELF1 due to the employed features inaccuracy . The results for the HMF which seems comparable to RELF1, depend on the tuning process of the fuzzy sets. Therefore it produces lower MSE values for several noise types for image PEPPERS - that was used in the training process - whereas for LENNA are a little worse.

Finally, Figure 9 presents the results of AF, MFF, HMF and RELF1 for LENNA, which was corrupted by normal distributed noise N(0,225) and impulsive noise of 0.01 probability on each channel. It is concluded that the three fuzzy filters presented preserve the details of the image and filter out several types of noise.

References

1. G. R. Arce and R. E. Foster, *Multi-level median filters: Properties and efficacy*, in Proc. ICASSP, New York, 1988, pp. 824-827.
2. J. Astola, P. Haavisto and Y. Neuvo, *Vector Median Filters*, Proceedings of the IEEE, vol. 78, No. 4, April 1990, pp. 678-689.
3. Y. Carron and P. Lambert, *Integration of linguistic knowledge for color image segmentation*, Proc. of EUSIPCO-96, Sept. 10-13 Trieste, Italy, 1996, pp. 1729-1732.
4. Chang-Shing Lee, Yau-Hwang Kuo and Pao-Ta Yu, *Weighted fuzzy mean filters for image processing*, Fuzzy Sets and Systems, vol. 89, no. 2, 1997, pp. 157-180.
5. Y. Choi and R. Krishnapuram, *A Robust Approach to Image Enhancement Based on Fuzzy Logic*, IEEE Trans. Image Proc., vol. 6, no. 6, June 1997, pp. 808-825.
6. M. N. Do and M. Vetterli, *Texture similarity measurement using Kullback-Leibler distance on wavelet subbands*, Proc. of IEEE International Conference on Image Processing (ICIP), Vancouver, Canada, September 2000, pp. 730-733.
7. A. Fotinos, N. Laskaris and S. Fotopoulos, *Fuzzy Color Filtering using Potential Functions*, Proc. Int. Conference on Digital Signal Processing (DSP 97), Santorini, vol. 2, 1997, pp. 987-990.
8. A. Fotinos, N. Laskaris and S. Fotopoulos, *Fuzzy Color Filtering using Relative Entropy*, Proc. European Signal Processing Conference (EUSIPCO 98), Rodos, vol. 2, 1998, pp. 813-816.
9. A. Fotinos, N. Laskaris, G. Economou and S. Fotopoulos, *Multidimensional Fuzzy Filtering using Statistical Parameters*, Multidimensional Systems and Signal Processing, vol. 10, 1999, pp. 415-424.
10. A. Fotinos, G. Economou, and S. Fotopoulos, *Use of relative entropy in colour edge detection*, Electronics Letters, vol. 35, no. 184, 1999, pp. 1532-153.
11. A. Fotinos, E. Zigouris, G. Economou and S. Fotopoulos, *A histogram base multichannel filter*, Proc. EUSIPCO 2000 Conference, September 13-15 2000, Tampere, Filand, 2000, pp. 1111-1114.
12. J.-N. Hwang, S.-R. Lay, and A. Lippman, *Nonparametric multivariate density estimation: a comparative study*, IEEE Trans. on Signal Processing, vol. 42, no. 10, 1994, pp. 2795-2810.
13. M.C. Jones and R.Sibson, *What is Projection Pursuit?*, Journal of Royal Statistical Society, Ser. A, vol 150, 1987, pp. 1-36.
14. J. M. Keller, *Fuzzy set theory in computer vision : a prospectus*, Fuzzy Sets and Systems, vol. 90, no. 2, 1997, pp. 177-182.
15. E. Kerre, M. Nachtegael, *Fuzzy Techniques in Image Processing*, Springer-Verlag Company, 2000.
16. S. Kullback, *Information Theory and Statistics*, Dover, New York, 1968.
17. S. Makrogiannis, G. Economou and S. Fotopoulos, *A Fuzzy Dissimilarity Function for Region Based Segmentation of Color Images*, Int. Journal of Pattern Recognition and Artificial Intelligence, vol. 15, no. 2, March 2001, pp. 255-267.
18. M. A. G. Mattoso and M. C. Fairhurst, *On the use of I-divergence for generating distribution approximations*, IEEE Trans. on Pattern Analysis and Machine Intelligence, vol. 5, no. 6, 1983, pp. 661-664.
19. M. Meguro, A. Taguchi, *Adaptive Weighted Filters by Using Fuzzy Techniques*, in Proc. ISCAS 96, pp. 9-12, Atlanta, USA, May 1996.

20. N. R. Pal and S. K. Pal, *A review on image segmentation techniques*, Pattern Recognition, vol. 26, no. 9, 1993, pp. 1277-1294.
21. E. Parzen, *On Estimation of a Probability Density Function and mode*, Ann. Math. Stat., vol. 33, 1962, pp. 1065-1076.
22. W. Pedrycz, *Fuzzy sets in Pattern recognition: methodology and methods*, Pattern Recognition, vol. 23, no. 1/2, 1990, pp. 121-156.
23. I. Pitas, A. N. Venetsanopoulos, *Nonlinear Digital Filters: Principles and Applications*, Kluwer Academic Publishers, Boston, 1990.
24. K. N. Plataniotis, D. Androutsos, A. N. Venetsanopoulos, *Adaptive Fuzzy Systems for Multichannel Signal Processing*, Proceedings of the IEEE, vol. 87, no. 9, September 1999, pp. 1601-1622.
25. T. Ross, *Fuzzy logic with engineering applications*, Mc-Graw Hill, New York, 1995.
26. F. Russo and G. Ramponi, *Nonlinear fuzzy operators for image processing*, Signal Processing, vol. 38, no. 3, 1994, pp. 429-440.
27. F. Russo, *Nonlinear Fuzzy Filters: An overview*, Proc. of EUSIPCO 96, Trieste 1996, pp. 1709-1712.
28. D. W. Scott, *Multivariate Density Estimation: Theory, Practice and Visualization*, John Wiley 1992.
29. D. Sindoukas, N. Laskaris and S. Fotopoulos, *Algorithms for color image enhancement using potential functions*, IEEE Signal Processing Letters, vol.4, no. 9, September 1997, pp. 269-272.
30. A. Taguchi, M. Meguro, *Adaptive L-filters Based on Fuzzy Rules*, in Proc. ISCAS-95, Seattle, Washington, USA, April 1995, pp. 961-964.
31. T. Terano, K. Asai and M. Sugeno, *Fuzzy systems theory and its applications*, Academic Press, 1995.
32. J. Tou and R. Conzalez, *Pattern recognition principles*, Addison-Wisley editions 1974.
33. P. E. Trahanias and A. N. Venetsanopoulos, *Vector Directional filters for color image processing*, ECCTD '93-Circuit Theory and Design, 1993, pp. 1645-1650.
34. H.-H. Tsai and P-T Yu, *Adaptive Fuzzy hybrid multichannel filters for removal of impulsive noise from color images*, Signal Processing, vol. 74, no. 2, 1999, pp. 127-151.
35. R. R. Yager and D. P. Filev, *Generation of fuzzy rules by mountain clustering*, Journal of Intelligent and Fuzzy Systems, vol. 2, no. 3, pp. 209-219, 1994.
36. X. Yang and P. S. Toh, *Adaptive Fuzzy Multilevel Median Filter*, IEEE Trans. on Image Processing, vol. 4, no. 52, 1995, pp. 680-68.
37. L. A. Zadeh, *Fuzzy Sets*, Information and Control vol. 8, 1965, pp. 338-353.

Part II

Fuzzy Filters for Edge Detection and Segmentation

Chapter 5

Fuzzy Based Image Segmentation

Otman Basir, Hongwei Zhu and Fakhri Karray

University of Waterloo
Department of Systems Design Engineering
Waterloo, ON, Canada, N2L 3G1
email: obasir@uoguelph.ca, h4zhu@engmail.uwaterloo.ca, karray@uwaterloo.ca

Summary. In this chapter, we introduce some recently developed fuzzy based techniques for image segmentation. They are fuzzy thresholding, fuzzy rule-based inferencing scheme, fuzzy c-mean clustering, and fuzzy integral-based decision making. A fuzzy integral based region merging algorithm for image segmentation, which combines both region and edge features of the image, is then used to merge regions recursively according to the criterion of the maximum fuzzy integral. The region merging process is regarded as a nonlinear process that fuses objective evidence, in the form of a fuzzy membership function that reflects the similarities between adjacent regions with respect to each feature, with the apriori system's expectation of the importance of that evidence provided by the corresponding feature. Using the maximum fuzzy integral criterion, the target number of regions can be reached by recursively merging regions. To handle the parameter initiation problem faced in the computation of fuzzy integral, an algorithm for automatically choosing such parameters is formulated as an optimization problem. A simulated annealing algorithm is designed to explore the value space of fuzzy densities and search the (near) optimal solution corresponding to a minimum cost value. This way, fuzzy densities are determined adaptively depending on the image at hand without requiring any human intervention. To evaluate the performance of the proposed approach, it is applied to magnetic resonance images (MRI) and natural images. The experimental results have demonstrated that the proposed approach brings out robust segmentation performance and outperforms other approaches efficiently.

1 Introduction

Image segmentation is the partitioning of an image into sections or regions. These partitions can be later associated with a set of objects or labels. Image segmentation, for many image processing applications, such as image description and recognition [1], medical imaging [2], image compression [3,4] and image retrieval [5], to name a few, is a fundamental task, and its quality reflects heavily on the overall system's performance. There is a number of factors that make image segmentation a difficult task, including: (1) lack of a priori knowledge about the number of segments in the image; (2) inherent limitations in the imaging sensors which often result in noisy and imperfect images; (3) the sensitivity of the image data when used to solve a given task is very high (like the case with SPECT imaging, where the high frequency information is often distorted resulting in fuzzy and unreliable edges); (4) the grey scale values and their distributions vary from image to another even in a single application such as the segmentation of structures from magnetic resonance images. In summary, vagueness in image processing is mainly due to grayness ambiguity, geometrical fuzziness and uncertain knowledge, and as such it is quite natural to explore the potential of fuzzy techniques to tackle the above problems [6].

Fuzzy approaches for image segmentation can be categorized into four classes [9]: segmentation via thresholding, segmentation via clustering, supervised segmentation and rule-based segmentation. Tizhoosh [6] categorizes such techniques into fuzzy clustering, fuzzy rule, fuzzy geometry, fuzzy thresholding, and fuzzy integral based segmentation techniques. A large amount of research work has been carried out in this area. In [7] a membership function (standard S function) of the image intensity value was designed to minimize a global fuzzy measure, and based on the optimal membership function, the object could be segmented from the background. A fuzzy thresholding scheme was proposed in [10]. This scheme minimizes the fuzziness in the thresholded description and allows for accommodating the variations in the gray values within each region. Examples of popular fuzzy clustering approaches are the fuzzy c-mean (FCM) [11] and the fuzzy region merging [12–14]. For example, a fuzzy region-growing algorithm was used to segment natural color images in [14]. In [15], the thresholds associated with membership functions were computed by means of fuzzy c-mean clustering (FCM) to perform fuzzy segmentation. Fuzzy rule-based approaches have been demonstrated to be successful in taking advantage of heuristic knowledge that is relevant to image segmentation [2,16] in different applications. In [2], an automated segmentation of human brain MR images was realized. Here, fuzzy if-then rules were used to express human knowledge and decompose the brain into the left cerebral hemispheres, the cerebellum and the brain stem. Another important fuzzy image segmentation technique is based on decision making by using fuzzy integrals [17]. This technique has demonstrated success in a wide range of applications [18,19]. Segmentation of gray level and color images using Sugeno integral was proposed in [20]. In [18,19] the fuzzy integral was proposed as a tool of information fusion for image segmentation.

Features provide some degree of fuzzy evidence that can assist in the region merging process. Nevertheless, this evidence tends to be fuzzy and therefore a simple and direct feature based analysis may not expose the real regional decomposition of the image. To address this problem, we propose a fuzzy integral based approach for merging watershed image regions [21]. The maximum fuzzy integral principle is then used as the merging criterion. We view the region merging as a fusion process, which can be implemented by the λ rule [17]. The fuzzy integral is a nonlinear function that is defined with respect to a fuzzy measure, which in turn is either a belief or a plausibility measure in the sense of evidence theory [22]. It also combines objective evidence for a hypothesis with the system's prior knowledge on the importance of the different features. In order to ensure high quality fuzzy integral based decision making this knowledge has to be assimilated effectively, often, in the form of fuzzy densities. In [23], a neural network was used to determine fuzzy densities needed in the computation of fuzzy integral. In [8] a genetic algorithm was proposed to obtain the fuzzy densities by introducing a proper fitness function. In this work we propose an alternative strategy to determine the fuzzy densities. We make use of the property that the fuzzy integral can fuse information to reach a unified decision. Image segmentation is first performed based on individual features and evaluated according to a cost function. Subsequently, a set of initial fuzzy densities are estimated from the relevant cost values, and tuned using the simulated annealing optimization technique.

The chapter is organized as follows. Section 2 introduces general fuzzy-based image segmentation approaches such as fuzzy thresholding, fuzzy rule-based inferencing scheme, fuzzy c-mean algorithm and fuzzy integral based decision making approach. A complete fuzzy integral based image segmentation algorithm is discussed in Sect. 3. Issues discussed in this section include: noise smoothing, edge preserving, the watershed transform for initial segmentation, feature selection, feature computation, fuzzy density determination using simulated annealing and the region merging implementation. Experimental results that demonstrate the performance of the proposed method are given in Sect. 4. Section 5 provides some concluding remarks.

2 Fuzzy approaches for image segmentation

2.1 Fuzzy thresholding for image segmentation

Thresholding is the simplest way to segment an image to its basic constituents. It has been used extensively in many image processing applications. Thresholding is based on the notion that regions corresponding to different objects can be distinguished by using a range function applied to the intensity values of image pixels. The fuzzy version has been studied in the literature based on the concept of fuzzy set theory [10,24,25].

A fuzzy subset of a set S is a mapping μ from S into $[0,1]$. For any p (a pixel, for example) belonging to S, $\mu(p)$ is called the degree of membership of p in S. An image X of $M \times N$ dimensions and L levels can be considered as an array of fuzzy singletons, each having a value of membership denoting its degree of brightness relative to some brightness level l, $l = 0, 1, \cdots L-1$. Image X can be expressed as

$$X = \{\mu_x(x_{mn}) = \mu_{mn}/x_{mn}; m = 1, \cdots M, n = 1, \cdots N\} ,$$

or $$X = \bigcup_m \bigcup_n \mu_{mn}/x_{mn}; m = 1, \cdots M, n = 1, \cdots N ,$$

where $\mu_x(x_{mn})$ or $\mu_{mn}/x_{mn}(0 \leq \mu_{mn} \leq 1)$ denotes the grade of possessing some property (e.g., brightness, edginess, smoothness, etc.) by the pixel located in (m, n).

Either global or local information of an image can be used to define a membership function characterizing properties such as pixel intensities, edge magnitudes and region geometrics. Membership functions can take various forms depending on the application. Examples of this function are: the standard S function, the π function, the trapezoidal function, the triangular function, and the Gaussian function.

Based on the membership function used, a number of fuzziness measures for image processing can be defined [7]. Indices of fuzziness, fuzzy entropy, compactness of fuzzy subset, and index of coverage of fuzzy subset, are all good candidates that have been used as objective criteria for determination of thresholds for image segmentation [7,26].

A simple and typical image segmentation using fuzzy thresholding technique is given as follows [27]:

- Step 1. Construct the standard S membership function $\mu(m, n)$ as

$$\mu_x(x_{mn}) = S(x_{mn}; a, b, c) = \begin{cases} 0 & \text{if } x_{mn} \leq a \ , \\ 2(x_{mn} - a)^2/(c-a)^2 & \text{if } a \leq x_{mn} \leq b \ , \\ 1 - 2(x_{mn} - c)^2/(c-a)^2 & \text{if } b \leq x_{mn} \leq c \ , \\ 1 & \text{if } x_{mn} \geq c \ . \end{cases}$$

  where $b = (a + c)/2$ is the cross-over point with $S(b; a, b, c) = 0.5$ and the window size is $w = c - a$. A standard S function is shown in Fig. 1.
- Step 2. Compute the area, length, breadth and the index of area coverage of a fuzzy set (IOAC) using the following definitions [7]:
  area: $a(\mu) = \sum_m \sum_n \mu_x(x_{mn})$;
  length: $l(\mu) = \max_m(\sum_n \mu_x(x_{mn}))$;
  breadth: $b(\mu) = \max_n(\sum_m \mu_x(x_{mn}))$;
  IOAC: $\text{IOAC}(\mu) = \frac{a(\mu)}{l(\mu) \times b(\mu)}$.
- Step 3. Vary b between 0 and L-1 and select planes for which $\text{IOAC}(\mu)$ has local minima. Among the local minima, let the global one have a cross-over point x, $\mu(x) = 0.5$ and $b = x$. The $\mu_x(x_{mn})$ plane, corresponding to the cross-over point x, can be viewed as a fuzzy segmented image of the image X.

Using the local fuzzy measure, the above approach can be extended to multiclass segmentation. The above algorithm was modified in [26], where the S function was replaced by a triangular function to locate edge pixels. A fuzzy clustering algorithm was used to extract the line segment edges to attain segmentation.

The basic idea behind fuzzy thresholding is to first transfer the selected image feature into a fuzzy subset by means of a proper membership function, and then to select and optimize a global or local fuzzy measure to attain the goal of image segmentation.

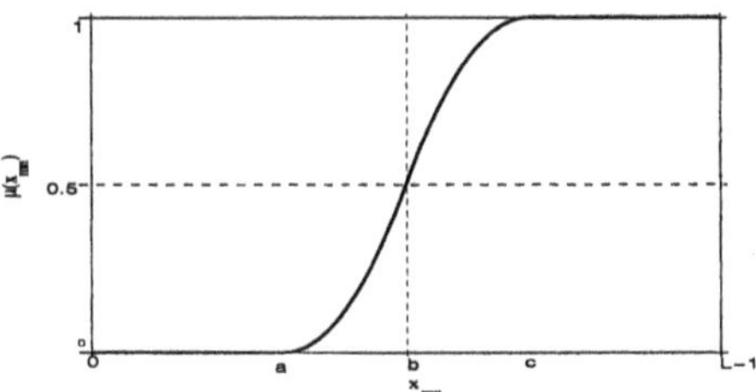

Fig. 1. Standard S function.

2.2 Fuzzy rule-based image segmentation

Another class of approaches for fuzzy logic based image segmentation is known as the fuzzy rule-based approach, where heuristic knowledge is modelled in the form of fuzzy rules. Fuzzy inference schemes make it possible to combine human heuristic knowledge such as heuristic rules in the form of linguistic terms [2,16]. To illustrate this approach let us consider how it deals with region merging. This is done under the assumption that initial regions have been obtained by some preprocessing (such as the watershed transform which will be introduced in Sect. 3.3). Based on the Mamdani method [28] the approach is described using the following five steps:

- Step 1. Determine the image features, each corresponding to a term set (linguistic labels). Each feature is then fuzzified by proper choice of a membership function. For instance, two fuzzy variables may be defined as follows between two adjacent regions R_i and R_j:
  region size: $\text{Size} = \frac{\text{Size}_i \times \text{Size}_j}{\text{Size}_i + \text{Size}_j}$,
  difference of mean intensities: $\text{Difu} = |\mu_i - \mu_j|$.
  where μ_i and μ_j denote the mean intensity of region R_i and region R_j, and Size_i and Size_j denote the size of region R_i and region R_j respectively. We can assign the corresponding fuzzy sets for these two fuzzy variables. For example, we may choose to have each variable take values in the three fuzzy sets "small", "medium" and "large" (in real-life situations more sophisticated considerations may be required). The corresponding membership function can be represented as in Fig. 2(a).
  Furthermore, we define a fuzzy variable, Sim, in the range [0,1], quantifying the similarity between adjacent regions: R_i and R_j under consideration. The larger Sim is, the similar is R_i to R_j are. Consequently, we

define some fuzzy sets associated with Sim, such as "vsim"(very similar), "psim"(possibly similar), "pnsim"(possibly not similar) and "nsim"(not similar), each of which have a membership function taking the form similar to the one depicted in Fig. 2(a).

- Step 2. We set up a knowledge-base that is composed of fuzzy if-then rules. For example, we may write two rules as:
  Rule 1: If Size "small" and Difu "small" then Sim "vsim",
  Rule 2: If Size "small" and Difu "medium" then Sim "psim".
- Step 3. For each rule, combine the fuzzified inputs and get the rule strength, say α, and combine the rule strength with the output membership function to find the consequence of the rule.
  Suppose we have the region size size0 and the difference of mean intensity difu0 for two adjacent regions. For Rule 1 its strength and consequence are computed by:
  strength: $\alpha_1 = \mu_{small}(\text{size0}) \wedge \mu_{small}(\text{difu0})$,
  consequence: $\mu'_{vsim}(\text{Sim}) = \alpha_1 \wedge \mu_{vsim}(\text{Sim})$.
  where $\wedge$ refers to the intersection operator (min), Sim ranges over the value that the rule conclusions can take. Consequently $\mu'_{vsim}(\text{Sim})$ is a membership derived by Rule 1 to express the degree of similarity for the two considered regions.
  Similarly for Rule 2, we have:
  strength: $\alpha_2 = \mu_{small}(\text{size0}) \wedge \mu_{med}(\text{difu0})$,
  consequence: $\mu'_{psim}(\text{Sim}) = \alpha_2 \wedge \mu_{psim}(\text{Sim})$.
- Step 4. Combine the consequences from variant rules and get an output distribution by the conflict-resolution process. For the above Rule 1 and Rule 2, the combined output is
  $$\mu_{Sim}(\text{Sim}) = \mu'_{vsim}(\text{Sim}) \vee \mu'_{psim}(\text{Sim}) ,$$
  where $\vee$ is the union operator (max) and $\mu_{Sim}(\text{Sim})$ is a pointwise membership function from the combined conclusion of Rule 1 and Rule 2.
- Step 5. Defuzzify the output distribution and get a crisp similarity measure for the current region pair. There is more than one approach for defuzzifing the output distribution. One of the popular approaches is the Center of Area (COA) approach. The COA approach calculates the center Sim* of gravity of the output distribution as:
  $$\text{Sim}^* = \frac{\sum_{j=1}^{q} \text{Sim}_j \times \mu_{Sim}(\text{Sim}_j)}{\sum_{j=1}^{q} \mu_{Sim}(\text{Sim}_j)} .$$
  where q is the number of quantization levels of the output. Sim_j is the output at the quantization level j, and $\mu_{Sim}(\text{Sim}_j)$ represents its membership value in Sim.
  The above fuzzy inference scheme is illustrated in Fig. 2(b).

Once we get all the pair-wise similarities of all possible regions, we can merge each pair with the largest Sim* value recursively, until we reach the expected number of regions in the final segmentation.

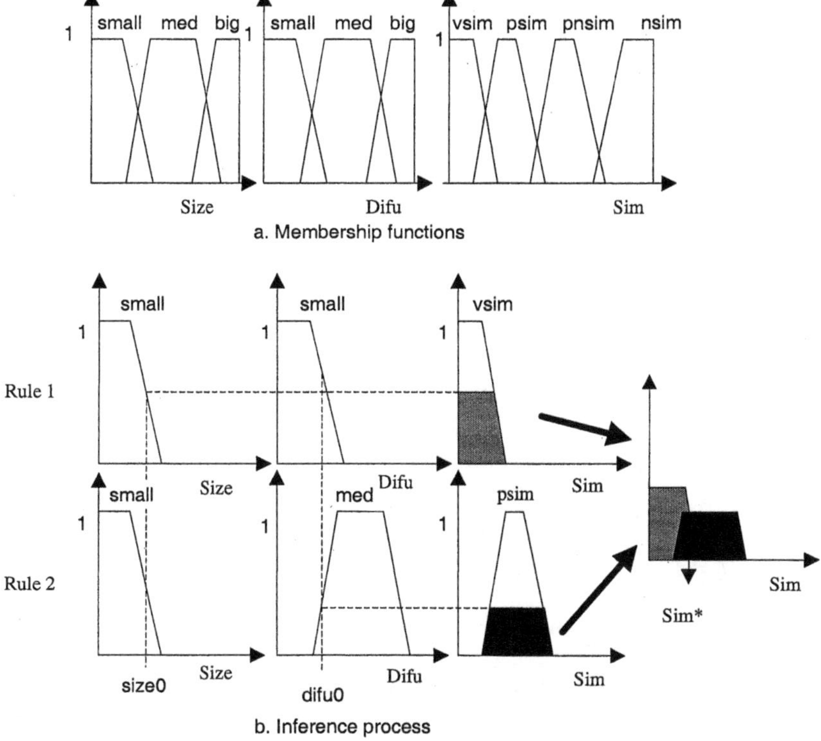

Fig. 2. Fuzzy rule-based inference

2.3 Fuzzy c-mean clustering

Fuzzy c-mean clustering (FCM) [11], also known as fuzzy ISODATA, is a popular data clustering algorithm in which each data point belongs to a cluster, to a degree specified by a membership grade.

FCM partitions a collection of n vectors, x_j, $j = 1, \cdots, n$, into c fuzzy groups, and finds a cluster center in each group, such that a cost function based on a distance measure is minimized. The fuzzy clustering takes into account the overlapping of the clusters and allows partial belongingness of the object to all clusters. In other words, a given data point can belong to any group with a degree of membership between 0 and 1. The results of a fuzzy clustering is a fuzzy c-partition represented by the following $(c \times n)$ matrix U:

$$M_{fc} = \left\{ U \in \Re^{c \times n} | \mu_{ik} \in [0,1], \forall i, k; \sum_{i=1}^{c} \mu_{ik} = 1, \forall k; n > \sum_{k=1}^{n} \mu_{ik} > 0, \forall i \right\} .$$

Column j of the $c \times n$ matrix U represents membership values of x_j in the c fuzzy subsets of X. Row i of U exhibits values of a membership function μ_i

on X where $\mu_{ik} = \mu_i(x_k)$ denotes the grade of membership of x_k in the ith fuzzy subset of X.

Let $J_m : M_{fc} \times \Re^{cp} \to \Re^+$ be $J_m(U,V) = \sum_{i=1}^{c} \sum_{k=1}^{n} \mu_{ik}^m d_{ik}^2$,

where,

$U \in M_{fc}$ is a fuzzy c-partitin of X;

$v_i \in \Re^p$ is the cluster center or prototype of μ_i, $i = 1, \cdots c$;

$V = [v_1, v_2, \cdots, v_c]$ with $v_i \in \Re^p$;

d_{ik}^2 is a measure of the distance between the x_k point and the v_i element, which can be considered as a central vector of the cluster i;

$m \in [1, \infty)$ is the weighting exponent that controls the fuzziness of the final partition.

$J_m(U,V)$ is a squared error clustering criterion. Let's define the sets $I_k = \{i | 1 \le i \le c, d_{ik} = \|x_k - v_i\| = 0\}$, and $\widetilde{I}_k = \{1, \cdots c\} - I_k$, then (U,V) may be globally minimal [11] in the following cases:

if $I_k = \emptyset \Rightarrow \mu_{ik} = 1 / \sum_{j=1}^{c} \left(\frac{d_{ik}}{d_{jk}} \right)^{2/(m-1)}$,

or if $I_k \neq \emptyset \Rightarrow \mu_{ik} = 0, \forall i \in \widetilde{I}_k$, and $\sum_{i \in I_k} \mu_{ik} = 1$,

where $v_i = \sum_{k=1}^{n} (\mu_{ik})^m x_k / \sum_{k=1}^{n} (\mu_{ik})^m, \forall\, i$.

To obtain the optimal solution in FCM, an iterative algorithm has been devised in [11].

The FCM is an algorithm implementing the fuzzy clustering paradigm. It updates the degrees of membership in a fashion that minimizes fuzziness of within-cluster variance. FCM provides the most popular method to assign multiclass membership values to pixels in image processing. Once the memberships of data points, such as pixels for each class (cluster), are available, we can assign pixels to the class with the highest membership.

In terms of generating membership functions for later processing, the fuzzy c-mean algorithm is unsupervised, as it requires no initial set of training data. When the conventional FCM is used for image processing, it does not use any spatial information. Additional considerations associated with spatial information are required in real applications. For example, rule-based approaches as introduced above can be applied to take into consideration the memberships of neighboring pixels according to image local context so that the spatial relationship between pixels can be considered for the segmentation.

2.4 Introduction to the fuzzy integral

Fuzzy integral is an important numerical-based approach which is useful for both image segmentation and object recognition. A particularly useful fuzzy measure is the Sugeno integral [29]. The fuzzy integral may be interpreted as an evaluation of object classes where the subjectivity is embedded in the fuzzy measure.

The notion of fuzzy integral has been used in multi-information fusion, and multi-information decision making, where the fuzzy integral has been used as either a belief or a plausibility measure in the sense of Dampster-Shafer evidence theory [22]. It overcomes some shortcomings of those traditional decision making criteria, for it differs from the Dampster-Shafer theory of evidence [22] and the probabilistic Bayesian reasoning [30], in that it considers both evidence supplied by each information source and the expected worth of each subset of sources in its decision making process. The fuzzy integral process is a nonlinear combination of the objective evidence, in the form of a fuzzy membership function. The fuzzy integral has proven to be a valuable tool in overcoming inherent ambiguities present in many decision making problems [23,31].

In this section we describe fuzzy measures and fuzzy integrals related to the segmentation task. Let $X = \{x_1, x_{2,} \cdots x_n\}$ be the set of information sources, a fuzzy measure g is a real-valued function defined on the power set of X with range [0,1], satisfying the following properties [17,29]:

1. $g(\emptyset) = 0$ and $g(X) = 1$;
2. $g(A) \leq g(B)$ if $A \subseteq B$;
3. If $\{A_i\}$is an increasing sequence of subsets of X, then $\lim\limits_{i \to \infty} (A_i) = g(\bigcup\limits_{i=1}^{\infty} A_i)$.

A fuzzy measure g is called a Sugeno measure (g_λ-fuzzy measure) if it additionally satisfies the following property [17,29]:

For all $A, B \subseteq X$ with $A \cap B = \emptyset$,

$$g_\lambda(A \cup B) = g_\lambda(A) + g_\lambda(B) + \lambda g_\lambda(A) g_\lambda(B), \qquad \text{for } \lambda > -1 \ . \tag{1}$$

The subscript λ will be omitted unless it is needed for clarity. Consider the above set of information source X, and let $g^i = g(\{x_i\})$. The mapping $x_i \to g^i$ will be referred to as a fuzzy density function. The fuzzy density value g^i is interpreted as the importance of the single information source x_i in determining the evaluation of a class hypothesis. A set of fuzzy density values can be constructed for the information sources in the set X.

The value of λ for any Sugeno fuzzy measure can be uniquely determined [17,29] for a finite set X using Eq. (1) and the fact that $X = \bigcup\limits_{i=1}^{n} \{x_i\}$ and $g_\lambda(X) = 1$, which leads to solving the following equation for λ:

$$\lambda + 1 = \prod_{i=1}^{n} (1 + \lambda g^i), \lambda \in (-1, +\infty) \ . \tag{2}$$

Let (X, Ω) be a measurable space and let $h: X \rightarrow [0, 1]$ be a Ω-measurable function. The fuzzy integral over $A \subseteq X$ of the function h with respect to a fuzzy measure g is defined as

$$\begin{aligned} \textstyle\int_A h(x) \circ g(\cdot) &= \sup_{E \subseteq X} \left[\min(\min_{x \in E} h(x), g(A \cap E)) \right] \\ &= \sup_{\alpha \in [0,1]} [\min(\alpha, g(A \cap F_\alpha))] \ , \end{aligned}$$

where $F_\alpha = \{x | h(x) \geq \alpha\}$.

For the finite case, suppose $h(x_1) \geq h(x_2) \geq \cdots \geq h(x_n)$ (if this is not the case, order the set of information sources X so that this relation holds). Then the fuzzy integral, e, with respect to a fuzzy measure g over x_i can be calculated by

$$e = \max_{i=1,\ldots,n} [\min(h(x_i), g(A_i))] \ , \tag{3}$$

where $A_i = \{x_1, x_2, \cdots x_i\}$.

When g is a fuzzy measure, the value of $g(A_i)$ can be determined recursively as

$$g(A_1) = g(x_1) = g^1, \tag{4}$$

$$g(A_i) = g^i + g(A_{i-1}) + \lambda g^i g(A_{i-1}) \ , \tag{5}$$

where $1 < i \leq n$.

When using the definition of the fuzzy integral for image segmentation, such as via region merging as will be introduced in the next section, we may then set a set of sources X as some image features, such as the region size, edge strength, etc. We can then use the fuzzy integral to guide region merging.

3 A robust fuzzy integral based image segmentation

3.1 Overview of the segmentation algorithm

The proposed texture segmentation scheme is shown in Fig. 3 and is specified next, using three phases: Phase 1- Preprocessing with the noise removal and edge preservation by fuzzy rule-based filtering; Phase 2- Initial segmentation by the watershed transform, and Phase 3- Maximum fuzzy integral based region merging combined with the optimization of fuzzy densities via the simulated annealing algorithm.

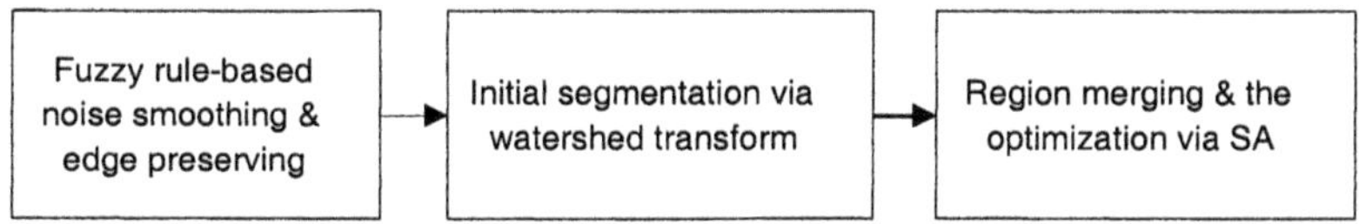

Fig. 3. Overview of the proposed algorithm.

3.2 Preprocessing with noise removal and edge preservation

The first common step in performing segmentation is to smooth the image so as to reduce its noise content, without degrading its edge content. Among a large number of smoothing techniques, some popular global smoothing techniques, such as Gaussian and Gaussian-like filters, while smoothing out the noise, they also remove genuine high frequency edge features and degrade localization. In some situations it is possible to resort to some local and adaptive smoothing filters based on image statistics. Recently, fuzzy rule-based noise filtering algorithms have been demonstrated to work well as local and adaptive filters [32,33].

Let F_G be a linear filter that can be used to remove additive Gaussian noise. F_G can be designed to average neighboring pixels' intensities within a proper window area, for example, a 3×3 window size.

$$I_G(m,n) = \frac{\sum_{i=-1}^{i=1} \sum_{j=-1}^{j=1} \mu(m+i,n+j) I(m+i,n+j)}{\sum_{i=-1}^{i=1} \sum_{j=-1}^{j=1} \mu(m+i,n+j)} ,$$

where $I_G(m,n)$ is the filtered result corresponding to the input original gray level image $I(m,n)$ and $\mu(m+i,n+j)$ is a weighting function. Here we can consider it as the membership function, denoting how close $I(m,n)$ is to $I(m+i,n+j)$, and it is implemented by the exponent function

$$\mu(m+i,n+j) = \exp\left(-\frac{\Delta I(m+i,n+j)^2}{K}\right) , \tag{6}$$

with $\Delta I(m+i,n+j) = I(m,n) - I(m+i,n+j)$, and where the positive constant K controls the spread of the membership function. This constant is application dependent.

Let F_M be the traditional median filter with its filtering result $I_M(m,n)$, and F_N be the operation where the input is kept unchanged.

Now we define a function, comp, to signify the compatibility, i.e., the degree to which the neighboring pixels are compatible to the center pixel in the 3×3 window as

$$\text{comp}(m,n) = \sum_{i=-1}^{i=1} \sum_{j=-1}^{j=1} \mu(m+i,n+j) - 1 ,$$

where $\text{comp}(m,n)$ ranges in the interval $[0,8]$. For an impulse pixel, due to the significant difference between its intensity and that of its neighbors, the corresponding comp is very small, thus the median filter, F_M, is proper for filtering this impulse noise. For a pixel in uniform regions, obviously it has a large comp value, and thus the linear filer F_G is suitable. For a pixel on strong edges, the corresponding comp is medium, and it may be acceptable to adopt the filter F_N, i.e., to keep the pixel intensity unchanged.

Using the theory of fuzzy logic, comp can be mapped into three fuzzy subsets say, "s"(small), "m"(medium) and "l"(large), with corresponding membership functions, $\mu_s(\cdot)$, $\mu_m(\cdot)$ and $\mu_l(\cdot)$. The relationships between the three fuzzy subsets are shown in Fig. 4, where the trapezoidal shape membership functions are used for fuzzification. The parameters $a \cdots h$ may be selected empirically or by training. Then fuzzy rule inference scheme is used to aggregate the separate decisions to attain the final decision as follows. Assuming M+1 rules, we adopt the if-then-else rule paradigm (similar to [32,33]) as follows:

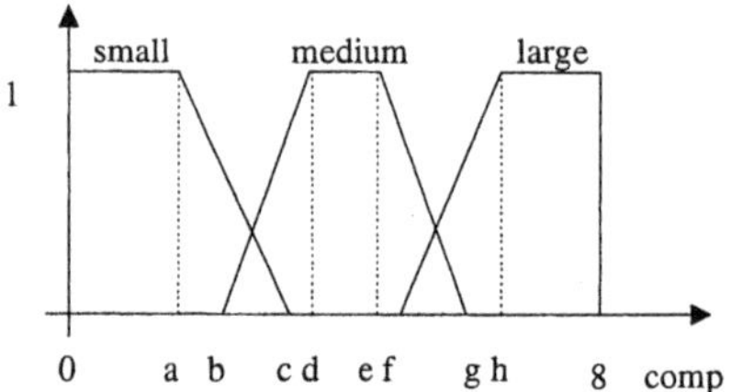

Fig. 4. Membership function of the comparability.

Rule 1: If X_{11} *is* $A_{11} \oplus \cdots \oplus X_{1i}$ *is* $A_{1i} \oplus \cdots \oplus X_{1N_1}$ *is* A_{1N_1} *then* F_1;

...

Rule k: If X_{k1} *is* $A_{k1} \oplus \cdots \oplus X_{ki}$ *is* $A_{ki} \oplus \cdots \oplus X_{kN_k}$ *is* A_{kN_k} *then* F_k;

...

Rule M: If X_{M1} *is* $A_{M1} \oplus \cdots \oplus X_{MN_M}$ *is* A_{MN_M} *then* F_M;

Rule M+1: else F_{M+1}.

A_{ki} is the linguistic label associated with the ith input variable X_{ki} in the kth rule, N_k is the number of input variables in the kth rule, and $\oplus$ is the aggregation operator [34]. F_k is the desired action in the kth rule. In our case F_k is an elementary filter, which may be the linear filter or median filter.

Similar to Sect. 2.2, we can get the strength for each rule in the above system as follows

strength of Rule k: $\alpha_k = \mu_{A_{k1}}(a_{k1}) \wedge \cdots \wedge \mu_{A_{ki}}(a_{ki}) \wedge \cdots \mu_{A_{kN_K}}(a_{kN_K})$,
where $\mu_{A_{ki}}$ is the membership function for fuzzy subset "A_{ki}", and $a_{k1} \cdots a_{kN_k}$ are values (input vector) corresponding to the linguistic labels respectively. We can then compute the output of the fuzzy rule-based system for an input vector as

$$Y = f[(\alpha_1 \circ F_1), \cdots, (\alpha_{M+1} \circ F_{M+1})],$$

where $\circ$ denotes the composition operator, and f is the defuzzification function. If we select the multiplication as the composition operator, the weighted average for the defuzzification function, and the filter output for the consequent, the overall output Y for noise removal can be expressed as:

$$Y = \frac{\sum_{k=1}^{M+1} \alpha_k F_k}{\sum_{k=1}^{M+1} \alpha_k}. \tag{7}$$

In our case, we can write the rules as:
Rule 1: If the compatibility comp is small, then use median filter F_M;
Rule 2: If the compatibility comp is large, then use linear filter F_G;
Rule 3: else use the filter F_N.
Corresponding to Eq. (7), the final filtering result is:

$$Y(m,n) = \frac{\mu_s(comp(m,n))I_M(m,n)+\mu_m(comp(m,n))I_N(m,n)+\mu_l(comp(m,n))I_G(m,n)}{\mu_s(comp(m,n))+\mu_m(comp(m,n))+\mu_l(comp(m,n))}$$

3.3 The watershed algorithm for initial segmentation

An image segmentation method based on the morphological watershed transform can be found in [21]. This transform overcomes the problem of disconnected contours and false edges. The main idea is depicted in Fig. 5 based on immersion simulations. Let I be a grey scale digital image, then the watershed algorithm can be viewed as puncturing holes at the local minima of I, if this image is regarded as a topographic surface immersed into water. Starting with the minima of lowest basin of I, each region filling with water is called a catchment basin. When the water from two separate catchment basins begins to merge, a dam is constructed at the encountering location to prevent water from one catchment basin to enter the other. At the end of this immersion procedure each minimum is surrounded by dams delimiting its associated catchment basins. All the dams correspond to the watersheds of the image I.

The watershed computation algorithm used here is based on the immersion simulations reported in [21]. This algorithm consists of two steps: sorting and flooding by recursive detection, and fast labelling of the different catchment basins. An advantage of the watershed algorithm is that it produces the closed and one pixel-wide contours, for it uses the region-based technique, which contrasts other pure edge-based segmentation algorithms that result in disconnected and false edges. For the watershed algorithm, a typical problem is oversegmentation, which is caused due the local variance of image pixel intensity, texture and/or noise. As a result, a region may be decomposed into many small regions after the watershed transform, more than what it should realistically be.

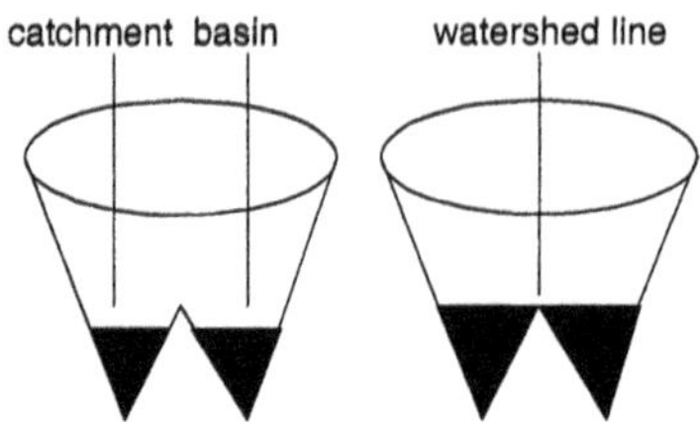

Fig. 5. Illustration of watershed algorithm.

3.4 Determination of features for image merging

Following the initial segmentation from the watershed transform, a region merging process is used to tune the image segmentation results from the watershed process. Here we employ the fuzzy integral introduced in Sect. 2.4 as the region merging criterion.

Let $R = \{R_1, \cdots R_n\}$, be a set of neighboring regions of R_0, and R_0 be the region under consideration. Let $f = \{f_1, f_2, f_3, f_4\}$ be a set of features introduced for region merging. Let f_{ik} denote the value of feature f_i between R_0 and R_k. Let $h : f \to [0,1]$ be a function where $h(f_{ik})$ denotes the confidence level of f_{ik} that indicates how certain we are in merging R_0 to R_k based on feature f_i. We associate with feature f_i, a degree of importance (or fuzzy density), denoted by g^i or $g(f_i)$ in Sect. 2.4.

Feature Selection

Let IM_0 denote a grey level image to be segmented (this can be the original image I, or the filtered image Y), with its pixel intensity denoted by $g_0(x,y)$ at point (x,y). Let IM_1 be the corresponding gradient magnitude image with pixel intensity $g_1(x,y)$ at point (x,y). Let IM_w be the watershed image, where only pixels at watersheds have a value 1; others are set to 0. Suppose that the current image region under consideration, R_0, has n neighboring regions R_k, $k = 1, \cdots n$, as shown in Fig. 6. Let Eg_k denote the shared edge, which consists of all the pixels on the edge between R_0 and R_k. Based on the above notations, let us define the following features as the information sources, as in Sect. 2.4. For each neighboring region R_k of R_0 we have:

$$\text{Feature } f_1: \quad f_{1k} = \begin{cases} 1 - \frac{\text{ComSize}(k)}{\tau_1} & \text{if ComSize}(k) < \tau_1 \,, \\ 0 & \text{otherwise} \,. \end{cases}$$

$$\text{Feature } f_2: \quad f_{2k} = \begin{cases} 0 & \text{if } G_k < \tau_2 \,, \\ G_k - \tau_2 & \text{otherwise} \,. \end{cases}$$

$$\text{Feature } f_3: \quad f_{3k} = \begin{cases} 0 & \text{if } |\mu_0 - \mu_k| < \tau_3 \,, \\ |\mu_0 - \mu_k| - \tau_3 & \text{otherwise} \,. \end{cases}$$

$$\text{Feature } f_4: \quad f_{4k} = |\sigma_0 - \sigma_k| \,.$$

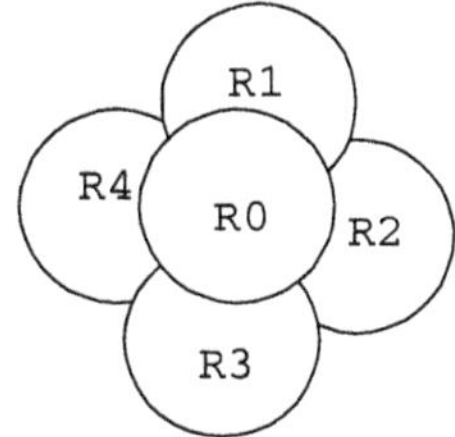

Fig. 6. Neighboring regions.

In the above definitions of features, the associated variables are determined as follows.

G_k is a measure of magnitude of Eg_k in image IM_1. It is obtained by applying the order statistic filter to magnitudes of Eg_k.

$\mu_k = \frac{1}{\text{Size}(R_k)} \sum_{(x,y)\in R_k} g_0(x,y)$ is the mean intensity of pixels in R_k. Size(R_k) denotes the total number of pixels in R_k.

$\sigma_k = \sqrt{\frac{1}{\text{Size}(R_k)} \sum_{(x,y)\in R_k} (g_0(x,y) - \mu_k)^2}$, is the standard deviation of the pixel intensities in R_k.

$\text{ComSize}(k) = \frac{\text{Size}(R_0)\times\text{Size}(R_k)}{\text{Size}(R_0)+\text{Size}(R_k)}$ is a measure integrating both the size of R_0 and the size of R_k.

τ_1 is an application-dependent threshold.

τ_2 and τ_3 are two thresholds, which are determined by the following scheme.

Since all edge pixels in the final segmentation are contained in the watershed pixels of IM_w, there exists a value of τ_2 that will yield a user-set number of regions (ExpNum). This way a feasible value of τ_2 can be determined based on the expected the number of regions.

To determine τ_3, a statistical test is applied by analyzing the histogram of the differential mean intensities between all adjacent region pairs. Let $\Delta\mu = |\mu_1 - \mu_2|$. μ_1 and μ_2 are the mean intensities of a pair of neighboring regions in IM_w. The probability density function of the occurrence of $\Delta\mu$, is expressed by $p(\Delta\mu)$ associated with which the distribution function, $P(\Delta\mu) = \int_{t=0}^{\Delta\mu} p(t)dt$, is a non decreasing function with respect to $\Delta\mu$. Furthermore, there is a limiting number of initial regions, and $\Delta\mu$ ranges in the interval $[\Delta\mu_{\min}, \Delta\mu_{\max}]$, for which $P(\Delta\mu)$ has a discrete form. An effective way to determine the value of τ_3 is to choose a ratio r_3 and find a value of τ_3 such that

$$\tau_3 = \min\{\Delta\mu \mid P(\Delta\mu) \geq r_3 \times P(\Delta\mu_{\max})\} .$$

Assuming there are IniNum regions in the initial segmentation, and that we expect ExpNum regions in the final segmentation, r_3 can be computed heuristically according to $r_3 = \frac{\text{IniNum}-\text{ExpNum}}{\text{IniNum}}$.

Computation of $h(f_{ik})$

To insure effective segmentation, we propose two types of fuzzy integrals, namely, the relative fuzzy integral and the absolute fuzzy integral.

1. For the relative fuzzy integral case the following formulas are used to compute features f_{ik}, $i = 2, 3, 4; \forall k$.
   $h(f_{ik}) = 1/\sum_{j=1}^{n} \left(\frac{f_{ik}}{f_{ij}}\right)^{2/(m-1)}$ for all $f_{ij} \neq 0$, $j = 1, \cdots n$.
   or $h(f_{ik}) = \begin{cases} 1/\mathrm{Num}(i) & \text{if } f_{ik} = 0 \\ 0 & \text{otherwise} \end{cases}$
   where $\mathrm{Num}(i)$ is the number of neighboring regions for which feature f_i is equal to zero. The fuzzifier m is set to 2.
   For feature f_1, $\forall k$, we define $h(f_{1k}) = f_{1k}/\sum_{j=1}^{n} f_{1j}$. $h(f_{1k})$ denotes the normalized relative size of regions R_0 and R_k. It is clear that regions with relatively small sizes are likely to be merged.
   Obviously $h(f_{ik})$ is equivalent to a class membership function denoting the similarity between a region and its adjacent regions.
2. Absolute fuzzy integral case
   Here we normalize feature values as follows:
   $h(f_{ik}) = 1 - f_{ik}/255, i = 2, 3, 4$ (assuming 0-255 gray levels). For $i = 1$, we set $h(f_{1k}) = 0$.

3.5 Implementation of region merging

Starting with the initial segmentation of the watershed transform, a hash table is allocated for all initial regions, where the ith entry corresponds to the region R_i. The table includes the maximum absolute/relative fuzzy integral values. In addition, the table includes the following information about its neighboring regions of R_i: region label, intensities of pixels on the shared boundary, the fuzzy integrals of the region, and three global arrays that record the region size, the sum of pixels' intensities, the sum of the square of pixel intensities. As the region merging process progresses, these arrays are updated recursively. The process of the region merging is described as follows:

- Step 1. For each region, a set of fuzzy integrals is computed corresponding to all its neighboring regions using Eqs. (3), (4) and (5), where x_i is replaced by f_i (f_i's serve as information sources). The neighboring region with the maximum relative fuzzy integral is then found. Suppose there are m regions with their labels ranging from 1 to m. Let the set $\{\mathrm{MaxR}(1), \cdots, \mathrm{MaxR}(m)\}$ be the label set of neighboring regions with the maximum relative fuzzy integrals associated with these m regions.

- Step 2. Based on the set $\{\text{MaxR}(1), \cdots, \text{MaxR}(m)\}$, do the following
  1. Find a region group $j_1, j_2 \cdots, j_l$ such that $\text{MaxR}(j_1) = j_2$ and $\text{MaxR}(j_2) = j_3, \cdots, \text{MaxR}(j_l) = j_1$. l is a preset value to control group size. When the preset value is 2, the group merging becomes the region pair merging.
  2. Merge region group $j_1, j_2 \cdots, j_l$. The basic operation is based on pair-wise region merging.
- Step 3. Repeat Steps 1 and 2 until no further region groups can be found that satisfy the conditions in Step 2.
- Step 4. If the number of regions in the current segmentation image is less than ExpNum, exit the loop; otherwise we find and merge the region pair (one region has a small size less than τ_1) with the maximum absolute fuzzy integral in the current cycle and return to Step 1.

3.6 The optimal implementation of image segmentation

Cost function

In region merging, different features do not contribute equally and therefore can not be considered as equally important. Therefore, determining the right fuzzy density value for each feature is quite important, and directly affects the region merging results.

As $g(f_i)$ ranges from 0 to 1, finding the optimal set of $g(f_i)$ can be stated as a problem of combinatorial optimization. The cost function is a multivariate, with a landscape that is fraught with local optima. This renders deterministic optimization techniques such as local hill-climbing and gradient descent ineffective, since they are prone to stalling in a local optimum. Stochastic optimization techniques, on the other hand, are more suitable since they are capable of forgoing the local optima in the solution space in favor of a global optimum. In this work, we use simulated annealing algorithm (SA) to explore the value space of fuzzy densities and look for the (near) optimal solution.

It is important to devise an appropriate cost function for qualifying the quality of a segmentation. This function provides a feedback for searching for optimal fuzzy densities. Generally speaking a "good" segmentation must satisfy the following two properties [35].

- A good region of the segmented image should contain one primitive (a texture or a constant gray level) characterized by the stability of the statistics inside it.
- Any pairs of adjacent regions should contain two different primitives to prevent oversegmentation, and hence emphasize the disparity of statistics between regions.

In addition to the above two points, we also take the edge strength in consideration. In this work we limit the segmentation to considering gray-level images, and define the cost function as

$$E = \frac{\text{STD}}{\text{DIFMEAN} \times \text{EDG}}, \tag{8}$$

where,

$\text{STD} = \sqrt{\frac{\sum_{i=1}^{N} \sum_{(x,y) \in R_i} (g_0(x,y) - \mu_i)^2}{\sum_{i=1}^{N} \text{Size}(R_i)}}$. STD denotes the overall standard deviation, and represents the overall uniformity. N is the number of regions in the current segmented image. The smaller STD, the more uniform the segmented regions.

$\text{DIFMEAN} = \frac{\sum_{i=1}^{N} \min|u_i - u_j|}{N}$. DIFMEAN denotes the overall average minimum contrast per region between adjacent regions. Obviously the bigger DIFMEAN, the better the segmentation.

$\text{EDG} = \frac{\sum_{i=1}^{N} \sum_{(x,y) \in Edg_i} g_1(x,y)}{\sum_{i=1}^{N} \text{EdgNum}(R_i)}$. EDG is an overall average measure of edge strength. Edg_i is the set of pixels on the boundary of region R_i. $\text{EdgNum}(R_i)$ is the number of pixels in Edg_i. The larger EDG is, the stronger boundaries the regions have. The optimal segmentation should correspond to the minimum cost of E. In what follows we show how the simulated annealing approach is used to optimize segmentation performance.

Determination of Initial Parameters of Fuzzy Density $g(f_i)$

As features are applied to compute the fuzzy integral and implement region merging, we should assign each feature a suitable initial value so as to enhance the computational effectiveness of the optimization task. Generally, there are two approaches to determine the initial parameters of $g(f_i)$. One approach is to generate them randomly. The other approach is to generate them empirically. Both approaches suffer some disadvantages. For example, the randomly generated initial parameters, no doubt, carry no application related information, and hence may cause a heavy computational load for searching optimal parameters and delay the convergence of search. On the other hand, using application dependent experience to determine the initial parameters is more effective, but nevertheless it is often influenced by subjective factors such as the problem-specific knowledge and experience. As a result, the assigned initial parameters may be biased, or may tend to be more suitable for one image than another. To overcome these problems, we introduce a statistical and automatic approach to determine the initial parameters adaptively.

As $g(f_i)$ expresses the importance of Feature f_i, intuitively, the feature with a larger fuzzy density should result in a good segmentation result. Based on this point of view, we measure the initial $g(f_i)$ corresponding to all 4 features using the following algorithm.

- Step 1. For Feature f_i, let $g(f_j) = \delta(j - i)$; $j = 1, \cdots, 4$. Implement region merging using the fuzzy integral calculated from Feature f_i, and stop when the expected number of regions is reached. Compute the cost E, denoted as $E(f_i)$ by Eq. (8) using the result image.
- Step 2. Repeat Step 1 for each feature, and get $E(f_i)$, $i = 1, \cdots, 4$.
- Step 3. Find the feature having the minimum cost among $E(f_i)'s$, say $f_{\min}$ with the minimum cost $E_{\min}$, then assign $f_{\min}$ a reasonable big fuzzy density from the interval $[0,1]$ (note the fuzzy density is limited to the range from 0 to 1), such as $g(f_{\min}) = 0.8$. So is to handle the feature $f_{\max}$ having the maximum cost $E_{\max}$, let $g(f_{\max}) = 0.1$.
- Step 4. For the other features the initial fuzzy densities are computed by $g(f_i) = g(f_{\min}) + (g(f_{\max}) - g(f_{\min})) \times \frac{E(f_i) - E_{\min}}{E_{\max} - E_{\min}}$.

Simulated Annealing for the Fuzzy Density Optimization and Region Merging

Simulated annealing [36,37] attempts to find optimal answers to a problem in a manner analogous to the formation of crystals in cooling solids to make the system energy minimum and reach a stable state.

The proposed algorithm starts with an initial point (a set of initial fuzzy densities $g(f_i)$) in the search space limited by the estimated intervals of all features. To evaluate how well a candidate set of fuzzy densities is, we use the cost function defined in Eq. (8) as the criterion. From a candidate density with cost E_i, a new density (neighboring point) with cost E_{i+1} is generated from a random perturbation according to the nature of the problem. The difference between the costs E_i and E_{i+1} is used in conjunction with the current system temperature T to compute the acceptance probability of the new candidate. The most common way to compute acceptance probability is based on the Boltzman distribution:

$$P_{accept} = \min\left(1, \exp(\frac{E_i - E_{i+1}}{cT})\right), \tag{9}$$

where c is a scale factor.

To clarify the proposed fuzzy integral based image segmentation and its (near) optimal implementation using the simulated annealing algorithm, we outline the entire procedure in 5 steps as follows:

- Step 1. Generate a gradient magnitude image by applying Canny edge operator [38] to an original gray-level (for a noise free image) or a filtered image after image preprocessing with noise removal and edge preserving (for a noisy image).

- Step 2. Perform the watershed transform based on the gradient magnitude image.
- Step 3. Build the region adjacency graph (RAG) carrying initial features information. In a region adjacency graph a node represents a region, and an arc denotes adjacency between nodes connected by this arc.
- Step 4. Initialize the parameters τ_2, τ_3, initial fuzzy densities, the start temperature *StartTemp*, the stop temperature *StopTemp*, the rate of decrease of the temperature *Ratio*, etc.
- Step 5. Simulated annealing for region merging.

    for($t=0$; $t<TempNum$;++t)
    {
        for($in=0$; $in<inLoopNum$;++in)
        {
            1. Randomly generating a new set of $g(f_i)$, say $g_n(f_i)$;
            2. Region merging (introduced in Sect. 3.5) using the computation of fuzzy integral associated with $g_n(f_i)$;
            3. Computing the new cost E_{i+1};
            4. Accept $g_n(f_i)$ and update E_i by E_{i+1} with probability, $P_{accept} = \min(1, \exp(\frac{E_i - E_{i+1}}{cT}))$.
        }
        $T = T \times Ratio.$
    }
    Output the (near) optimal segmented image.

    Where $Ratio = (\frac{StopTemp}{StartTemp})^{\frac{1}{TempNum}}$, $TempNum$ is the number of temperature iterations, $inLoopNum$ is the number of iterations for each temperature, and the initial T is set as $StartTemp$.

4 Experimental results

The segmentation algorithm based on the fuzzy integral computation and the SA optimization is implemented and tested on some gray-scale images. Some of the experimental results on three images (with image size 256×256, and 256 gray levels) are reported in this chapter. This image set consists of two MRI images (a head image and a knee image) and a natural image (peppers).

For each image, the proposed algorithm is applied to both the noise free version and the noisy version with mixed noise (with zero mean, 0.01 variance Gaussian additive noise and the salt & pepper noise with 0.05 density). We manually set the expected number of regions, ExpNum, in the final segmentation. Automatic selection of the target number of regions has been reported in the literature, see for example approaches reported in [39].

In the noisy cases, to remove the mixed noise, the parameter K in Eq. (6) is set to 800 for the computation of $\mu(m+i, n+j)$ in Sect. 3.2, and the parameters involved in the membership function in Fig. 4 are set empirically as:

$a = 2.5,\ b = 2.5,\ c = 4.0,\ d = 4.0,\ e = 4.0,\ f = 4.0,\ g = 5.5,\ h = 5.5$

In the simulated annealing process, the cooling schedule is set as follows. The start temperature $StartTemp = E_{\max}$, where $E_{\max}$ is the maximum cost when computing the initial fuzzy densities. The stop temperature $StopTemp = 0.5E_{\min}$, where $E_{\min}$ is the minimum cost when computing the initial fuzzy densities. c is set to 0.5 in Eq. (9), $TempNum$ to 30, and $inLoopNum$ to 30.

To demonstrate the performance and effectiveness of the proposed algorithm, we compare it to two other image segmentation algorithms, namely, the FCM clustering algorithm, and the algorithm in [40]. Fuzzy c-mean method [11] is applied to compute the membership for each pixel intensity. We assign each pixel to the class corresponding to its maximum membership for the final crisp segmentation. The number of classes are set manually, which are shown in the corresponding figures. In [40] the region merging criterion is to find and merge the adjacent region pair iteratively, say R_i and R_j, with the global minimum value for the term

$$\frac{\text{Size}(R_i)\times\text{Size}(R_j)}{\text{Size}(R_i)+\text{Size}(R_j)} \times (\mu_i - \mu_j)^2 ,$$

where $\text{Size}(R_i)$, μ_i, respectively, denote the number of pixels in region and the average pixel intensity in R_i.

The experimental results are arranged as follows. For each image, the noise free original image is placed in Fig. (a), with its mixed noisy version in Fig. (b). The image is overlapped by segmented boundaries of the image in Fig. (e). The resulted noise free image is shown in Fig. (c). The filtered image (after noise removal) is overlapped by the segmented boundaries of Fig. (f); the result is shown in Fig. (d). Figures (g) and (h), respectively, show the results when the FCM approach and the approach in [40] are applied to the noisy images.

The noise filtering scheme introduced in Sect. 3.2 can be applied to noisy images more than one time successively, depending on the application requirements and the quality of input images. In our experiments, one iteration of noise filtering is implemented as a preprocessing step for both noisy MR images in Fig. 7, and Fig. 8. Five iterations of noise filtering are used for noisy pepper image in Fig. 9.

Figure 7(e), Fig. 7(f), Fig. 8(e), Fig. 8(f), Fig. 9(e), Fig. 9(f) show that the final segmentation results have closed one pixel-wide contours. By observing the overlapped images in Fig. 7(c), Fig. 7(d), Fig. 8(c), Fig. 8(d), Fig. 9(c), Fig. 9(d), it is easy to find that the majority of important image regions are successfully extracted including some detailed regions even in the noisy versions. Furthermore, the extracted boundaries with closed and one pixel-wide contours, conform the human visual understanding and are exactly consistent with the true ones in the original gray level images. Also the results in (d) and (f) of each figure demonstrate good noise removal performance of the proposed noise filtering scheme, which greatly benefits the final segmentation to noisy images.

The obtained fuzzy densities and the corresponding λ value's are shown in Table 1. In the final segmentation and after using the SA algorithm, the 4 features were assigned different fuzzy densities (i.e, features vary in importance). Generally speaking, feature f_2 (edge strength), and Feature f_3 (the contrast of adjacent regions), have significant importance for both the noise free version and the noisy version of the 3 images. Feature f_1 (region size) has medium importance, while Feature f_4 (difference of stand deviations of adjacent regions) has the least importance. Consider row 3 and row 5 of Table 1, and compare them with the noise free experiments depicted in row 2 and row 4 of Table 1. Feature f_2 becomes the most important feature and hence $g(f_2)$ ranks first. This demonstrates that the proposed noise filtering scheme is able to enhance edges when one iteration is run. When 5 iterations of noise filtering were applied to the noisy pepper image, feature f_2 ranked second-next to feature f_3 in the sense of the importance. This can be attributed to extensive noise filtering which degraded the edges to some extent. This edge degradation is not significant as the difference between $g(f_2)$ and $g(f_3)$ in the last row of Table 1 is small.

Table 1. Fuzzy densities and the corresponding λ's.

Experiment	$g(f_1)$	$g(f_2)$	$g(f_3)$	$g(f_4)$	λ
Noise free head image, Fig. 7(e)	0.445398	0.665100	0.80	0.10	-0.955341
Noisy head image, Fig. 7(f)	0.303874	0.742808	0.673513	0.240606	-0.938260
Noise free knee image, Fig. 8(e)	0.630455	0.728249	0.800	0.10	-0.978410
Noisy head image, Fig. 8(f)	0.563230	0.728934	0.715146	0.133990	-0.963026
Noise free pepper image, Fig. 9(e)	0.498063	0.80	0.779040	0.10	-0.975551
Noisy pepper image, Fig. 9(f)	0.348141	0.672192	0.724387	0.145540	-0.927702

Comparing the proposed approach with the FCM approach, it is obvious that the FCM approach results in many trivial and small regions, especially, when a few iterations of noise filtering are used (see Fig. 7(g) and Fig. 8(g)). This can be attributed to the fact that the FCM approach does not utilize any information other than pixel intensities.

Considering the results if the approach reported in [40], it is clear that some regions are merged even though they have strong boundaries. This can be verified from the left-hand boundary of the head in Fig. 7(h). Furthermore, compared to approach proposed in this chapter, it is evident that there are more artifact edges produced by the algorithm in [40] (see more artifact edges in the head of Fig. 7(h), and more regions in the body of peppers, such as in the left long pepper and the big pepper in bottom-middle part of Fig. 9(h)).

These limitations are understandable as only the region size and the pixel average intensity are considered in the approach of [40].

However, to achieve such better segmentation quality and accuracy, more computational power and more memory space are required for our algorithm compared with the FCM and the approach in [40]. This is due to the fact that more computationally complex features are to be computed, in addition to computing the fuzzy integral and determining the optimal fuzzy densities by SA.

5 Conclusions

In this chapter, we have introduced some general fuzzy based approaches for image segmentation from different viewpoints of fuzzy logic. We have presented an unsupervised fuzzy integral based image segmentation approach. In this approach, the fuzzy rule-based inference scheme is applied to aggregate multiple filters for noise smoothing and edge preserving. After edge extraction and watershed transform-based initial segmentation, the proposed approach incorporates 4 image features to implement region merging based on a nonlinear computation. This multi-feature fusion scheme is more likely to simulate the human decision making process. Experimental results have shown that this approach outperforms other reported approaches in terms of segmentation quality and efficiency. A simulated annealing algorithm is developed to optimize the fuzzy densities based on information deduced from the images. This makes the approach more intelligent in obtaining knowledge from the image to be segmented so as to maximize the reliability of the segmentation results.

The approach employs relatively small constraints on parameter setting. Some crucial parameters, such as the fuzzy densities, are computed adaptively from the image under consideration. As a result, the approach lends itself to a wider range of applications.

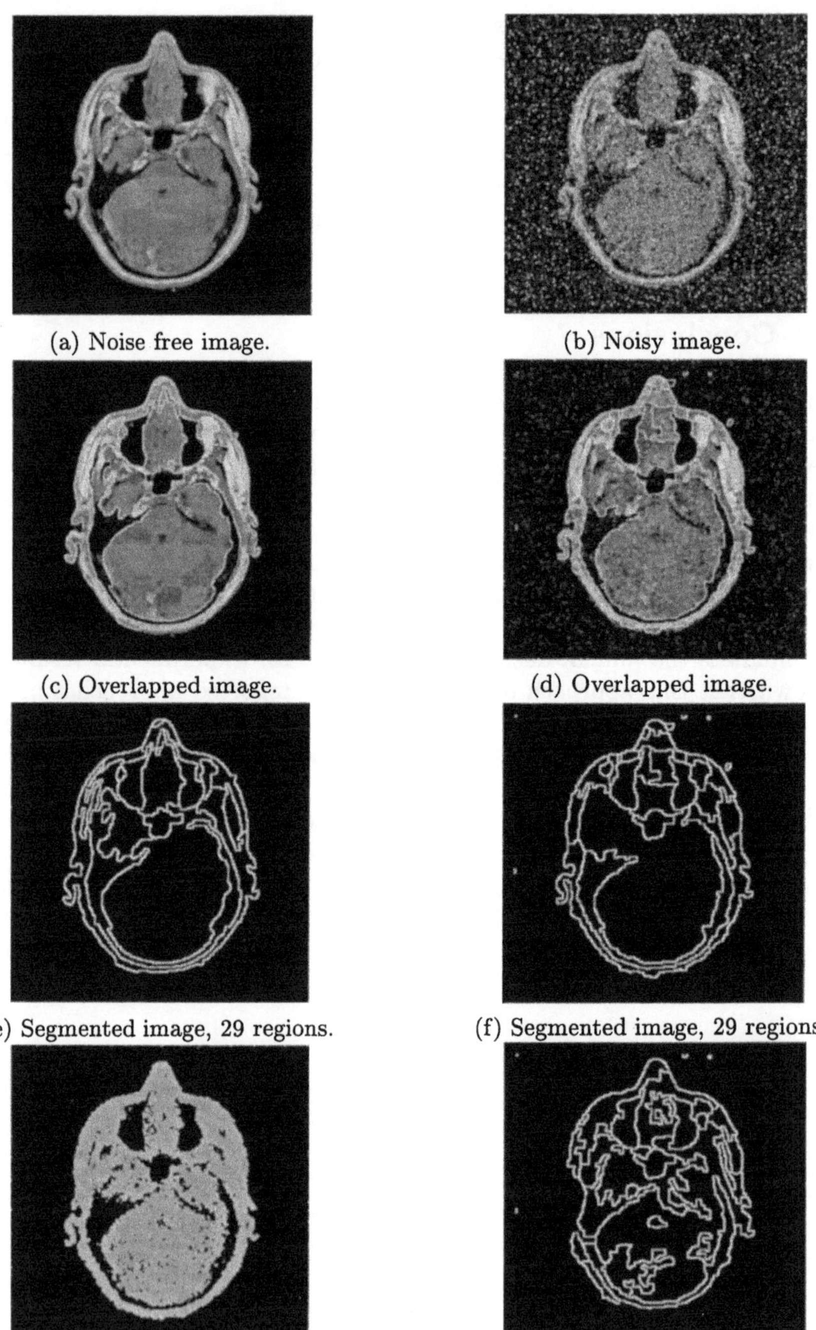

(a) Noise free image. (b) Noisy image.

(c) Overlapped image. (d) Overlapped image.

(e) Segmented image, 29 regions. (f) Segmented image, 29 regions.

(g) Segmented image, FCM, 2 classes. (h) Segmentation by the method in [40], 29 regions

Fig. 7. The experimental results for head MRI.

References

1. Suetens P., Fua P., and Hanson A. J., *Computational strategies for object recognition*, in: ACM Comput. Surv., Volume 24, 1992, 5-61.
2. Hata Y., Kobashi S., Hirano S., Kitagaki H. and Mori E., *Automated segmentation of human brain MR images aided by fuzzy information granulation and fuzzy inference*, in: IEEE Trans. Syst. Man Cybern., Part C: Applications and Reviews, Volume 30, 2000, 381-395.
3. Willemin P., Reed T. R. and Kunt M., *Image sequence coding by split and merge*, in: IEEE Trans. Commun., Volume 39, 1991, 1845-1855.
4. De Natale F.G.B., Desoli G.S., Giusto D.D., Vernazza G., *Polynomial approximation and vector quantization: a region-based integration*, in: IEEE Trans. Commun., Volume 43, 1995, 198-206.
5. Frigui H., *Adaptive image retrieval using the fuzzy integral*, in: Proc. 18th Intern. Conf. North American, 1999, 575-579.
6. Tizhoosh H. R., *Fuzzy image processing: introduction in theory and practice*, Springer-Verlag, Gemany, 1997.
7. Yager R. R., Zadeh L. A. (editors), *An introduction to fuzzy logic applications in intelligent systems*, Kluwer Academic Publishers, Boston, 1992, 147-183.
8. Zhu H., Basir O., and Karray F., *Fuzzy integral based image segmentation and the optimal implementation using genetic algorithm*, accepted by CVPRIP'2002, in: Intern. Conf. Computer Vision, Pattern Recognition and Image Processing, 2002.
9. Bezdek J. C., Keller J., Raghu K. and Pal N. R., *Fuzzy models and algorithms for pattern recognition and image processing*, Kluwer Academic Pubkishers, Boston, 1999.
10. L. K. Huang, M. J. J. wang, *Image thresholding by minimizing the measure of fuzziness*, in: Patt. Recog., Volume 28, 1995, 41-51.
11. Bezdek J. C., *Pattern recognition with fuzzy objective function algorithms*, in: Plenum Press, New York and London, 1981, 65-85.
12. Moghaddamzadeh A., Bourbakis N., *A fuzzy region growing approach for segmentation of color images*, in: Patt. Recog., Volume 30, 1997, 867-881.
13. Hong Peow Ong, Rajapakse J. C., *Fuzzy-region-segmentation*, in: Proc. Intern. Joint Conf. Neural Networks, Volume 2, 2001, 1374-1379.
14. Maeda J., Ishikawa C., Novianto S., Tadehara N. and Suzuki Y., *Rough and accurate segmentation of natural color images using fuzzy region-growing algorithm*, in: Proc. 15th Intern. Conf. Patt. Recog., Volume 3, 2000, 638-641.
15. Jawahar C. V., Biswas P. K. and Ray A. K., *Analysis of fuzzy thresholding schemes*, in: Patt. Recog., Volume 33, 2000, 1339-1349.
16. Makrogiannis S., Economou G. and Fotopoulos S., *A fuzzy dissimilarity function for region based segmentation of color images*, in: Intern. Journ. Pattern Recognition and Artificial Intelligence, Volume 15, 2001, 255-267.
17. Sugeno M., *Theory of fuzzy integrals and its applications*, in: thesis, Toko Institute of Technology, 1974.
18. Pham T. D., Yan H., *Color image segmentation using fuzzy integral and mountain clustering*, in: Fuzzy Sets and Systems, Volume 107, 1999, 121-130.
19. Zhu H., Basir O. and Karray F., *Fuzzy integral based region merging for watershed image segmentation*, in: Proc. FUZZ-IEEE'2001, the 10th Intern. Conference on Fuzzy Systems, 2001.

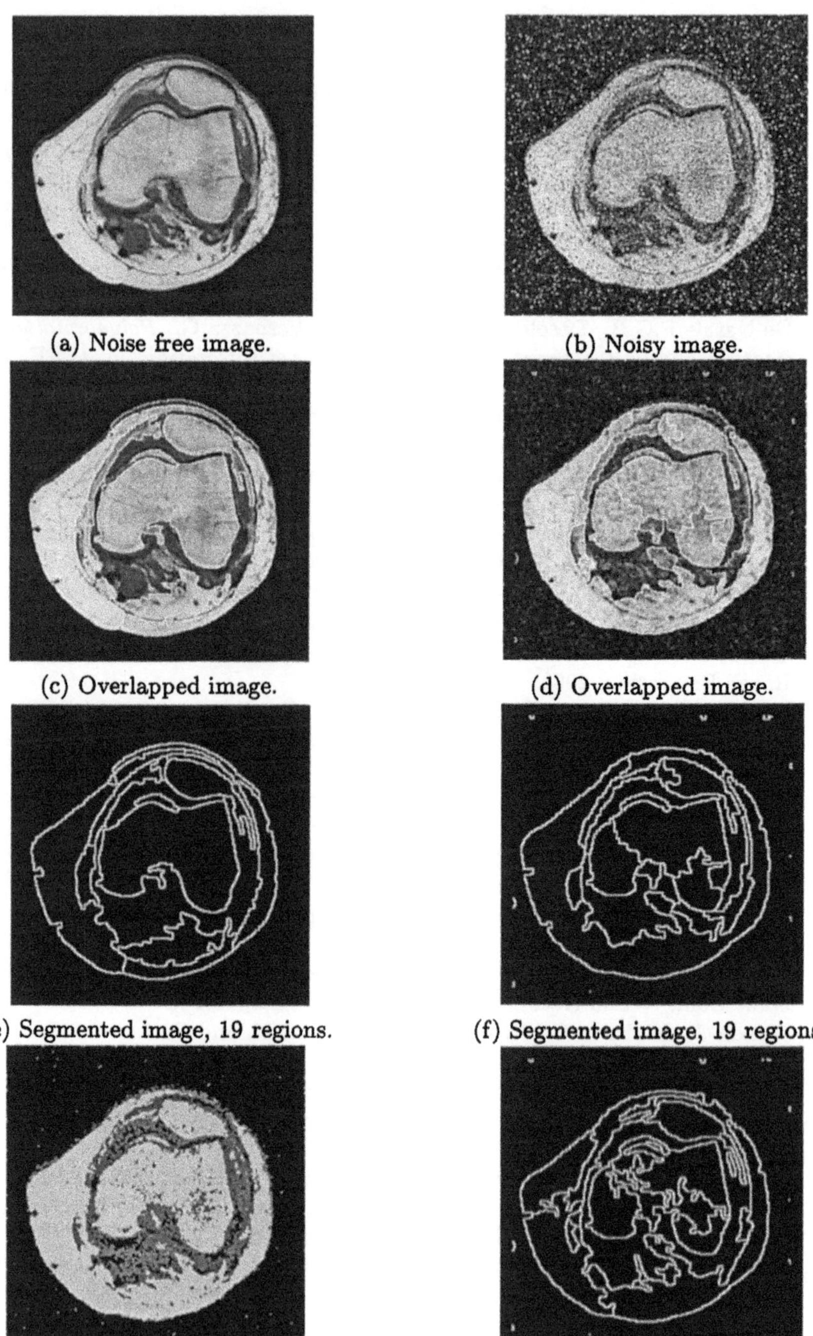

(a) Noise free image. (b) Noisy image.

(c) Overlapped image. (d) Overlapped image.

(e) Segmented image, 19 regions. (f) Segmented image, 19 regions.

(g) Segmented image, FCM, 3 classes. (h) Segmentation by the method in [40], 19 regior

Fig. 8. The experimental results for knee MRI.

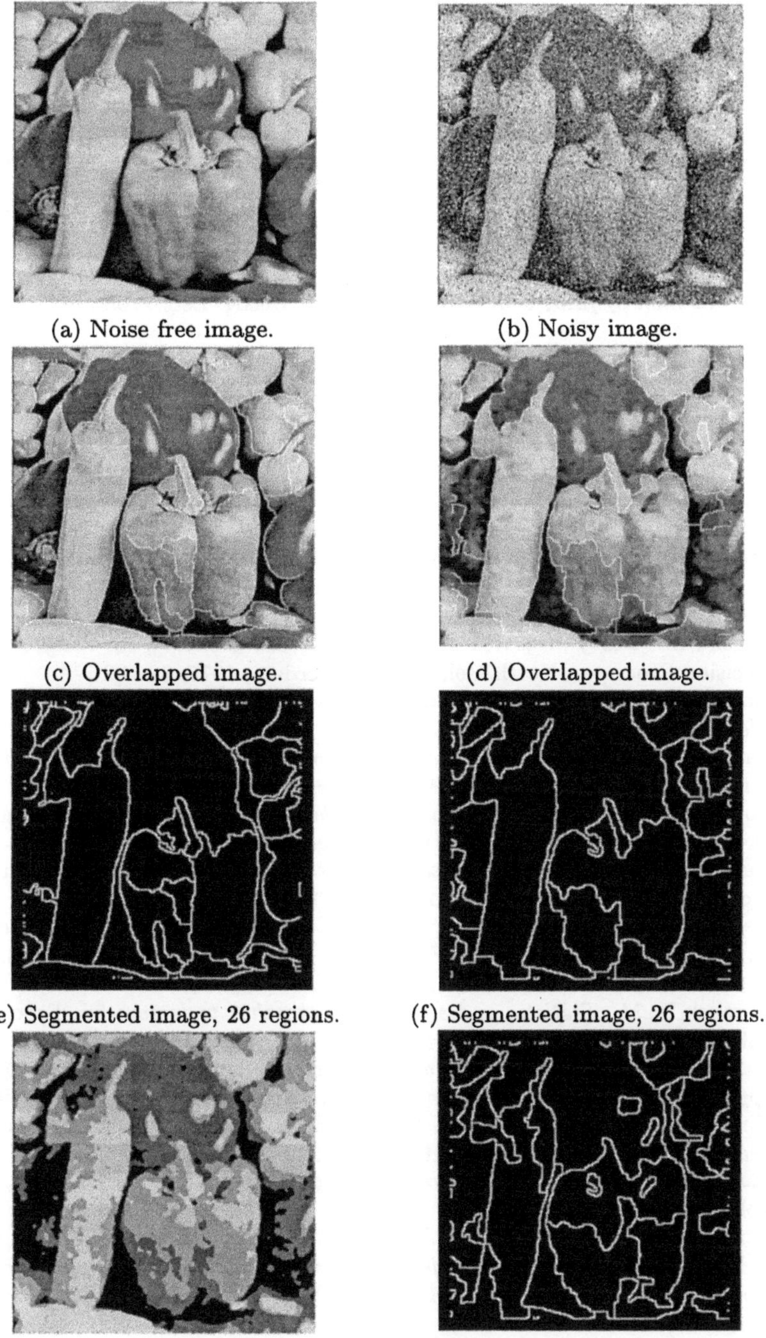

(a) Noise free image. (b) Noisy image.

(c) Overlapped image. (d) Overlapped image.

(e) Segmented image, 26 regions. (f) Segmented image, 26 regions.

(g) Segmented image, FCM, 4 classes. (h) Segmentation as in [40], 26 regions.

Fig. 9. The experimental results of peppers.

20. Grabisch M., Nguyen H. T. and Walker E. A., *Fundamentals of uncertainty calculi with applications to fuzzy inference*, Kluwer Academic Publishers, Dordrecht, 1995, 283-284.
21. Vincent L., Soille P., *Watersheds in digital space: an efficient algorithm based on immersion simulations*, in: IEEE Trans. PAMI, Volume 13, 1991, 583-598.
22. Shafer G. A., *A mathematical theory of evidence*, Princeton Universyty Press, Princeton, N.J., 1976.
23. Keller J. M., Osborn J., *Training the fuzzy integral*, in: Intern. Journ. Approximate Reasoning, Volume 15, 1996, 1-24.
24. Pal S. K., Resonfeld A., *Image enhancement and thresholding algorithm by optimization of fuzzy compactness*, in: Patt. Recog. Lett. Volume 7, 1988, 77-86.
25. Murthy C. A., Pal S. K.,*Fuzzy thresholding: mathematical framework, bound functions and weighted moving average technique*, in: Patt. Recog. Lett. Volume 11, 1990, 197-206.
26. Bigand A., Bouwmans T. and Dubus J. P., *Extraction of line segments from fuzzy images*, in: Patt. Recog. Lett., Volume 22, 2001, 1405-1418.
27. Pal S. K., *Fuzzy skeletonization of an image*, in: Patt. Recog. Lett., Volume 10, 1989, 17-23.
28. Mamdani E. H., Assilian S., *An experiment in linguistic synthesis with a fuzzy logic controller*, in: Intern. Journ. of Man-Machine Studies, Volume 7, 1975, 1-13.
29. Sugeno M., *Fuzzy measures and fuzzy integrals: a survey*, in: Automatic and Decision Processes, North Holland, Amesterdam, 1977, 89-102.
30. Charnik E., McDermott D., *Introduction to artificial intelligence*, Addison-Wesley, Reading, Mass., 1986.
31. Tahani H., Keller J. M., *Information fusion in computer vision using the fuzzy integral*, in: IEEE Trans. Syst. Man Cybern., Volume 20, 1990, 733-741.
32. Young Sik Choi, Krishnapuram R., *A robust approach to image enhancement based on fuzzy logic*, in: IEEE Trans. Image Proc., Volume 6, 1997, 808-825.
33. Russo F., Ramponi G., *Combined FIRE filters for image enhancement*, in: Proc. 3rd IEEE Conf. Fuzzy Syst., Volume 1, 1994, 260-264.
34. Keller J. M., Krishnapuram R., Rhee F. C.-H., *Evidence aggregation networks for fuzzy logic inference*, in: IEEE Trans. Neural Network, Volume 3, 1992, 761-769.
35. Haralick R. M., *Statistic and structure approaches to texture*, in: Proc. of the IEEE, Volume 69, 1979, 786-804.
36. Atarts E., Korst J., *Simulated annealing and boltzmann machine-a stochastic approach to combinatorial optimization and neural computing*, Wiley, 1989.
37. Kirkpatrick S., Gelatt Jr C. D., Veccchi M. P., *Optimization by simulated annealing*, IBM Research Report RC 9355, 1982.
38. Canny J., *A computational approach to edge detection*, in: IEEE Trans. PAMI, Volume 8, 1986, 679-698.
39. Rezaee M.R., van der Zwet P.M.J., Lelieveldt B.P.E., van der Geest R.J., Reiber J.H.C., *A multiresolution image segmentation technique based on pyramidal segmentation and fuzzy clustering*, in: IEEE. Tran. Image Proc., Volume 9, 2000, 1238-1248.
40. Haris K., Efstratiadis S. N., Maglaveras N. and Katsaggelo A. K., *Hybrid image segmentation using watersheds and fast region merging*, in: IEEE Trans. Image Proc., Volume 7, 1998, 1684-1699.

Chapter 6

Fuzzy Thresholding and Histogram Analysis

Manuel Guillermo Forero-Vargas

Universidad Nacional de Colombia
System Engineering Department
OHWAHA Research Group
Ciudad Universitaria, Bogotá, Colombia
email: mforero@ing.unal.edu.co

Summary. This chapter provides a comprehensive discussion of several thresholding techniques that employ the concept of the measure of fuzziness. The basic concepts and ideas of the measures of fuzziness are introduced. A unified description of the fuzzy thresholding methods based on the measure of fuzziness is given. A segmentation procedure for chromatic images based on fuzzy rules is presented. Finally, the application of the methods in image processing in order to obtain optimal threshold values for segmenting images and the conclusion are shown.

1 Introduction

The segmentation of digital images is a process that divides an image into parts, zones, or disjointed classes that are similar with respect to one or more attributes in order to identify the structures or objects present in the image. One of the most popular tools is thresholding or gray level segmentation. Supposing that the objects in the image are characterized by different gray levels, the process of thresholding consists of attempting to classify the pixels as belonging to either the set of background pixels or the set of object pixels as a function of the level of intensity. The threshold is determined by the pixel values found in the image. A very common way to select a threshold is by using the histogram of the gray levels in the image.

Let (x, y) be the spatial coordinates of each pixel in an image Q, $q(x, y)$ be the gray level or intensity of the pixel in the position (x, y), where $0 \leq q \leq L - 1$ and L is the maximum number of positive integers that can represent the gray levels in the image. Consider the histogram of an image composed of a clear object on a dark background or viceversa. To extract the object of the image a threshold value t is chosen, converting the gray level image $q(x, y)$ into a binary one $b(x, y)$, where the pixels that have a gray level intensity superior to t which belong to the object, are represented with a color q_1, while those belonging to the background are represented with another q_0, or in other terms:

$$b(x,y) = \begin{cases} q_0 & \text{if } q(x,y) \leq t \\ q_1 & \text{if } q(x,y) > t \end{cases}$$

In practice, it could be necessary to choose more than one threshold to partition the image.

The optimal choice of the threshold value t^* of t based on a certain criterion is a difficult process due to the presence of noise, as well as the vagueness and ambiguity among the classes, because of the fact that they overlap in the histogram. Several surveys and evaluations of thresholding techniques have been made [5,11,16].

New progress in the fuzzy set theory provides different possibilities for developing new image segmentation techniques. Fuzzy models have the capacity to work with ambiguity and noisy data and the applicability of the fuzzy theory as an alternative to improve the selection of the optimal threshold in order to obtain correct segmentation is evident [1,6].

Different attempts have been made to use fuzzy sets in image segmentation, but the gray level technique described here is based on the concept of a measure of fuzziness. That is defined as the distance between the original gray level image and the thresholded image [6,14]. By minimizing the fuzziness, an optimal threshold value should be obtained.

Color image segmentation can also be improved by taking into account the information provided by the different channels. The color method described here apply the gray level thresholding technique to the different color channels and the knowledge about the regions in the image to establish fuzzy rules that are employed in the segmentation.

2 Membership function

According to the classical set theory, an element either belongs to a set or it does not. In the fuzzy set theory, the belonging of element x to set A is gradual and characterized by a particular plausibility, expressed by a membership or belonging function. Let X be the universal set of elements x, then a fuzzy set A in X is defined by:

$$A = \{(x, \mu_A(x)) \,|x \in X\}$$

where μ_A is called the membership function of the fuzzy set A.

To threshold an image, a membership function associated with each gray level of the histogram $\mu_Q(q)$ must be determined, that is the plausibility associated with the classification of each pixel to one of the two classes, e.g. object or background. The membership function, which should be a value between 0.5 and 1, defines the degree of belonging of any pixel q either to the background or the object set depending on the relationship between its intensity or gray level and the threshold t.

Different forms can be used to define the membership function. Whichever measure of fuzziness is used, an estimate of the mean gray level of the background m_b and that of the objects m_o is needed, where both values depend on the threshold t:

$$m_b(t) = \frac{\sum_{q=0}^{t} qh(q)}{\sum_{q=0}^{t} h(q)}$$

$$m_o(t) = \frac{\sum_{q=t+1}^{L-1} qh(q)}{\sum_{q=t+1}^{L-1} h(q)}$$

The smaller the difference between the gray level of any pixel q and the mean for its class, the greater will be the value of the membership function $\mu_Q(q)$. Such a good membership function can be defined that it will show how the gray levels are grouped around the mean gray level representative of each class [1,6]. If the membership values near the mean of the class are equal to it, the classification is not precise. A generalized bell membership function, based on the Cauchy distribution, is a good option [7]:

$$\mu_Q(q,t) = \begin{cases} \frac{1}{1+\frac{|q-m_b(t)|^{2b}}{D}} & \text{if } q \leq t \\ \frac{1}{1+\frac{|q-m_o(t)|^{2b}}{D}} & \text{if } q > t \end{cases}$$

This function is specified by three parameters: D, $m(t)$ and b that determine the behaviour of the membership function. Figure 1 shows how the membership function changes with the variation in D, b and $m(t)$.

D is a constant that adjusts the width of the function. Because the measure of fuzziness can be used only if $\mu(q)$ is in the interval $0.5 \leq \mu(q) \leq 1$, D is used to satisfy this condition. $D = L - 1$ or can be made equal to the difference between the maximum and minimum gray levels present in the image $D = q_{max} - q_{min}$ [1,14].

$m(t)$ is associated with the center of the class and represents the weighted average of the gray levels.

b is associated with the sharpness of a fuzzy set. b is usually positive and used to control the slopes at the crossover point.

The crossover is defined by [7]:

$$\text{crossover}(A) = \{x|\mu_A(x) = 0.5\}$$

When b changes from 1 to 0 the shape of μ becomes more abrupt and $m(t)$ sharper. In this case, the classification is stricter, so the pixels belonging to the class are closer to the mean and those not belonging have a very small membership value. If b tends to zero then μ tends to be a fuzzy singleton. A fuzzy singleton is a fuzzy set A whose support is a single point in the universal set with $\mu_A(x) = 1$ [1,3], that is A only has a membership grade for a single value. The support of a fuzzy set A is:

$$\text{support}(A) = \{x \in X | \mu_A(x) > 0\}$$

If $b = 1$, the shape of μ is less sharp and the curve is smoother. If $b > 1$ the core of the fuzzy set rises.

The core is defined by [7] as:

$$\text{core}(A) = \{x | \mu_A(x) = 1\}$$

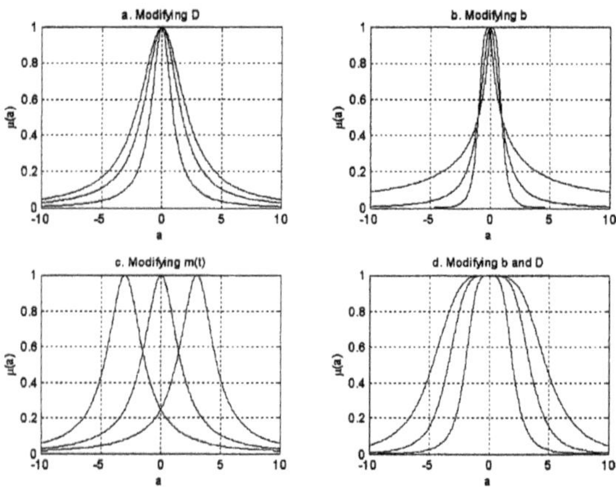

Fig. 1. Variation of the shape of the membership function μ changing D, b and $m(T)$ [7].

Other options can be employed to define the membership functions. The Gaussian and bell functions are popular because of their smoothness. The bell function is preferred because it has three parameters instead of the two possessed by the Gaussian one and represents one more degree of freedom [7]. This parameter makes it possible to control the steepness of the function at the crossover point.

Once the membership function is defined, it is necessary to find the degree of fuzziness of the segmentation measured for each threshold t.

3 Degree of fuzziness

There are two types of uncertainty: vagueness and ambiguity: [12].

- The vagueness corresponds to the uncertainty associated with the problem of finding well defined borders or distinctions between the objects to be segmented.
- The ambiguity is the uncertainty related to the difficulty of making the correct choice among two or more alternative objects.

A measure of fuzziness is taken as a measure of the uncertainty associated with vagueness and is employed to quantify the degree of fuzziness of a fuzzy set. The concept of fuzzy measures is related to the concept of ambiguity.

De Luca and Termini introduced the concept of measure of fuzziness as a function

$$f : P(x) \to \Re$$

where $P(x)$ denotes the set that includes all fuzzy subsets of the universal set X and $\Re$ is the real number domain. f assigns a measure value $f(A)$ to each fuzzy subset A of X associated to the degree of fuzziness of A.

f satisfies the following axioms:

Given two sets A and B:

- Axiom 1: Boundary conditions. $f(A) = 0 \Leftrightarrow A$ is a crisp set, that is $\mu_A(x) \in \{0, 1\}, \forall\ x$. Then the degree of fuzziness of a crisp set is zero.
- Axiom 2: Monotonicity. If $A \prec B \Rightarrow f(A) \leq f(B)$ where $A \prec B$ denotes that A is less fuzzy or sharper than B. In other words, if the uncertainty of A is less than that of B, the measured value $f(A)$ must be less than $f(B)$ and the membership function μ_A should rise more abruptly than μ_B.
- Axiom 3: $f(A)$ is maximum $\Leftrightarrow A$ is maximally fuzzy.
- Axiom 4: $f(A) = f(\bar{A})$ where $\bar{A}$ is the complement of A.

Three measures are commonly used to measure the fuzziness. According to axioms 2 and 3 different definitions have been adopted for the concepts of sharper and maximally fuzzy [12]. The first two measures, Shannon and Kaufmann, are based on the next two definitions.

Let A and B be two fuzzy sets.

- Definition 1: According De Luca and Termini, the "less fuzzy" relation can be defined by:

$$A \prec B = \begin{cases} \mu_A(x) \leq \mu_B(x) & \text{for } \mu_B(x) \leq \frac{1}{2} \\ \mu_A(x) \geq \mu_B(x) & \text{for } \mu_B(x) \geq \frac{1}{2} \end{cases} \qquad \forall\ x \in X$$

  Figure 2 shows an example where the membership function μ_1 is sharper than μ_2.
- Definition 2: A is maximally fuzzy if $\mu_A(x) = 0.5\ \forall,\ x \in X$.

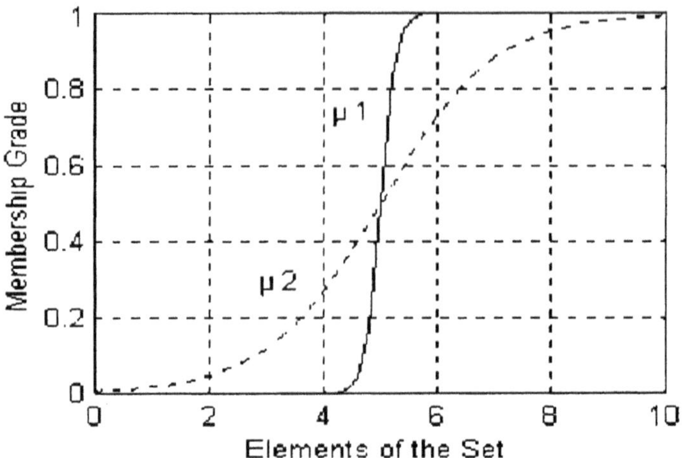

Fig. 2. Sharpness of a fuzzy set. Membership functions μ_1 and μ_2. It can be observed than μ_1 rises more abruptly than μ_2.

3.1 Shannon's entropy measure of fuzziness

De Luca and Termini [2] introduced the concept of entropy measure , the first function used to measure the degree of fuzziness of a fuzzy set, which is based on Shannon's function.

Shannon's entropy measure is considered to be the fundamental base in the information theory and its function is employed in the context of measuring information and uncertainty. It is given by:

$$H[p(x)] = -\sum_{x \in X} p(x) \log_2 p(x) \qquad \forall x \in X$$

where $p(x)$ denotes the distribution of probability in the universal set X.

Motivated by Shannon's entropy measure, De Luca and Termini defined, as a measure of fuzziness, the following non-probabilistic entropy function $f(A)$:

$$f(A) = \sum_{x \in X} S\left[\mu_A(x)\right]$$

where $S\left[\mu_A(x)\right]$ is Shannon's function given by:

$$S\left[\mu_A(x)\right] = -\mu_A(x) \log_2 \mu_A(x) - \left[1 - \mu_A(x)\right] \log_2 \left[1 - \mu_A(x)\right].$$

This is the sum of the uncertainties of the fuzzy set A defined by the membership function $\mu_A(x)$ and its complement $\bar{A}$ defined by $[1 - \mu_A(x)]$. $f(A)$ is regarded as the entropy of the fuzzy set A. Notice that the product $0 \log_2 0$ is supposed to be equal to zero.

The normalized measure $\hat{f}(A)$ of $f(A)$ is defined as:

$$\hat{f}(A) = \frac{f(A)}{|X|}$$

where $|X|$ is the cardinality of the universal set X and the normalized measure observes this relation:

$$0 \leq \hat{f}(A) \leq 1.$$

3.2 Kaufmann's measure of fuzziness

The second measure of fuzziness, referred to as an index of fuzziness [12] , was defined by Kaufmann as [8]:

$$f(A) = \left\{ \sum_{x \in X} |\mu_A(x) - \mu_C(x)|^d \right\}^{\frac{1}{d}} \qquad d \in [1, \infty)$$

This measure is expressed in terms of a metric distance (e.g. the Minkowsky distance) between the fuzzy set A and the nearest crisp set C where μ_C is defined as [12]:

$$\mu_C(x) = \begin{cases} 0 & \text{if } \mu_A(x) \leq 0.5 \\ 1 & \text{if } \mu_A(x) > 0.5 \end{cases}$$

When $d = 1$ Minkowski's measure becomes the Hamming distance, also denominated City-block or Manhattan distance, and when $d = 2$, it is the Euclidean one.

3.3 Yager's measure of fuzziness

The third measure is based on the following definitions 3 and 4 instead of definitions 1 and 2 [12].

- Definition 3:

$$A \prec B \Leftrightarrow |\mu_A(x) - \mu_{\bar{A}}(x)| \geq |\mu_B(x) - \mu_{\bar{B}}(x)| \;, \forall \; x \in X$$

where $\bar{A}$ and $\bar{B}$ are the complements of A and B respectively, or in other terms,

$$\mu_{\bar{A}}(x) = 1 - \mu_A(x).$$

- Definition 4: If $\mu_A(x) = e_c \ \forall \ x \in X \Rightarrow A$ is maximally fuzzy, provided that $\bar{A}$ has an equilibrium point e_c.

Yager [18] has proposed a measure of fuzziness based on the idea that a crisp set C does not have elements in common with its complement $\bar{C}$, and that in a fuzzy set A each element may belong to A and to $\bar{A}$ with certain plausibility. The fuzziness of a fuzzy set can be expressed by Minkowsky's distances between the fuzzy set A and its complement $\bar{A}$.

This measure is calculated using the following expression:

$$f(A) = 1 - \frac{D_d(A, \bar{A})}{|X|^{\frac{1}{d}}}$$

where:

$$D_d(A, \bar{A}) = \left\{ \sum_{x \in X} |\mu_A(x) - \mu_{\bar{A}}(x)|^d \right\}^{\frac{1}{d}} \qquad d \in (0, \infty)$$

and d is an integer used to define a distance measure; $d = 1$ corresponds to the Hamming distance and $d = 2$ to the Euclidean one.

4 Unified formulation

All three measures of fuzziness shown can be expressed in a unified formulation suggested by Knopfmacher [10], Loo [13], Forero and Rojas [3]. $f(A)$ is expressed in terms of $\sum_{x \in X} F_x [\mu_A(x)]$ [9,17].

In this way, a function $f(A)$ can be used as a measure of fuzziness if:

- F_x is defined in the domain $0 \leq \mu(x) \leq 1 \ \forall \ x \in X$.
- F_x is monotonic and symmetric around 0.5.
- F_x is increasing in the interval $0 \leq \mu(x) \leq 0.5$.
- $F_x(0.5)$ is the unique maximum. If F_x is normalized then $F_x(0.5) = 1$.
- $F_x(0) = 0$ and $F_x(1) = 0$ give the following cases:

If $\mu_A(x) = 1, \ \forall \ x \in X \Rightarrow x \in A$, then A is a crisp set and $F_x(1) = 0$.
If $\mu_A(x) = 0, \ \forall \ x \in X \Rightarrow x \in \bar{A}$, then A is a crisp set and $F_x(0) = 0$.
$F_x[\mu_A(x)]$ can be seen as a cost function. The cost function is maximal when $\mu_A(x) = 0.5, \ \forall \ x \in X$ and is minimal when $\mu_A(x) = 1, \ \forall \ x \in X$. Since the cost F_x must decrease when the membership value μ rises, that is, when the fuzziness becomes less [1], F_x is restrained to the interval $0.5 \leq \mu_A(x) \leq 1$. Notice that in the interval $0 \leq \mu_A(x) \leq 0.5$, F_x corresponds to the complement of the set A.

The rules established for this formulation can be used to define new measures.

4.1 Shannon's measure of fuzziness

Shannon's measures of fuzziness can be expressed in the unified formulation as seen previously:

$$f(A) = \begin{cases} \sum_{x \in X} S\left[\mu_A(x)\right] & \text{if } 0 < \mu_A(x) < 1 \\ 0 & \text{if } \mu_A = 0 \vee \mu_A = 1 \end{cases}$$

In this case, the cost function corresponds to the Shannons entropy function, that is, $F_x[\mu(x)] = S[\mu(x)]$.

Figure 3 shows the cost $S\left[\mu_A(x)\right]$ as a function of the degree of membership of a fuzzy set. In proportion to the increase in the plausibility of an event, the entropy decreases. In the same way, when the degree of membership μ_A of an element of a fuzzy set rises, the fuzziness and the cost decrease.

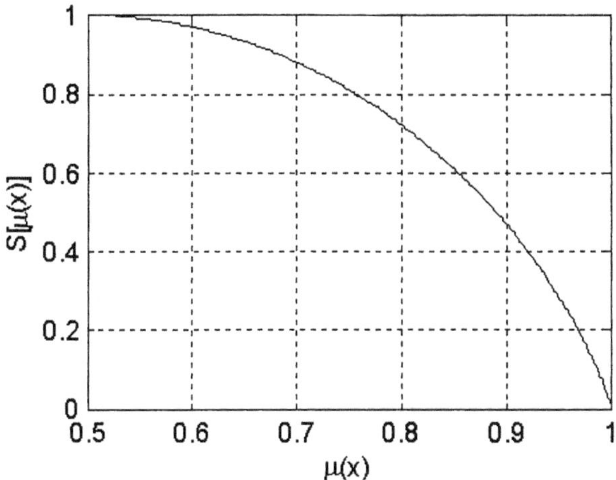

Fig. 3. Cost function $S\left[\mu_A(x)\right]$.

4.2 Kaufmann's measure of fuzziness

Kaufmann's measure is expressed in the unified formulation as:

$$f(A) = \left\{ \sum_{x \in X} K\left[\mu_A(x)\right] \right\}^{\frac{1}{d}} \qquad d \in [1, \infty)$$

where K is the cost function F_x equal to:

$$K\left[\mu_A(x)\right] = \left|\mu_A(x) - \mu_C(x)\right|^d$$

or

$$K\left[\mu_A(x)\right] = \left\{\min\left[\mu_A(x), 1 - \mu_A(x)\right]\right\}^d$$

$f(A)$ can be normalized:

$$\hat{f}(A) = \frac{f(A)}{|X|^{\frac{1}{d}}}$$

Minkowski's measure holds only for $d \in [1, \infty)$. However, it must be noticed that a cost function K can be equally obtained in the interval $d \in [0, 1)$ where d is a quasi-metric distance.

Figure 4 shows the normalized cost function $\hat{K}\left[\mu_A(x)\right]$, given by $\hat{K} = 2^d K$, for different values of d.

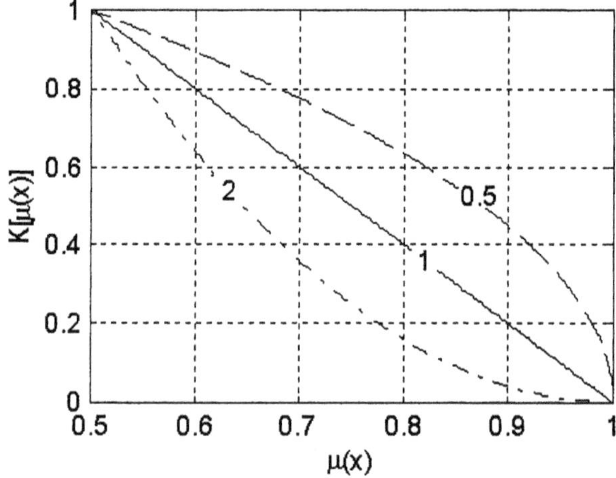

Fig. 4. Normalized cost function $\hat{K}\left[\mu_A(x)\right]$ for $d = 0.5, d = 1$ and $d = 2$.

4.3 Yager's measure of fuzziness

Yager's measure of fuzziness can be normalized and expresed in the unified formulation by:

$$f(A) = \frac{\left\{\sum Y\left[\mu_A(x)\right]\right\}^{\frac{1}{d}}}{|X|^{\frac{1}{d}}}$$

where the cost function F_x is expressed as:

$$Y[\mu_A(x)] = 1 - |2\mu_A(x) - 1|^d \qquad d \in [1, \infty).$$

As was explained before in relation to Kaufmann's cost function, it must be noticed that the cost function Y can be equally expanded to the interval $d \in [0, 1)$ where d is a quasi-metric distance.

Figure 5 shows the cost function $Y[\mu_A(x)]$ for different values of d. Notice that $Y[\mu_A(x)] = \hat{K}[\mu_A(x)]$ for $d = 1$.

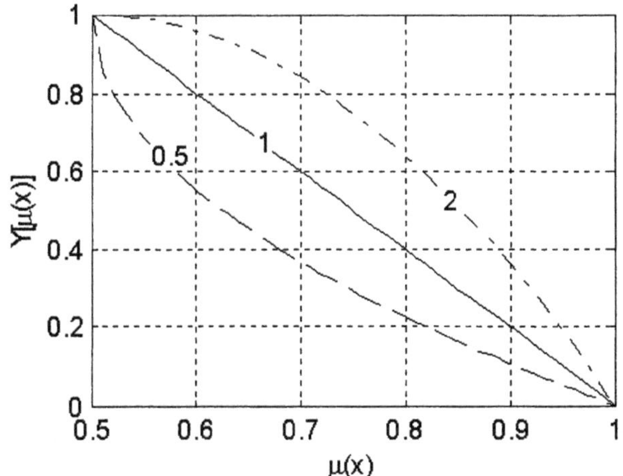

Fig. 5. Cost function $Y[\mu_A(x)]$ for $d = 0.5, d = 1$ and $d = 2$.

4.4 Gaussian measure of fuzziness

To test the unified formulation, Forero and Rojas proposed the Gaussian measure, based on the definition of the normal distribution of probability [3]. If the measure of fuzziness is normally distributed, the Gaussian measure is defined by:

$$f(A) = \frac{\sum G[\mu_A(x)]}{|X|}$$

where:

$$G[\mu_A(x)] = \begin{cases} e^{-\frac{1}{2}\left[\frac{\mu_A(x)-c}{\sigma}\right]^2} & \text{if } 0 < \mu_A(x) < 1 \\ 0 & \text{if } \mu_A(x) = 0 \vee \mu_A = 1 \end{cases}$$

In order to make $G[\mu_A(x)]$ a cost functionF_x, a mean of $c = 0.5$ and a standard deviation of $\sigma = 0.15$ is chosen, so $\mu_A(0.5) = 1$ but $\mu_A(0) = \mu_A(1) = 0.0039 \approx 0$. Therefore, it is necessary to impose values for the borders of the interval [0,1].

Figure 6 shows the cost function $G[\mu_A(x)]$. The maximal cost and fuzziness is obtained when the membership function is 0.5 and zero when it is 1.

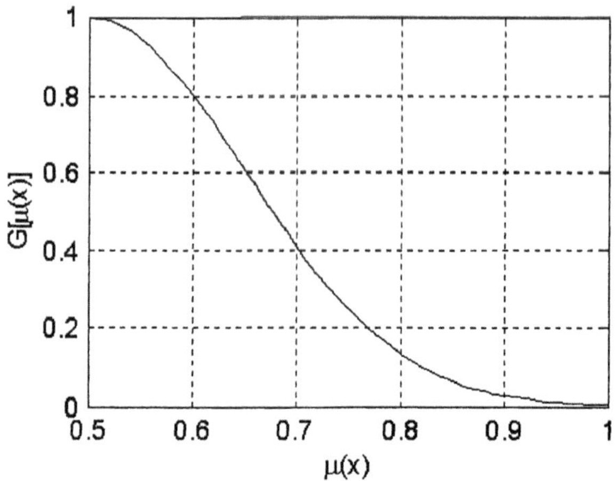

Fig. 6. Cost function $G[\mu_A(x)]$.

5 Threshold determination

Huang and Wang use the entropy measure as a criterion function for selecting an optimal threshold [6]. This technique can be extended for any measure of fuzziness [3]. An image can be considered as an array of fuzzy singletons of size $N \times M$, where each singleton corresponds to an image pixel.

Each pixel has a membership value associated with it. As a result an image can be defined as a function $Q : N \times M \overset{q(x,y)=q}{\longrightarrow} q \overset{\mu_Q(q,t)}{\longrightarrow} [0,1]$ equal to:

$$Q = \{(q(x,y), \mu_Q[q(x,y),t])| \ 0 \le x \le N-1, \ 0 \le y \le M-1, \\ 0 \le q,t \le L-1, \ 0 \le \mu \le 1\}$$

The normalized fuzziness function of an image $f[q(x,y)]$ can be calculated by:

$$f[q(x,y),t] = \frac{1}{NM} \sum_{x=0}^{N-1} \sum_{y=0}^{M-1} F[\mu(q(x,y),t)]$$

Because the histogram of the image $h(q)$ shows the frequency distribution of the gray levels, the bidimensional function Q can be simplified to a unidimensional one using the histogram. Therefore $Q : q \xrightarrow{h(q)} \mathbf{N} \xrightarrow{h(q)\mu_Q(q)} [0,1]$, where $\mathbf{N}$ is the set of the natural numbers. Q is then expressed as:

$$Q = \{(h(q), \mu_Q[h(q),t]) \mid 0 \leq q,t \leq L-1, \wedge, 0 \leq \mu \leq 1\}$$

where $\mu_Q(h(q),t) = h(q)\mu_Q(q,t)$.

The normalized fuzziness function of the image is simplified by:

$$f[q(x,y),t] = f(q,t) = \sum_{q=0}^{L-1} h(q)F[\mu_Q(q,t)]$$

So the measures of fuzziness can be expressed as a function of the gray level histogram and the cost. This can be summarized for each measure seen as follows:

- Shannon:

$$S[\mu_Q(q,t)] = -\mu_Q(q,t)\log_2 \mu_I(q,t) - [1-\mu_Q(q,t)]\log_2[1-\mu_Q(q,t)]$$

$$f(q,t) = \begin{cases} \sum_{q=0}^{L-1} h(q)S[\mu_Q(q,t)] & \text{if } 0 < \mu_Q(q,t) < 1 \\ & \text{for } 0 \leq t \leq L-1 \\ 0 & \text{if } \mu_Q(q,t) = 0 \vee \mu_Q(q,t) = 1 \end{cases}$$

- Kaufmann:

$$K[\mu_Q(q,t)] = |\mu_Q(q,t) - \mu_C(q,t)|^d$$

or

$$K[\mu_Q(q,t)] = \{\min[\mu_Q(q,t), 1-\mu_Q(q,t)]\}^d \qquad d \in (0,\infty)$$

$$f(q,t) = \left\{\sum_{q=0}^{L-1} h(q)K[\mu_Q(q,t)]\right\}^{\frac{1}{d}} \quad \text{for } 0 \leq t \leq L-1$$

- Yager:

$$Y\left[\mu_Q(q,t)\right] = 1 - |2\mu_Q(q,t) - 1|^d \qquad d \in (0,\infty)$$

$$f(q,t) = \left\{ \sum_{q=0}^{L-1} h(q) Y\left[\mu_Q(q,t)\right] \right\}^{\frac{1}{d}} \text{ for } 0 \le t \le L-1$$

- Gauss:

$$G\left[\mu_Q(q,t)\right] = \begin{cases} e^{-\frac{1}{2}\left[\frac{\mu_Q(q,t)-c}{\sigma}\right]^2} & \text{if } 0 < \mu_Q(q,t) < 1 \\ 0 & \text{if } \mu_Q(q,t) = 0 \vee \mu_Q(q,t) = 1 \end{cases}$$

$$f(q,t) = \sum_{q=0}^{L-1} h(q) G\left[\mu_Q(q,t)\right] \text{ for } 0 \le t \le L-1$$

The concept of the measure of fuzziness associated with the histogram is then employed for the optimal selection of the threshold in order to binarize the image. Considering that the objects are composed of pixels having similar gray levels, they should have a similar membership value μ and fuzziness function f. f is employed as a criterion for classifying the pixels into two classes: $[C_0, C_1]$ (e.g. background and objects), such that each class is represented by the mean gray level with a minimal standard deviation. f measures whether or not each pixel has a degree of belonging close to the mean of a class.

The fuzziness function $f(q,t)$ is calculated for all the values of t. The optimal threshold t^* is the one that minimizes $f(q,t)$:

$$t^* = \arg[\min f(q,T)] \qquad 0 \le t \le L-1$$

or better,

$$t^* = \arg[\min f(q,T)] \qquad q_{\min} \le t \le q_{\max}$$

The optimal threshold t^* is then defined as the gray level that maximizes as much as possible the separation between the two class means. That is, when the fuzziness is minimal. In proportion as $f(q,t^*)$ becomes closer to zero the gray levels of the pixels belonging to each class draw near the mean gray level representative of the class and the two class variances become as small as possible. If $f(q,t^*)$ is bigger, the gray levels are further from the mean gray level representative of each class and the classification is difficult.

6 Applications

In this section, the results obtained by applying the technique of thresholding seen for the different measures of fuzziness are presented. The above mentioned methods are applied to one image "head" shown in Fig. 7. The normalized cost functions were employed in each case. The threshold values determined from the four measures of fuzziness are shown in Table 1. The shape of the cost functions, the measures of fuzziness and the membership functions for each gray level, and the binary images obtained by the thresholding methods considered in this chapter are shown in Figs. 8 to 13. Figures 9 to 11 show the results obtained with Kaufmann's measure of fuzziness for different values of b and d.

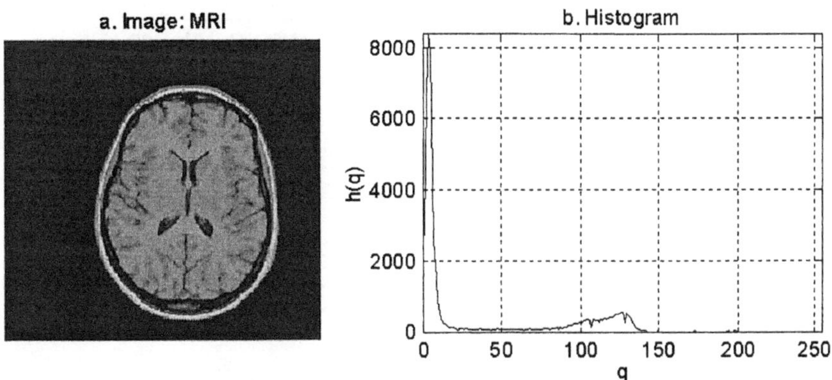

Fig. 7. a. MRI of a real head. b. Histogram.

Regarding the figures, it can be observed that the values of the cost and the fuzziness for each gray level decrease to a minimum, corresponding to the optimal threshold value, and then rise when the intensity increases.

The membership value is maximum at each one of the two class means and decreases in proportion to the increase between the gray level and its class mean. It can be seen that the membership values are always greater than 0.5.

The shape of the cost function and the measure of fuzziness changes for different values of b and d. When b and d decrease, their shapes become sharper the closer they are to the minimum. If these parameters rise, the shape of the curves is smoother and the threshold value could change because the shape becomes flat around the minimum.

Measure	Shannon	Kaufmann	Yager	Gauss
b	0.5	0.5	0.5	0.5
d	-	0.5	0.5	-
t^*	42	27	48	59
b	0.5	0.5	0.5	0.5
d	-	1	1	-
t^*	42	46	46	59
b	0.5	0.5	0.5	0.5
d	-	2	2	-
t^*	42	59	46	59
b	1	1	1	1
d	-	2	2	-
t^*	58	64	59	64
b	3	3	3	3
d	-	2	2	-
t^*	69	99	73	75

Table 1. Optimal threshold values obtained for the "head" image.

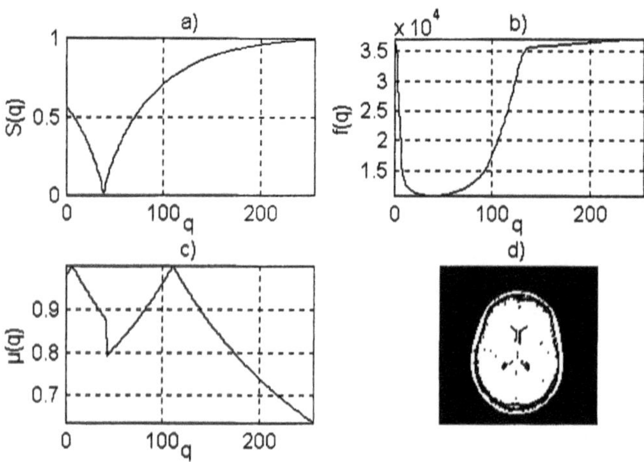

Fig. 8. Thresholding results of MRI using Shannon's measure of fuzziness a. Cost function. b. Measure of fuzziness. c. Membership function for $t^* = 42, m_b = 5.78, m_o = 110.03, D = 255, b = 0.5$. d. Thresholded image.

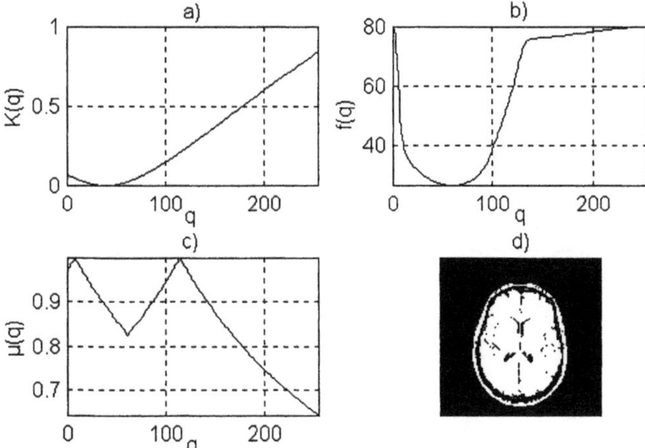

Fig. 9. Thresholding results of MRI using Kaufmann's measure of fuzziness a. Cost function. b. Measure of fuzziness. c. Membership function for $t^* = 27, m_b = 5.03, m_o = 106.04, D = 255, b = 0.5, d = 0.5$. d. Thresholded image.

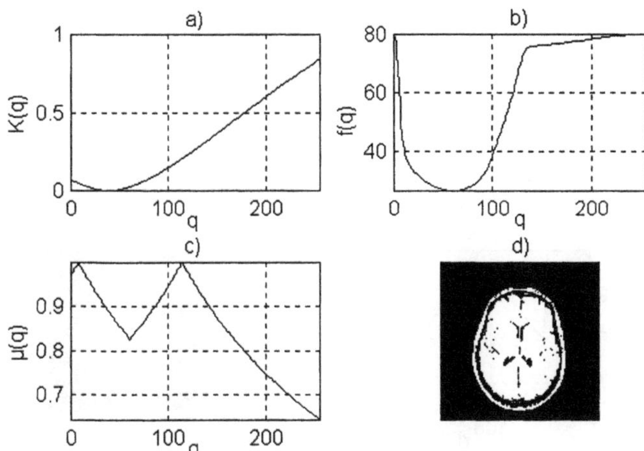

Fig. 10. Thresholding results of MRI using Kaufmann's measure of fuzziness a. Cost function. b. Measure of fuzziness. c. Membership function for $t^* = 59, m_b = 7.05, m_o = 113.95, D = 255, b = 0.5, d = 2$. d. Thresholded image.

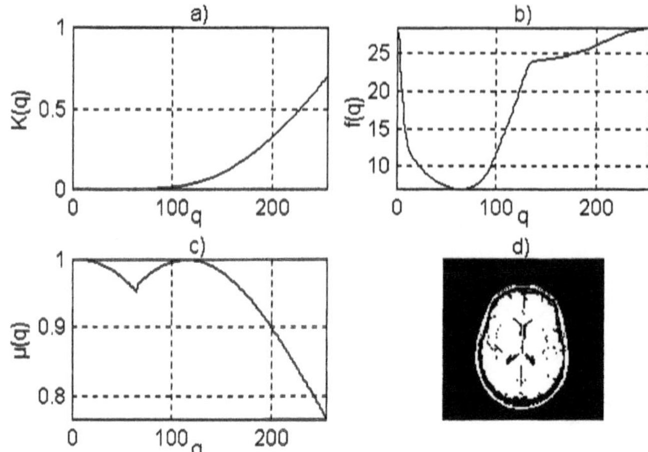

Fig. 11. Thresholding results of MRI using Kaufmann's measure of fuzziness a. Cost function. b. Measure of fuzziness. c. Membership function for $t^* = 64, m_b = 7.47, m_o = 114.93, D = 255, b = 1, d = 2$. d. Thresholded image.

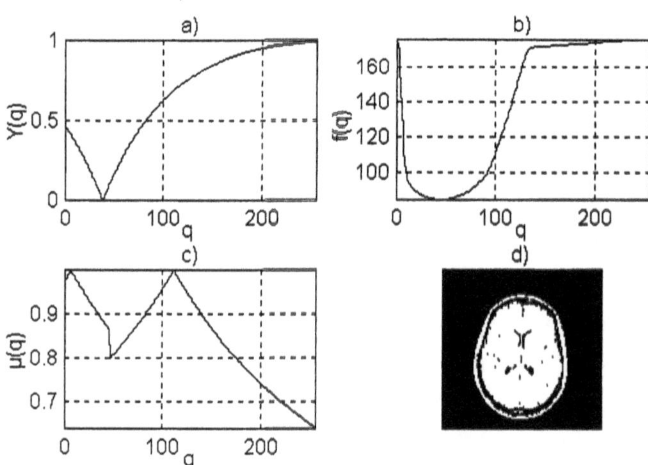

Fig. 12. Thresholding results of MRI using Yager's measure of fuzziness a. Cost function. b. Measure of fuzziness. c. Membership function for $t^* = 46, m_b = 6.03, m_o = 110.94, D = 255, b = 0.5, d = 2$. d. Thresholded image.

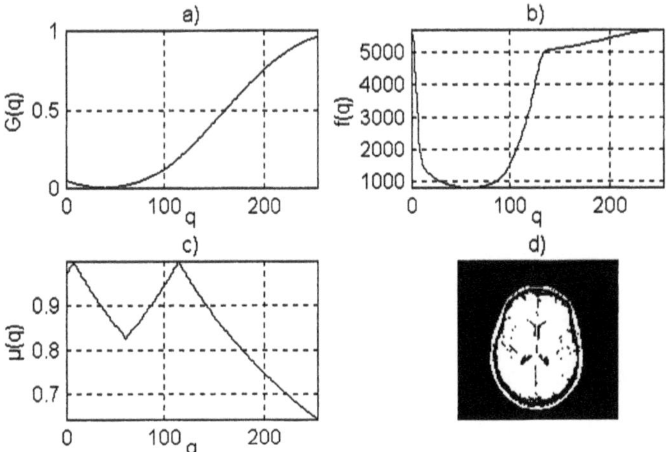

Fig. 13. Thresholding results of MRI using Gauss's measure of fuzziness a. Cost function. b. Measure of fuzziness. c. Membership function for $t^* = 59, m_b = 7.05, m_o = 113.95, D = 255, b = 0.5$. d. Thresholded image.

7 Multilevel luminance thresholding

Until now, the two-class classification problem only has been taken into account. However, a multilevel segmentation can be made. The methods seen can be extended to multithresholding.

A first approximation consists of extending the method seen above to multilevel thresholding [1]. It is assumed that the histogram of the image is multimodal. It is necessary to obtain k different gray level regions of an image, therefore $k-1$ optimal threshold values $t_1, \ldots, t_k$ must be selected.

The normalized fuzziness function of the image is simplified by:

$$f(q, t_1, \ldots, t_{k-1}) = \sum_{a=1}^{k-1} \sum_{q=t_{a-1}}^{L-1} h(q) F_a[\mu_Q(q, t_1, \ldots, t_{k-1})]$$

where $t_0 = 0$ and the membership function is expressed as:

$$\mu_Q(q, t_1, \ldots, t_{k-1}) = \begin{cases} \frac{1}{1+\frac{|q-m_1(t)|^{2b}}{D}} & \text{if } q \leq t_1 \\ \frac{1}{1+\frac{|q-m_2(t)|^{2b}}{D}} & \text{if } t_1 < q \leq t_2 \\ \vdots & \\ \frac{1}{1+\frac{|q-m_k(t)|^{2b}}{D}} & \text{if } q > t_{k-1} \end{cases}$$

The optimal thresholds are the t_{k-1} values of t that minimize the fuzziness function $f(q, t_1, \ldots, t_{k-1})$.

A second approximation is made using a recursive thresholding. The process consists of separating the darker regions from the brighter ones during the first stage by applying the previous technique. Then, new histograms are obtained of each segmented region and a new segmentation is made using the new histograms. The process continues until each new histogram becomes unimodal [15].

8 Color processing

The histogram of the images has been used previously for image segmentation by using thresholding techniques. However, they are not directly adapted for color image segmentation because in this case the histogram is multidimensional. A gray scale image $r(x, y$ can be obtained from a color image $\mathbf{q}(x, y)$ by using different transformations. Supposing that an image is in RGB color format, the simplest and most commonly used transformation is:

$$r(x, y) = \frac{\mathbf{q}_r(x, y) + \mathbf{q}_g(x, y) + \mathbf{q}_b(x, y)}{3}$$

where q_r, q_g, q_b are the red, green and blue components. Then, the resulting image is segmented as before.

The concept of multithresholding for gray level segmentation can be used for employment in multichannel image segmentation [15]. The image is segmented after a threshold value is selected in any of the color channels. The process is repeated until the histograms for each region become unimodal or the resulting segmentation is sufficient. Forero et al. [4] propose another approach for fuzzy segmentation of color images based on the information that is entailed in each separate chromatic histogram and the proportion of one color channel in relation to another.

A group of fuzzy rules for segmenting images are built from the information of the color images. The first step in developing a set of fuzzy rules to segment an image is to determine the antecedent fuzzy sets. Hence, it is necessary to study the spectral behavior of the images by looking for the conditions that could separate regions in each of the color image channels. In order to build fuzzy rules, the regional conditions which determine the antecedents of the rules of inference, are found via histogram analysis applied to each color component as was seen before in order to establish a particular language rule. Sometimes a channel does not give sufficient information to construct the fuzzy rules and can be ignored. Also, the information obtained from the histograms of each color component is not always enough for segmenting an image. In these cases, the information given by the relationship between the ratio of two channels can make it possible to improve the results.

9 Color application

The above mentioned color technique was developed on a database of 30 "bacilli" images. An example of an image is shown in Fig. 14.a. The main characteristics of these images are:

- Each region in the background and bacilli have similar colors, intensity and texture.
- The colors shown by bacilli fluctuate between fluorescent green and yellow and in some cases become white.

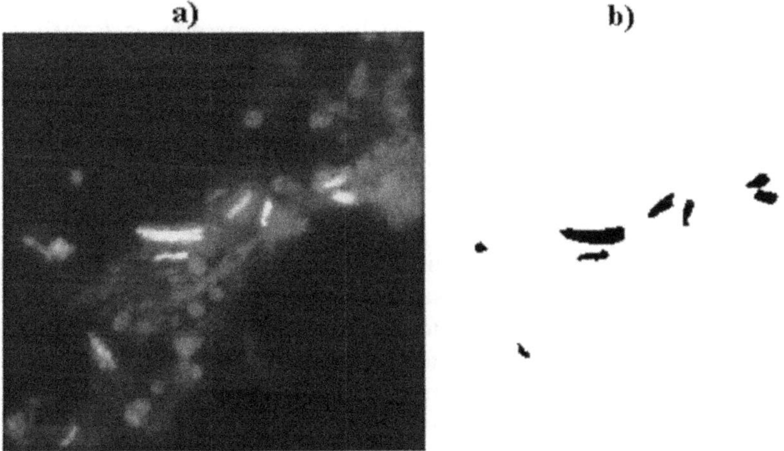

Fig. 14. Bacilli analysis. a) Original bacilli color image. b)Result after the fuzzy rules were applied.

For each image, color channel histograms were obtained and several background and object pixels were picked up for image analysis. Given that images are in RGB color format, it was observed in the histograms that the bacillus that appears in green shows a high value in the green channel and very low intensity in the others. The yellow bacillus shows a high and similar intensity in the green and red channels. Finally, when the bacillus appears white, the intensity in all three channels is similar and presents a high value. On the other hand, the background has a color that can change between black, red or blue. When it appears red, it is due mainly to artifacts. When the background is black, it was observed that intensities are very low in all channels and when the background is red, it shows a higher red intensity than the green one although that value can be high too. The blue background appears when the blue channel has a higher value compared with the other channels. Based on these characteristics, the membership functions of the background and bacilli fuzzy sets were constructed. The gray level techniques seen above were employed to select threshold values. The techniques were applied for each channel and the resulting binary images were employed to choose the thresholds to be employed. Finally, a L membership function was taken for the background set and a Γ membership function for the bacilli set.

A L membership function is specified by two parameters a, b with $0 < a < b$ as follows:

$$\mu_Q(q) = \begin{cases} 1 & \text{if } q < a \\ \frac{b-q}{b-a} & \text{if } a > q > b \\ 0 & \text{if } q > b \end{cases}$$

A Γ membership function is specified by two parameters a, b with $0 < a < b$ as follows:

$$\mu_Q(q) = \begin{cases} 0 & \text{if } q < a \\ \frac{q-a}{b-a} & \text{if } a > q > b \\ 1 & \text{if } q > b \end{cases}$$

Figure 15 presents the membership functions established for each fuzzy set in the red and green channels. It was found that the blue channel does not give sufficient information to construct the fuzzy rules and can be ignored. However, it was also found that the information obtained from the histograms was not enough for segmenting the images. So the ratio between green and red ratio was employed. Taking into account the fact that the bacilli fluctuate between green, yellow and white, this ratio makes it possible to establish the membership function of each pixel as bacilli or background.

Beginning with the membership functions, simple fuzzy rules were established to determine whether each pixel belongs to the bacilli O or to the background B sets:

If $\left(\mathbf{q}_r(x,y) \in B \vee \mathbf{q}_g(x,y) \in B\right) \wedge \left(\frac{\mathbf{q}_g(x,y)}{\mathbf{q}_r(x,y)} \in B\right)$ then $\mathbf{q}(\mathbf{x},\mathbf{y}) \in B$

If $\left(\mathbf{q}_r(x,y) \in O \vee \mathbf{q}_g(x,y) \in O \vee \frac{\mathbf{q}_g(x,y)}{\mathbf{q}_r(x,y)} \in O\right)$ then $\mathbf{q}(x,y) \in O$

Figure 14.b shows the resulting image of the application.

From the visual aspect of the binary images, it has been found that this technique allows very good results. The advantage of this technique is its flexibility, i.e. by changing the rules it is easy to modify the membership of the region. By analyzing the color of each region in the image, a relationship between the colors can be established in order to segment the image.

10 Conclusion

In summary, this chapter has presented a survey of various fuzzy thresholding methods employing the concept of measures of fuzziness applied to gray level

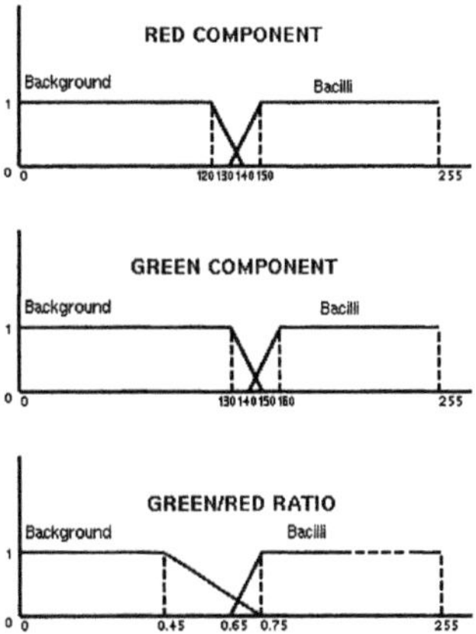

Fig. 15. Membership functions associated with the bacilli and background sets for the red and green channels and the green to red ratio.

images. The measures of fuzziness can be expressed as a function of cost. The requirements that measures of fuzziness and cost functions must accomplish have been presented. New measures can be proposed as long as they fit these requirements. Three of the cost methods described here optimize a criterion function. The membership functions can be modified or changed and some of the measures of fuzziness are flexible and can be modified by changing the parameters they have in order to adjust the threshold value. This method is not directly adapted for color image segmentation because in this case the histogram is multidimensional.

A technique has been presented for color image segmentation. Fuzzy rules are established by using sample pixels and analyzing the color histograms. The technique is developed by applying the thresholding technique to adjust the rules and to determine the membership functions for each set and threshold values for each color channel and color ratio.

References

1. Chi Z., Yan H. and Pham T., *Optimal image thresholding*, Fuzzy algorithms: with applications to image processing and pattern recognition, World Scientfic, Singapore, pp. 45-84, 1998.
2. De Luca A. and Termini S., *A definition of a non-probabilistic entropy in the setting of fuzzy sets theory*, Information and control, Vol. 20, pp. 301-312, 1972.
3. Forero M.G. and Rojas O., *New formulation in image thresholding using fuzzy logic*, 11th portuguese conference on pattern recognition RECPAD2000, pp. 117-124, 2000.

4. Forero M.G., Sierra E.L., Alvarez J., Pech J., Cristóbal G.,Alcalá L. and Desco M., *Automatic sputum color image segmentation for tuberculosis diagnosis*, Proceedings of SPIE: Algoriths and systems for optical information processing V. Javidi Bahram and Psaltis Demetri Eds. Vol. 4471, pp 251-261, 2001.
5. Glasbey C.A., *An analysis of histogram-based thresholding algorithms*, CVGIP: Graphical models and image processing, Vol. 55, No. 6, pp. 532-537, 1993.
6. Huang L.K. and Wang M.J., *Image thresholding by minimimizing the measure of fuzziness*, Pattern recognition. Vol. 28, No. 1, pp. 41-51, 1995.
7. Jang J.S.R., Sun C.T. and Mizutani E., *Fuzzy sets*, Neuro-fuzzy and soft computing, pp. 13-46, 1997.
8. Kaufmann A., Introduction to the theory of fuzzy subsets, 1975.
9. Klir G.J. and Folger T.A., Fuzzy sets, uncertainty and information, Prentice Hall, Englewood Cliffs, 1988.
10. Knopfmacher J., *On measures of fuzziness*, Journal of mathematical analysis and applications, Vol. 49, pp. 529-534, 1975.
11. Lee S.U. and Chung S.Y., *A comparative study of several global thresholding techniques for segmentation*, Computer vison, graphics and image processing, Vol. 52, pp. 171-190, 1990.
12. Lin C.T. and Lee G., *Fuzzy measures*, Neural fuzzy systems: A neuro-fuzzy synergism to intelligent systems, Prentice Hall, Upper Saddle River, pp. 63-88, 1996.
13. Loo S.G., *Measures of fuzziness*, Cybernetica, Vol. 20, pp. 201-210, 1977.
14. Parker J.R., *Advances methods in grey-level segmentation*, Algorithms for image processing and computer vision, John Wiley and Sons, New York, pp. 116-149, 1997.
15. Pratt W.K., *Image segmentation*, Digital image processing, John Wiley and Sons, New York, pp. 597-627, 1991.
16. Sahoo P.K., Soltani S., Wong A.K.C. and Chen Y.C., *A survey of thresholding techniques*, Computer vision, graphics and image processing, Vol. 41, pp. 233-260, 1988
17. Sztandera L., Bock C., Trachtman M. and Velga J., *Soft computing and molecular modeling in chemical design*, National textile center annual report, pp. 1-10, 1999.
18. Yager R.R., *On the measure of fuzziness and negation. Part 1: Membership in the unit interval*, International journal of general systems, Vol.5, pp. 221-229, 1979.

Chapter 7

Color Image Segmentation by Analysis of $3D$ Histogram with Fuzzy Morphological Filters

Aymeric Gillet, Ludovic Macaire, Claudine Botte-Lecocq and Jack-Gerard Postaire

Laboratoire Automatique I3D
Universite Sciences et Technologies Lille 1
Cite scientifique - Bat P2 59 655 Villeneuve d'Ascq
France Tel : 00 33 - (0)3 20 43 41 69 Fax : 00 33 - (0)3 20 43 65 67
Ludovic.Macaire@univ-lille1.fr

Summary. This chapter presents a new color image segmentation method using fuzzy mathematical morphological filters applied to the $3D$ color histogram. Segmentation consists in detecting the different modes which are present in the $3D$ color histogram and associated with homogeneous regions. In order to detect these modes, we show how a color image can be considered as a fuzzy subset of pixels characterized by its membership function to a mode. This function is evaluated thanks to a concavity analysis of the $3D$ color histogram. Then, a fuzzy morphological transformation is applied to this membership function in order to enhance the modes. The effectiveness of our proposed fuzzy morphological approach is then illustrated with color images.

1 Introduction

A color is an additive combination of the three primary colors : red, green and blue. In color image processing, color features are evaluated from the stimuli R (red), G (green) and B (blue) [1]. The color of a pixel is usually given as three values corresponding to the (R, G, B) space.

Color image segmentation is defined as the partition of a color image into regions which are sets of connected pixels whose colors are homogeneous [2]. We consider that pixels, whose color feature levels are equal to R, G and B, are associated with points or observations in the three-dimensional color space (R, G, B). We assume that all the pixels belonging to one region give rise in the color space to one cluster of points which can be considered as a pixel class. Furthermore, pixels which form several distinct regions of the same color, are assigned to the same pixel class. The color image segmentation method proposed in this paper is based on an original unsupervised pixel classification scheme by means of mode detection.

Many statistical procedures aim at mode detection by estimating the probability density function (*pdf*) underlying the distribution of the observations [5]. The modes are characterized by high local concentration of observations in the color space separated by valleys with low local concentration of observations. Under the assumption that each mode of the *pdf* corresponds to a pixel class, color image segmentation can be viewed as a *pdf* mode detection problem.

The most widely used tools for the approximation of the *pdf* are the 1D histograms of the color features. Pioneer works are based on marginal analysis of the distribution of the pixel colors in the three-dimensional color space, considering one-dimensional color feature histograms [11]. Tominaga partitions the image into regions by a recursive analysis of these histograms [18]. Schettini proposes to apply a "scale space filter "process to each of the one-dimensional color feature histograms to determine modal intervals [15]. Then, the three-dimensional color space is partitioned into hexahedra which are defined by the end points of the modal intervals. The non-empty hexahedra correspond to pixel classes. However, well separated clusters in the three-dimensional color space may correspond to only a single mode in each of the one-dimensional color feature histograms.

As color is a three-dimensional information, we choose to use the $3D$ histogram for approximating the *pdf*. The color histogram is composed of bins whose coordinates are the three color features. Each bin indicates the number of pixels in the image which have this particular color. As the colors of pixels are usually quantified with 256 levels, a $3D$ histogram requires a considerable amount of computer memory. This explains why only few authors analyze the $3D$ histogram for color image segmentation [9]. A simple solution consists in color quantization by only considering the highest bits for each feature in order to reduce the computational complexity. This solution yields poor results, because this quantization phase leads to a very crude segmentation. Another approach consists in mode detection by thresholding the values of the bins. Nevertheless, the adjustment of the threshold is very difficult.

To avoid this problem, Shafarenko proposes to apply the watershed algorithm to the chromaticity $2D$ histogram in the CIE (L, u^*, v^*) color space [17]. It consists in reducing the acquisition noise thanks to an adapted size gaussian filtering. The smoothed histogram is then analyzed by the watershed algorithm to detect the modes which correspond to the pixel classes.

Park proposes to detect the clusters of the binary $3D$ histogram thanks to morphological filters [12]. First, the difference between two gaussian smoothed $3D$ histograms is thresholded in order to provide the binary $3D$ histogram. Then the initial seeds of clusters are detected from this binary $3D$ histogram thanks to a binary closing operation. Finally, an adaptive dilation is applied to the binary $3D$ histogram in order to extract the clusters. The aim of this adaptive dilation scheme is to make two neighboring clusters meet each other at the valley of the binary $3D$ histogram while preventing the clusters to expand toward empty bins. If two neighboring clusters meet during dilation, the dilation is stopped in the direction in which they tend to merge. The process is completed either after a pre-specified maximum number of iterations or when any cluster can be dilated any more. The results depend both on the size of the structuring element used for morphological operations and on the choice of the color space.

Basic binary morphological filters have proved to be an efficient tool for cluster detection when the cluster shapes are rather spherical [4]. But when there is no a priori knowledge about the shapes of the clusters to be detected, it is interesting to introduce some adaptivity in the morphological filters [19].

One problem with traditional clustering techniques is that there are only two values, either 1 or 0, to indicate the membership degree of each observation to a cluster. This requires well-defined boundaries between the clusters, which is not the usual case for real images. The uncertainties arising in image segmentation can be solved by using the fuzzy set theory.

Fuzzy C-Means is a method which allows ambiguous boundaries between clusters, and has received much attention [8]. This iterative optimization method evaluates the membership degree of an observation to each of the clusters by means of a fuzzy distance between this observation and each of the cluster centers. A each iteration, these cluster centers are updated and an objective criterion is used to minimize the fuzzy intra-cluster distance.

During the coarse stage of a "coarse to fine "segmentation algorithm, Lim analyses the $1D$ color feature histograms in order to detect the cluster centers [10]. During the fine stage, he uses the Fuzzy C-Means method to assign the pixels which were not classified during the coarse stage to the closest detected cluster. The applicability of this technique, which tries to solve the problem of cluster validity by using a coarse to fine concept, is limited by the 1D histograms thresholding. The fuzzy C-Means method requires the a priori knowledge of the number of clusters to be detected, and its computational cost is quite high with large data sets.

For color image segmentation purpose, Pham proposes the mountain clustering which uses the fuzzy integral as a distance measure [13]. This method does not require initial estimates of cluster centers. Nevertheless, segmentation results depend on two main parameters which are difficult to adjust, namely the spatial resolution of the image, which specifies the degree of detail in the segmentation process, and an integration threshold.

In this paper, we propose to define fuzzy morphological erosion and dilation filters specifically adapted to the detection of the modes of the $3D$ color histogram. For this purpose, we describe in the second section how to extract, from the $3D$ histogram, a fuzzy subset defined by its membership function. In the third section, we present a fuzzy morphological transformation which is applied to this membership function in order to enhance the different modes. In the last section, we describe the pixel classification scheme which analyses the detected modes in order to yield the segmented image. Extensive experiments are carried out on color images in this section and the effectiveness of our color image segmentation method is assessed.

2 Fuzzy subset extracted from the $3D$ color histogram

2.1 *Pdf* approximation

As we have mentioned in the first section, many clustering procedures are based on the detection of the modes of the approximated underlying *pdf*. In our context, we propose to approximate this *pdf* by the $3D$ color histogram.

We consider that a pixel, whose color features are (R, G, B), is associated with an observation X, whose coordinates are (R, G, B) in the $3D$ color space. All the observations X, which represent the different colors in the image, form the set Δ.

The color histogram is composed of bins which indicate the number $HISTO(X)$ of pixels with the corresponding (R, G, B) color features. This histogram, which is normalized between 0 to 1, can be considered as an approximation of the underlying *pdf*. For the segmentation purpose, the key problem is to assign the observations of Δ to the different classes identified by means of the detected modes.

This clustering problem is considered as a fuzzy decision process in order to cope with uncertainties in color quantization. Indeed, in the case of a real color image, the boundaries between the color clusters associated to the different regions constituting the image cannot be considered as well defined surfaces as when using the crisp decision theory.

Hence, we propose to implement a fuzzification phase. This step quantifies the degree with which each observation X belongs to a class associated with a mode. It allows to evaluate the confidence degree in the event "X belongs to a class associated to a detected mode ". For the sake of simplicity, this event will be referred as "X belongs to a mode "in what follows. This fuzzification requires the transformation of the crisp observation set Δ into a fuzzy subset M characterized by a membership function. This function, denoted μ_M, is evaluated thanks to a convexity analysis of the color histogram.

2.2 Membership function evaluation

The membership function defining the fuzzy subset M can be evaluated by considering that an observation X belongs to a mode if the *pdf* approximated at X is concave [14]. A convexity test locally assigns a "concave "or "convex"label to each observation of the color space. The local convexity of the approximated *pdf* at an observation X is evaluated by analyzing the variation of this *pdf* estimator when an observation domain grows around X. So, we first evaluate the estimator $p^1(X)$ of the *pdf* using a cubic observation domain D_X^1 which contains neighbors of X, and then, we compute another estimator $p^2(X)$ using the cubic observation domain D_X^2, slightly larger than D_X^1, such as

$$p^1(X) = \frac{\sum_{Y \epsilon D_X^1} HISTO(Y)}{(2.m+1)^3} \tag{1}$$

and

$$p^2(X) = \frac{\sum_{Y \epsilon D_X^2} HISTO(Y)}{(2.n+1)^3}. \tag{2}$$

The parameters $(2.m+1)^3$ and $(2.n+1)^3$ are respectively the volumes of the first and second observation domains, with $n > m$. By experiment, the sizes $(2.m+1)$ and $(2.n+1)$ are respectively taken equal to 3 and 5. In these conditions, if $p^2(X) < p^1(X)$, the approximated *pdf* is considered as locally concave and the observation X is labelled as "concave ", otherwise X is labelled as "convex ".

Now, we introduce a Π-fuzzification function which provides the confidence degree of the event "X is a modal observation ". This degree depends on $\frac{k^1(X)}{(2.m+1)^3}$, where $k^1(X)$ is the number of concave observations which are included in D_X^1. If the labels of most neighbors of X included in D_X^1 are concave (respectively convex), X is considered as a concave (respectively convex) observation, i.e as a modal (respectively non-modal) observation whose membership degree is close to 1 (respectively close to 0). So, the membership function μ_M is defined as:

$$\begin{array}{ll} \text{if } \frac{k^1(X)}{(2.m+1)^3} \leq a, & \text{then } \mu_M(X) = 0, \\ \text{if } \frac{k^1(X)}{(2.m+1)^3} > b, & \text{then } \mu_M(X) = 1, \\ \text{if } a < \frac{k^1(X)}{(2.m+1)^3} < b, & \text{then} \end{array}$$

$$\mu_M(X) = \frac{1}{2} + \frac{1}{2}.sin(\frac{\pi}{b-a}.(\frac{k^1(X)}{(2.m+1)^3} - \frac{a+b}{2})). \tag{3}$$

In this chapter, the values of the parameters a and b are respectively taken equal to 0.2 and 0.8, so that we consider that an observation belongs to a mode if 80% of the neighboring observations in D_X^1 are labelled as "concave ". After this fuzzification phase, M is well defined by its membership function μ_M. However, even with this fuzzification scheme, experiments have shown that some clusters, associated with weak modes, tend to be ignored in the final clustering scheme. To cope with this limitation of the procedure, we propose to apply fuzzy morphological filters to the membership function μ_M in order to enhance the modes of M.

3 Mode detection by fuzzy morphological filters

The basic morphological filters are erosion and dilation [7]. The main effect of erosion is to enlarge the valleys, by eliminating the irregularities of the distribution, but this transformation also tends to shrink the modes. On the other hand, dilation is used to enlarge the modes, but it also tends to fill the valleys [16]. So, we propose to transform the membership function μ_M by means of fuzzy morphological filters which take advantage of these two basic operations without their drawbacks.

The aim of this procedure is to increase the contrast between modes and valleys. High values of μ_M correspond to observations X belonging to the modes of M while low values are associated with observations standing in the valleys between the modes. If an observation X is close to a mode, it is relevant to increase μ_M, i.e to enhance the dilation effect at the location of this observation and to limit the erosion effect in order to preserve the mode. Conversely, if the observation is far from all modes, it is probably located in a valley. In this case, it is useful to decrease μ_M, i.e to erode the membership function at the location of this observation and not to dilate it. So it would be pertinent to erode or to dilate the membership function more or less depending on the location of each observation in the color space. This location can be introduced in the definition of the structuring functions used by both fuzzy erosion and dilation filters.

In this section, after presenting the classical definition of fuzzy erosion and dilation filters, we describe new fuzzy morphological filters which include structuring functions specifically adapted to mode enhancement. Finally, we describe the morphological transformation which enhances the modes of M.

3.1 Classical fuzzy erosion and dilation filters

Among the available fuzzy erosion and dilation filters, we have selected the most classical one, which according to Bloch, corresponds to the greater fuzzy effect of dilation and erosion [3].

The chosen classical fuzzy erosion and fuzzy dilation filters at an observation X is defined as :

$$E_\gamma[\mu_M(X)] = \min_{Y \epsilon S_X} (\max[\mu_M(Y), 1 - \gamma(X - Y)]) \tag{4}$$

$$D_\gamma[\mu_M(X)] = \max_{Y \epsilon S_X} (\min[\mu_M(Y), \gamma(X - Y)]), \tag{5}$$

where S_X is the fuzzy structuring element centered at the observation X and defined by its membership function γ called the structuring function. This structuring function used by the erosion and dilation depends on the distances between X and each of the neighboring observations Y in S_X.

For the purpose of illustration, let us consider the $1D$ membership function μ_M of Fig. 1 and the $1D$ structuring function γ which is set to 1, for all the neighboring observations Y in S_X, the structuring element of size 3. The eroded and dilated membership functions $E_\gamma[\mu_M]$ and $D_\gamma[\mu_M]$ are presented in Figs. 2 and 3 respectively,where μ_M is represented by the dotted graph. This example shows that the fuzzy erosion tends to enlarge the valleys but tends to shrink the modes, whereas the fuzzy dilation tends to enlarge the modes while filling the valleys.

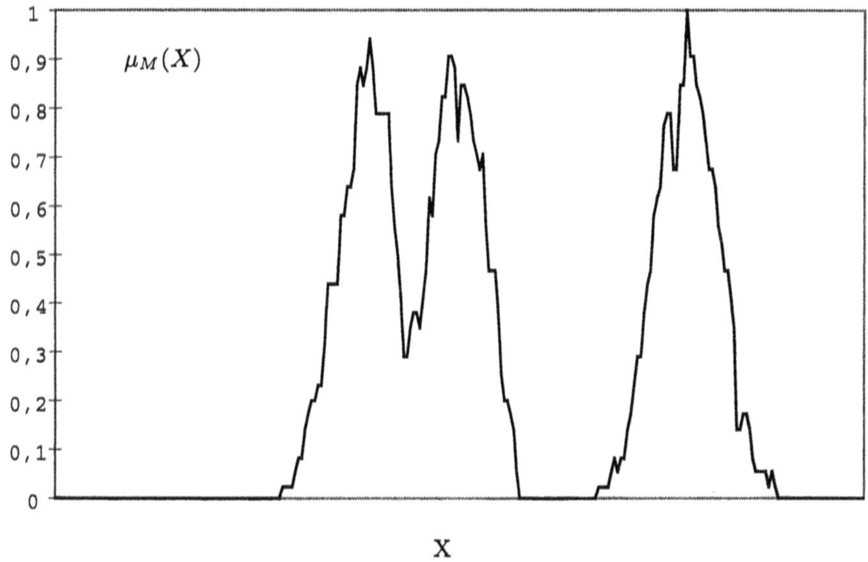

Fig. 1. Original $1D$ membership function μ_M.

For the purpose of mode detection, it would be pertinent to define new fuzzy morphological filters using different structuring functions for erosion and dilation.

3.2 Fuzzy erosion and dilation filters for mode detection

In the classical definitions Eqs. (4) and (5), the structuring function γ depends on the distance between the considered observation X and each of its neighbors Y, but does not depend on the membership degree of the neighbors. If a neighboring observation is characterized by an high membership degree, it would be interesting to take it into account only for the dilation purpose at X, but not for the erosion. On the other hand, if the membership degree of a neighbor is low, its influence should be higher for erosion than for dilation at X.

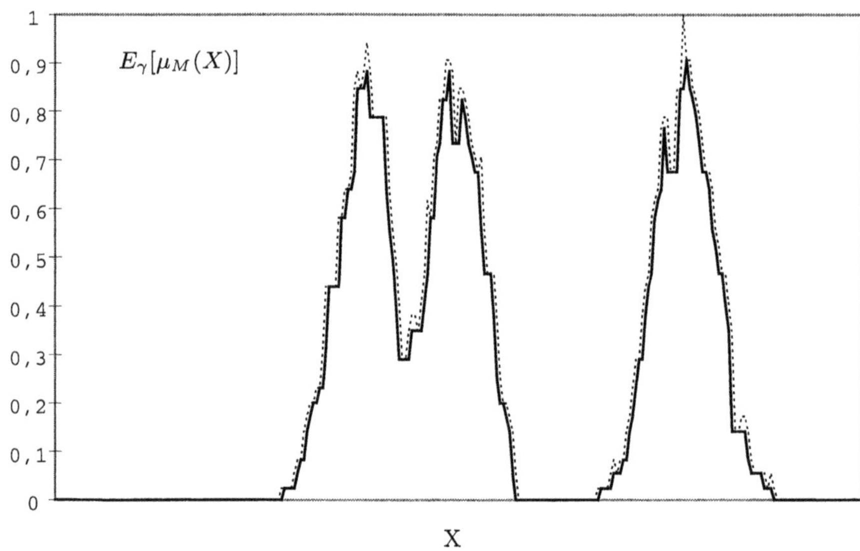

Fig. 2. Result of the $1D$ fuzzy erosion $E_\gamma[\mu_M]$.
Size of $S_X = 3$.

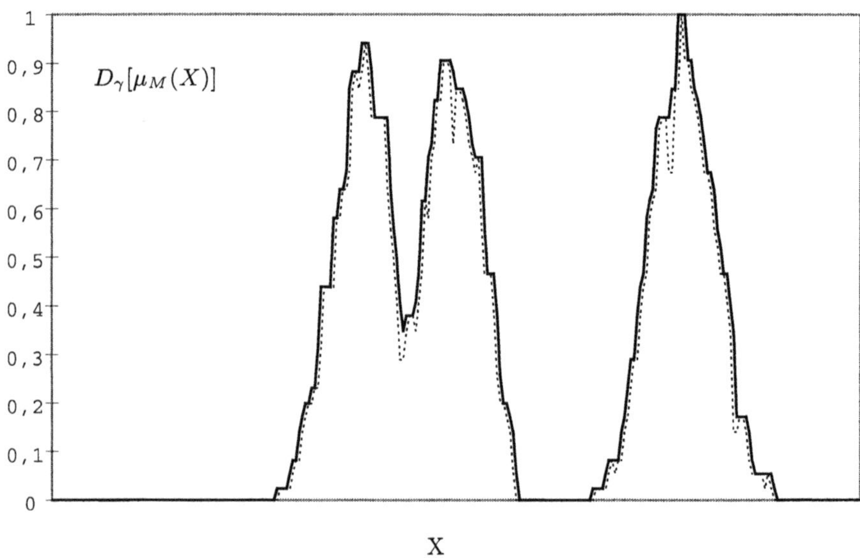

Fig. 3. Result of the $1D$ fuzzy dilation $D_\gamma[\mu_M]$.
Size of $S_X = 3$.

In order to erode and to dilate depending on the membership degrees of the neighbors in S_X, we define new morphological filters associated with the fuzzy structuring element S_X:

$$E'_{\gamma_E}[\mu_M(X)] = \min_{Y \epsilon S_X} (\max[\mu_M(Y), 1 - \gamma_E(Y)]) \tag{6}$$

$$D'_{\gamma_D}[\mu_M(X)] = \max_{Y \epsilon S_X} (\min[\mu_M(Y), \gamma_D(Y)]), \tag{7}$$

where γ_E and γ_D are the structuring functions respectively defined for fuzzy erosion and dilation and described in Sect. 3.3. These functions do not depend on the distance between between the observation X and each of its neighbors Y in S_X but on the membership degrees of these neighbors.

In order to detect the modes, the structuring functions γ_E and γ_D have to be defined so that the erosion enlarges the valleys without shrinking the modes, while the dilation enlarges the modes without filling the valleys.

3.3 Structuring functions adapted to mode detection

In the color space, a neighboring observation Y belonging to the structuring element S_X can be located in a mode, in a valley, or in the border between a mode and a valley. If this observation is close to a mode, its membership degree $\mu_M(Y)$ is high, whereas if Y is close to a valley, its membership degree is low. Furthermore, a neighbor Y which is close to the border between a valley and a mode, is characterized by significant local spatial variations between its membership degree and the membership degrees of its neighbors. The combination of these two criteria is well suited to decide if an observation is located in a mode, a valley or a border. We present now how to introduce these criteria into the structuring functions for mode enhancement.

Structuring function associated with the fuzzy erosion. In order to detect the modes, we have to erode the valleys while not deepening the modes. This fuzzy erosion is performed with a structuring element S_X, centered at each observation X and using a structuring function γ_E defined for each of the neighboring observations Y. An observation X is considered all the more as belonging to a valley (respectively a mode) as its neighbors are also considered as belonging to a valley (respectively a mode). In this case, the value of the structuring function at each neighbor Y belonging to a valley will be set to 1, so that its effect will be high for the erosion operation. If the neighbor Y is located in a mode, its effect on the erosion will have to be very low.

To meet these requirements, we propose for fuzzy erosion expressed by Eq. (6), the structuring function defined as:

$$\begin{aligned}&\gamma_E(Y) = 1 \qquad \text{if}[\mu_M(Y).(1 - g_{\mu_M}(Y))] \leq TH,\\ &\gamma_E(Y) = \mu_M(X) \text{ otherwise,}\end{aligned} \tag{8}$$

where TH is a threshold and $g_{\mu_M}(Y)$ is an approximation of the local spatial variations of $\mu_M(Y)$, evaluated as the mean of the local differences between the membership degree of Y and the membership degrees of its neighbors. To be more specific, $g_{\mu_M}(Y)$ is the normalized mean absolute difference between the membership degree of Y and the membership degrees of the neighboring observations which belong to the domain D_Y^1 centered at Y:

$$g_{\mu_M}(Y) = \frac{\sum_{Z \epsilon D_Y^1} |\mu_M(Y) - \mu_M(Z)|}{(2.m + 1)^3}. \tag{9}$$

If the expression $[\mu_M(Y).(1 - g_{\mu_M}(Y))]$ is lower than the threshold TH, a neighbor Y is considered as being located in a valley or in a border close to a valley. In this case, $\gamma_E(Y)$ is set to 1 so that the effect of the neighbor Y on the fuzzy erosion at the observation X is the highest one. If $[\mu_M(Y).(1 - g_{\mu_M}(Y))]$ is higher than the threshold TH, Y is considered as being located in a modal observation or in a border close to a mode. In this case, $\gamma_E(Y)$ is set to $\mu_M(X)$, so that this neighboring observation Y has no effect on the fuzzy erosion at the observation X.

When considering all the neighboring observations Y in the structuring element S_X, the erosion of the membership function μ_M is performed with only the neighboring observations which are located in a valley or in a border close to a valley. This erosion using the structuring function γ_E tends to deepen the valleys and to preserve the modes.

Structuring function associated with the fuzzy dilation. In the same way, the structuring function associated with the fuzzy dilation expressed by Eq. (7) can be defined as :

$$\begin{aligned}&\gamma_D(Y) = 1 \qquad \text{if}[\mu_M(Y).(1 - g_{\mu_M}(Y))] > TH,\\ &\gamma_D(Y) = \mu_M(X) \text{ otherwise.}\end{aligned} \tag{10}$$

If $[\mu_M(Y).(1 - g_{\mu_M}(Y))]$ is higher than TH, a neighbor Y, which belongs to the structuring element S_X, can be considered as being located in a mode or in a border close to a mode. Its effect on the fuzzy dilation at the observation X is the highest one. Conversely, if $[\mu_M(Y).(1 - g_{\mu_M}(Y))]$ is lower than TH, the observation Y is considered as being located in a valley or in a border close to a valley, and has no effect on the fuzzy dilation.

When considering all the neighbors Y in the structuring element S_X, the fuzzy dilation takes only into account the neighbors Y which are located in or near a mode. Dilating the membership function μ_M using this structuring function γ_D tends to enhance the modes without filling the valleys.

Figures 4 and 5 display the fuzzy eroded and dilated membership functions of Fig. 1 respectively, when the size of S_X is set to 3. The threshold TH is set to $0.5(1-0.1) = 0.45$, as to consider that an observation Y is located in a valley or in a border close to a valley if its membership degree is less than 0.5 and its local membership degree variation is higher than 0.1. Figure 4 shows that the membership function is only eroded at the observations X which are located in the valleys. Conversely, Fig. 5 shows that μ_M is dilated only at the observations which are located near the modes.

These results can be compared with those obtained by means of the classical morphological filters and illustrated by Figs. 2 and 3. We can see that the fuzzy classical erosion tends to shrink the modes while the proposed fuzzy erosion tends to only deepen the valleys. Furthermore, the classical fuzzy dilation tends to fill the valleys while the proposed fuzzy dilation tends to only enhance the modes. This example shows the improvement achieved with the proposed fuzzy morphological filters with respect to the using of the classical ones for mode enhancement.

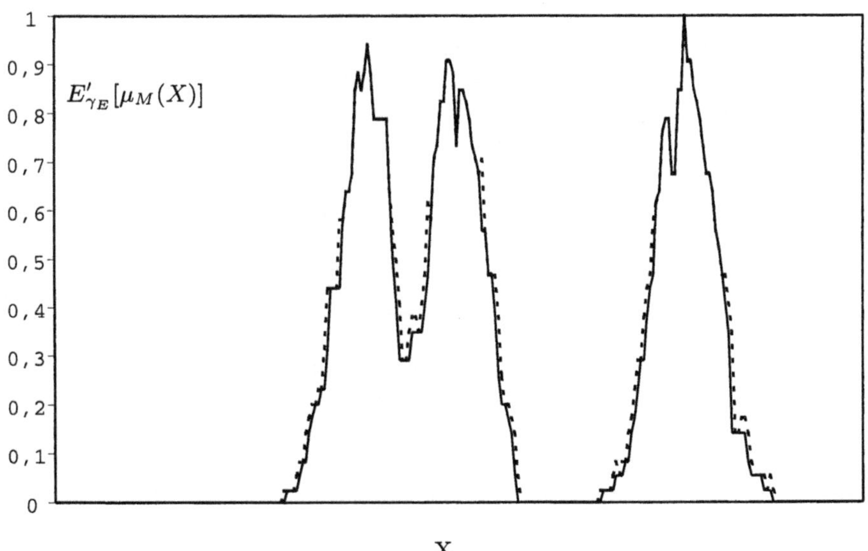

Fig. 4. Result of the $1D$ fuzzy erosion $E'_{\gamma_E}[\mu_M]$.
Size of $S_X = 3$ and $TH = 0.5(1-0.1) = 0.45$.

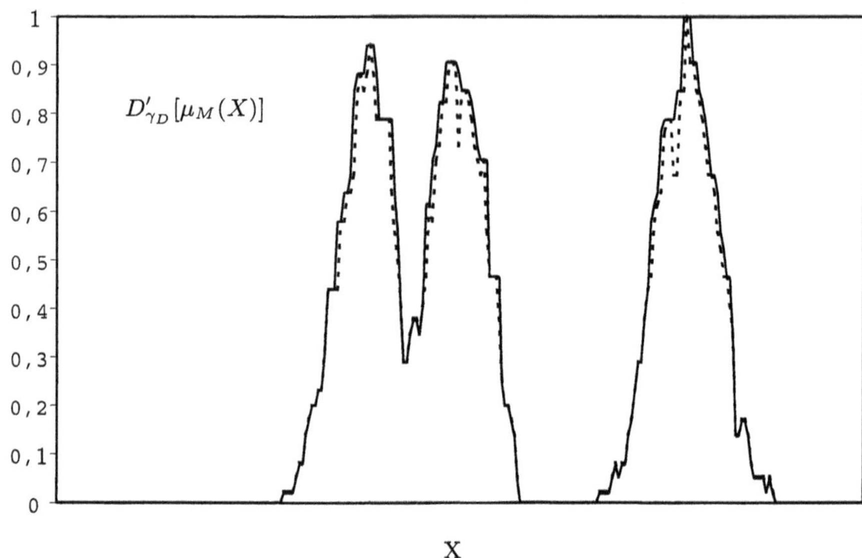

Fig. 5. Result of the $1D$ fuzzy dilation $D'_{\gamma_D}[\mu_M]$.
Size of $S_X = 3$ and $TH = 0.5(1 - 0.1) = 0.45$.

3.4 Fuzzy morphological transformation for mode extraction

In order to take advantage of the two fuzzy filters defined above, we propose to combine them into a fuzzy morphological transformation, denoted t, which performs a fuzzy erosion of the fuzzy subset M using the structuring function γ_E, followed by a fuzzy dilation of the resulting fuzzy subset using the structuring function γ_D [6].

This transformation yields the fuzzy subset M^t, characterized by its mode membership function μ_M^t, defined as :

$$\mu_M^t = t(\mu_M) = D'_{\gamma_D}[E'_{\gamma_E}(\mu_M)]. \quad (11)$$

Figure 6 presents the resulting membership function μ_M^t, displayed as a solid graph, of the function μ_M presented in Fig. 1 and displayed as a dotted graph. As required, the modes are enhanced while the main valleys are not filled. However, the effect of this transformation is rather weak. Hence, we propose to iterate it until the resulting mode membership function μ_M^T is stabilized, such as:

$$\mu_M^T = (\mu_M^t)^\infty \text{ with } (\mu_M^t)^j = t[(\mu_M^t)^{j-1}] = D'_{\gamma_D}[E'_{\gamma_E}[(\mu_M^t)^{j-1}]] \text{ , } j = 1, ..., \infty. \quad (12)$$

The initial conditions are $(\mu_M^t)^0 = \mu_M$.

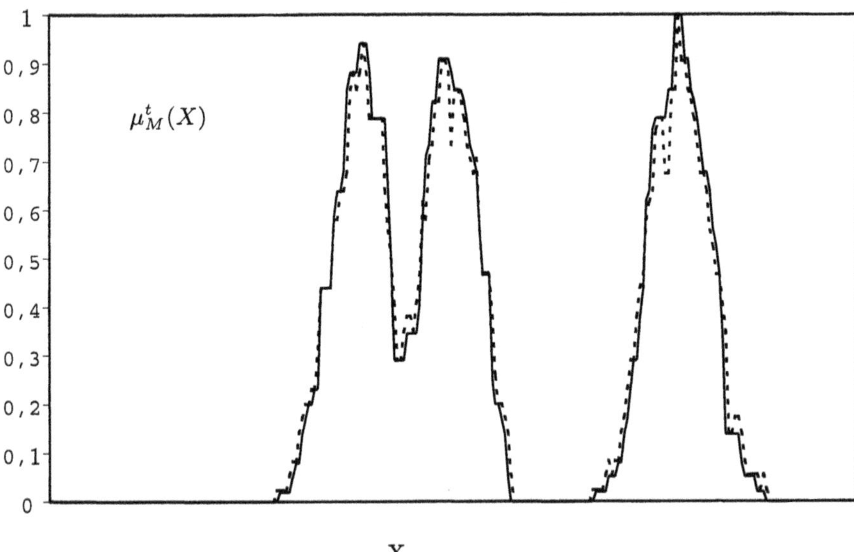

Fig. 6. Result of the transformation μ_M^t.
Size of $S_X = 3$ and $TH = 0.5(1 - 0.1) = 0.45$.

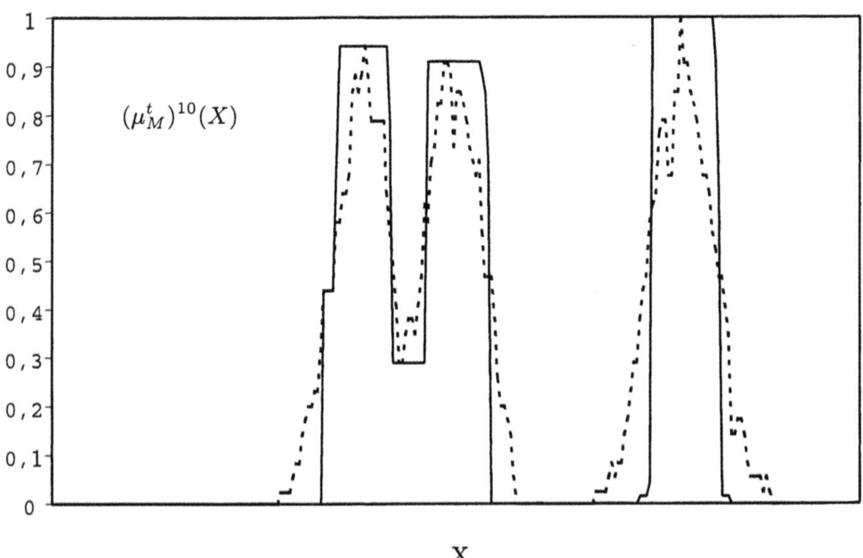

Fig. 7. Result of the transformation $(\mu_M^t)^{10}$.
Size of $S_X = 3$ and $TH = 0.5(1 - 0.1) = 0.45$.

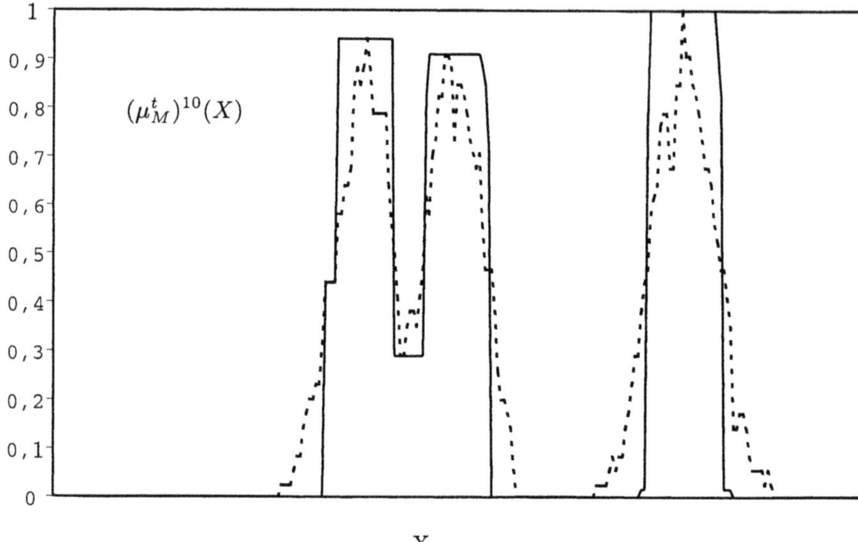

Fig. 8. Result of the transformation $(\mu_M^t)^{10}$.
Size of $S_X = 3$ and $TH = 0.5(1 - 0.2) = 0.40$.

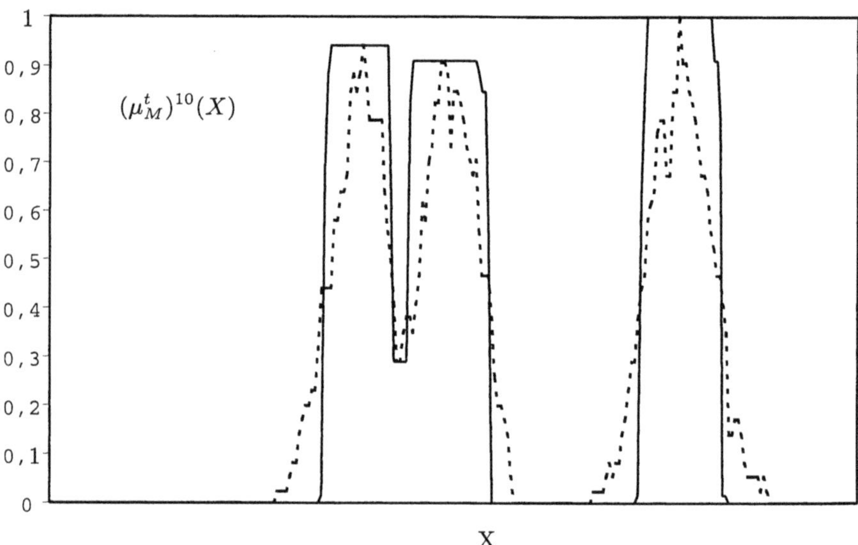

Fig. 9. Result of the transformation $(\mu_M^t)^{10}$.
Size of $S_X = 3$ and $TH = 0.5(1 - 0.3) = 0.35$.

Figure 7 shows the resulting membership function $(\mu_M^t)^{10}$ after 10 iterations. The three modes are well enhanced and it is important to notice that the higher and lower membership degrees associated to the modes and the valleys are preserved. Indeed the fuzzy dilation performed at the observation whose membership degree is the highest, propagates this high degree to its neighbors which belong to the structuring element centered in this observation and which are considered as being located in this mode or a border. Conversely, the proposed erosion performed at the observation whose degree is the lowest, propagates this low degree to the neighbors which belong to the structuring element centered in this observation and which are considered as being located in this valley or in a border.

The mode enhancement procedure is sensitive to the choice of the parameter TH used to decide if the neighboring observations act on the morphological filters.

Figures 7, 8, 9 display the transformed membership function $(\mu_M^t)^{10}$ with different values of TH, which are $0.5(1-0.1) = 0.45$, $0.5(1-0.2) = 0.40$ and $0.5(1-0.3) = 0.35$ respectively. These figures illustrate that lower the parameter TH is, the wider the modes are and the narrower the valleys are. This behavior of the procedure results from a relationship between the value of TH and the number of neighbors of X taken into account by the filtering process. The lower TH is, the stronger the effect of the dilation is.

The modes are easily detected thanks to the defuzzification of the transformed membership function μ_M^T, which is based on the convexity test described in the second section.

The local convexity of the transformed membership function μ_M^T at an observation X is evaluated by analyzing its variation when an observation domain grows around X. So, we first evaluate the estimator $p_*^1(X)$ of μ_M^T using the cubic observation domain D_X^1, and then, we compute another estimator $p_*^2(X)$ using the cubic observation domain D_X^2, slightly larger than D_X^1, such as:

$$p_*^1(X) = \frac{\sum_{Y \in D_X^1} \mu_M^T(Y)}{(2.m+1)^3} \tag{13}$$

and

$$p_*^2(X) = \frac{\sum_{Y \in D_X^2} \mu_M^T(Y)}{(2.n+1)^3}. \tag{14}$$

As for the fuzzification phase by means of convexity test, we set the sizes $(2.m+1)$ and $(2.n+1)$ to 3 and 5 respectively. In these conditions, if $p_*^2(X) < p_*^1(X)$, the transformed membership function μ_M^T is considered as locally concave and so we decide that the observation X belongs to a mode.

4 Experimental results

In order to illustrate the behavior and the efficiency of the above mode detection procedure, we propose to segment the classical real image of the "house" (see Fig. 10) and one synthesis image (see Fig. 16). The segmentation of the house image is quite challenging because it is difficult to extract regions which take into account the shadow effect. Furthermore, some regions such as the roof and the walls present local color non-homogeneities.

Fig. 10. The "house" color test image.

As it is difficult to display the values of the bins of the $3D$ color histogram, we present the three $1D$ color feature histograms (see Fig. 11). They show that it is not easy to discriminate the pixel classes which correspond to the different regions.

In order to extract the modes from the analysis of the $3D$ color histogram of the image of Fig. 10, we evaluate the membership function associated with the fuzzy subset M. Figures 12 and 13 present the support of M (defined by $\mu_M(X) \neq 0$), and the kernel of M (defined by $\mu_M(X) = 1$), respectively.

We can see, in Fig. 12, that the different modes corresponding to the pixel classes cannot be easily discriminated using only the support of M. Furthermore, the kernel of M only contains three main modes and some insignificant ones (see Fig. 13). These two figures show that for this image, the fuzzification step is not sufficiently discriminating and does not provide well defined and separated modes.

The mode enhancement is achieved thanks to the iteration of the fuzzy morphological transformation t defined by Eq. 11. The threshold TH is set to $[0.5(1-0.1)] = 0.45$ and the size of the structuring element is set to 3. After 3 iterations, the resulting membership function μ_M^T is stabilized and the modes

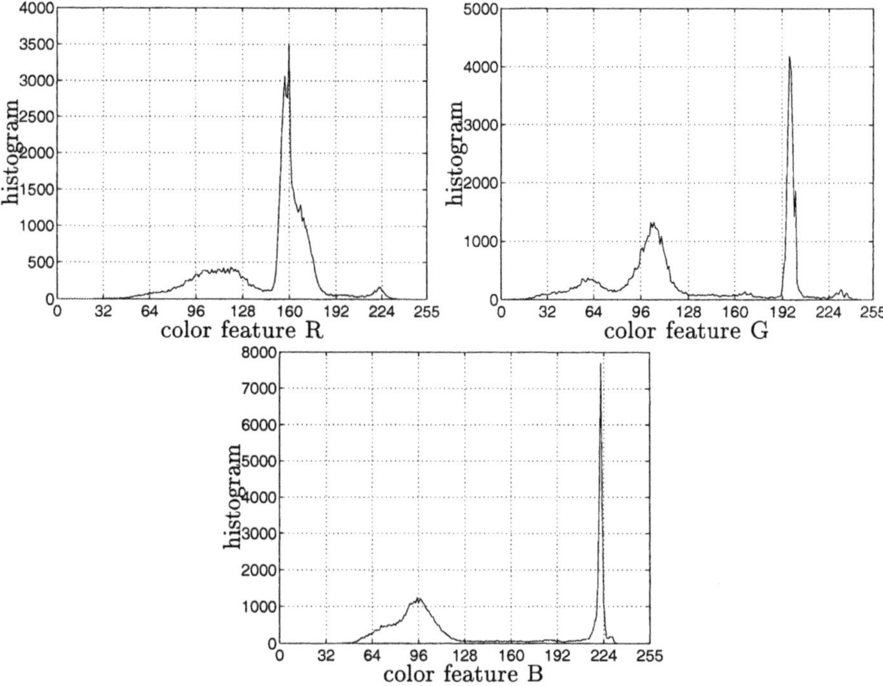

Fig. 11. $1D$ Color feature histograms of the image of Fig. 10.

are well enhanced. We propose to extract them through the defuzzification step. Figure 14 presents the six different modes which are extracted by means of our method. The modes are displayed as the subsets of observations that belong to each of them. A class of pixels is associated to each detected mode and is characterized by the gravity center of all the observations that belong to this mode. Each pixel of the analyzed color image is assigned to the class whose center is the nearest of the associated observation.

The pixels of the image of Fig. 15 are labelled with the colors defined by the coordinates of the centers of the classes to which they are assigned. This figure shows that the detected modes correspond to pixel classes which are not equiprobable. Indeed, the class of pixels which represents the wall of the house has a much larger population than the class which represents one of the windows. We can say that the number of classes determined by our scheme is relevant because the regions of the image of Fig. 10, such as the wall without the shadow, the wall with the shadow, the roof and the windows are well identified in the segmented image.

In order to evaluate the robustness of our approach, we propose to experiment a performance evaluation. It is based on the measure of the difference between the result of the segmentation and a reference segmentation. In order to make this reference segmentation available, we use the synthetic image

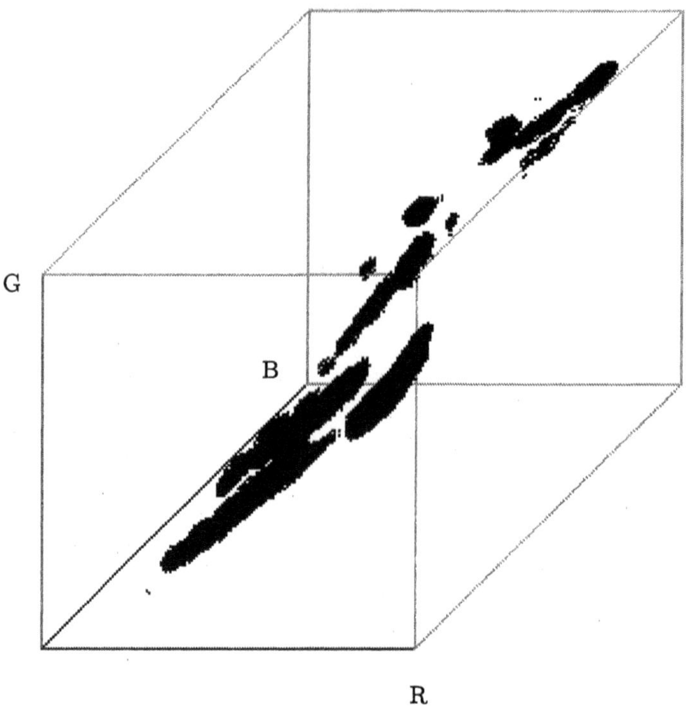

Fig. 12. Support of M extracted from the image of Fig. 10.

of Fig. 16. This synthetic image is composed of the 5 following regions with different shapes:

- a brown background R_1,
- an orange large square R_2,
- a red irregular shape R_3,
- a disc which is composed of 2 concentric regions:
  - a peripheral green ring R_4 and
  - a slightly darker green disc R_5.

The regions R_4 and R_5 are built so that their red mean values are equal and the differences between their green and blue mean values are equal to 10 and 20 levels, respectively.

In order to analyze the reliability of the mode detection procedure, we evaluate the effects of the noise on the number of detected modes. For this purpose, the synthetic image of Fig. 16 is corrupted by a non-correlated gaussian noise with a standard deviation σ which is independently added to each of the three color features. Figure 17 shows the number of detected

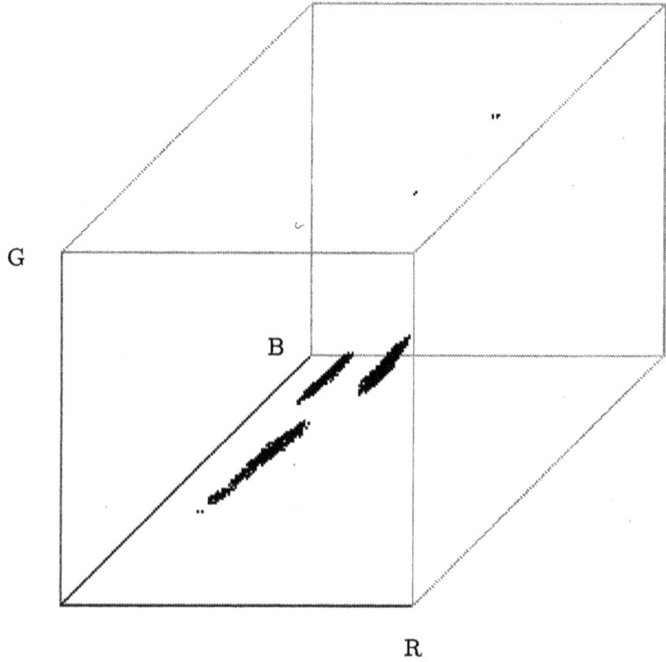

Fig. 13. Kernel of M extracted from the image of Fig. 10.

modes when the standard deviation ranges from 3 to 13, for a structuring element size set to 3. When σ is lower than 6, the number of detected modes corresponds exactly to the number of classes present in the pixel population of the image of Fig. 16. When σ is higher than 6 and lower than 13, only four modes are detected because the two modes corresponding to the classes associated to the regions R_4 and R_5, i.e the two circular concentric regions, are identified as a single one. In this case, the pixels of the two circular regions are assigned to the same class. When σ is higher than 12, the noise is so strong that the classes corresponding to R_2, R_4 and R_5 are merged into one detected mode.

In order to provide some insight into the behavior of the procedure, we consider the image of Fig. 16 corrupted for σ equal to 6 (see image of Fig. 18). Furthermore, by examining the three $1D$ color feature histograms presented by Fig. 19, we see that the distributions of the color features overlap. In order to detect the modes, we apply our approach with the same parameter values than the previous example. Figure 20 shows the five detected modes for σ equal to 6, which corresponds to the highest noise level for which the

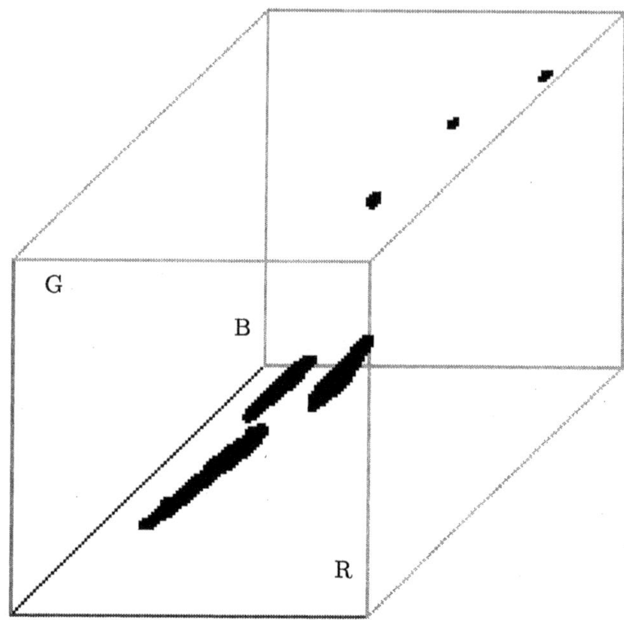

Fig. 14. Modes detected from the image of Fig. 10. Size of S_X = 3x3x3 and $TH = 0.5(1 - 0.1) = 0.45$.

Fig. 15. Image of Fig. 10 segmented with the detected modes of Fig. 14.

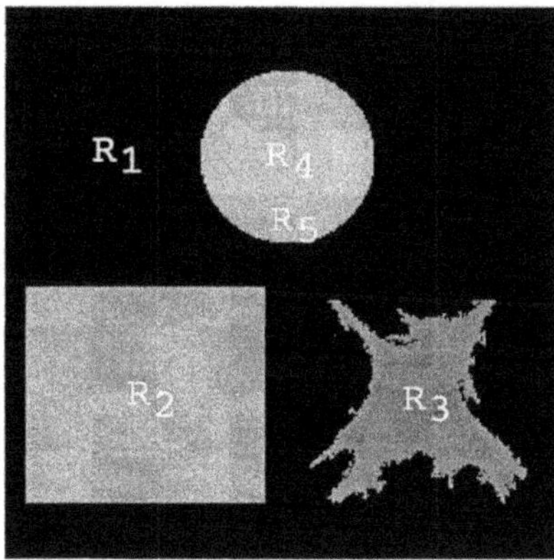

Fig. 16. Original synthetic image.

five modes are correctly identified. As for the real image, these modes are displayed as the subset of observations that belong to each of them.

The image of Fig. 21 shows how the pixels of the image of Fig. 16 are assigned to the five classes that correspond to the five regions of the original image. This result shows that our method is able to handle unequiprobable classes of pixels.

All the pixels of the regions R_1, R_2 and R_3 have been correctly classified. Some misclassified pixels appear into regions R_4 and R_5. However the error rate computed as the ratio of misclassified pixels to the total number constituting the two circular regions is equal to 3%.

The preceding evaluation of the mode stability with respect to the noise level has been performed with a structuring element size set to 3. We now evaluate the influence of this parameter.

Figure 17 shows that when the size of S_X is set to 5, any mode is detected until σ reaches 5. In this case, the size of the structuring element is so large that the erosion filtering erases all the modes of the membership function.

In conclusion, we have seen that the mode detection depends on two main parameters, namely the threshold TH and the size of the structuring element. TH is used for deciding if each neighbor of the center of the structuring element is taken or not into account for the erosion or dilation purpose. The examples of the third section (see Figs. 7, 8 and 9) show that the lower TH is, the stronger the effect of dilation is. If TH is set to a too lower value, the dilation filtering tends to merge neighboring modes. Finally, this section shows that the size of the structuring element deserves a great attention, because the erosion filtering with a too large structuring element whose size is large tends to erase the modes of the membership function.

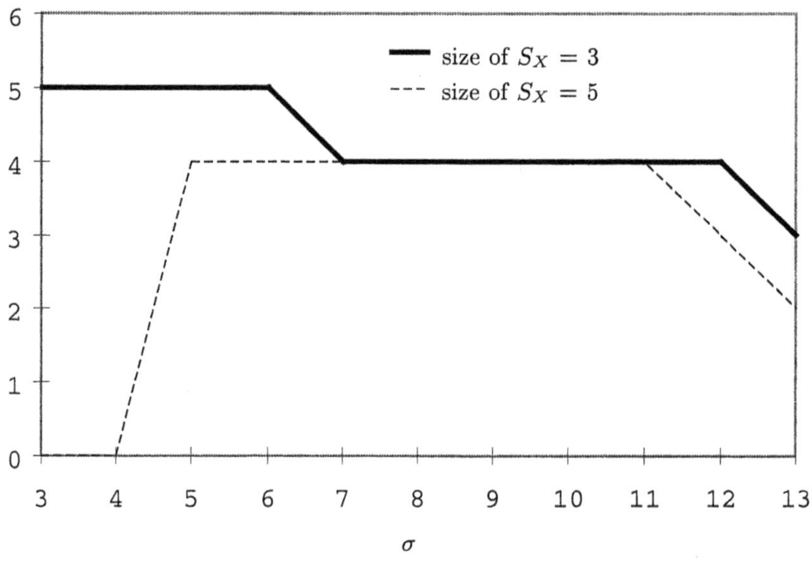

Fig. 17. Detected mode number with different noise standard deviations and structuring element sizes.

$$TH = 0.5(1 - 0.1) = 0.45.$$

5 Conclusion

In this paper, color image segmentation has been considered as a pixel classification problem, based on the detection of the modes of the $3D$ color histogram. Our scheme consists in associating each homogeneous region with a mode of the color histogram. This histogram is used to approximate the

Fig. 18. Image of Fig. 16 corrupted with a gaussian noise for $\sigma = 6$.

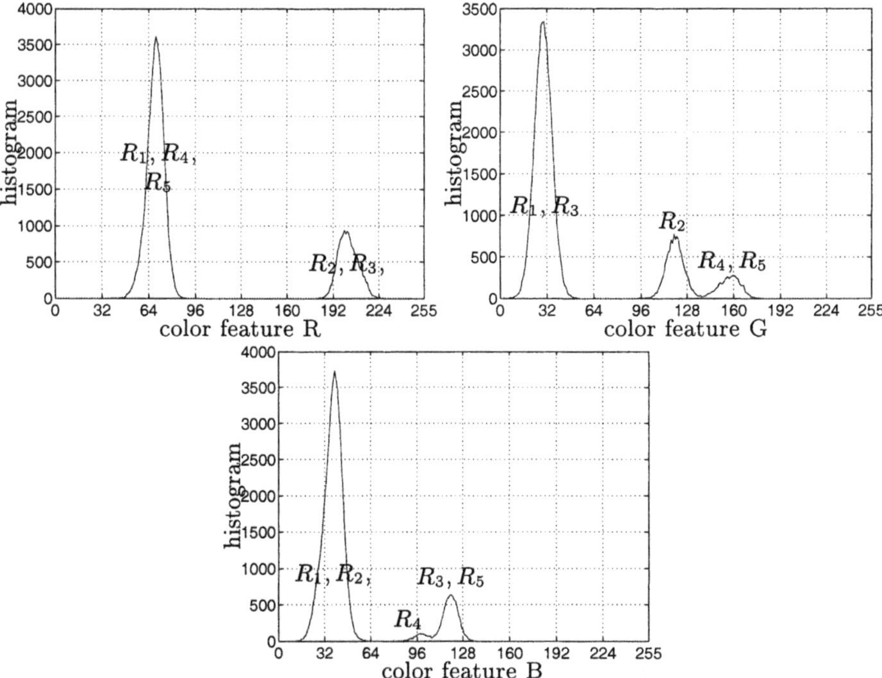

Fig. 19. $1D$ Color feature histograms of the image of Fig. 18.

underlying *pdf* whose concavity analysis yields a fuzzy subset defined by its membership function. A fuzzy morphological transformation, based on a combination of specific fuzzy erosions and dilations, is iteratively applied to this membership function. The pixels of the original color image are finally assigned to the classes associated with the detected modes.

This work leaves several possibilities of further improvement. First, the color histogram is a tool for approximating the *pdf*, which considers only the color properties of the pixels. The approximation of the *pdf* should be more pertinent in the context of image segmentation, if it was based on the co-occurrence matrices which take into account the colorimetric properties as well as the spatial interaction between the pixels. The second point concerns the evaluation of the structuring functions which should integrate a measure of the compatibility between the localization of the considered observation and those of its neighboring observations since the confidence in the assumption that an observation stands in a mode (respectively a valley) is an increasing function of the number of its neighbors that also stand in a mode (respectively a valley).

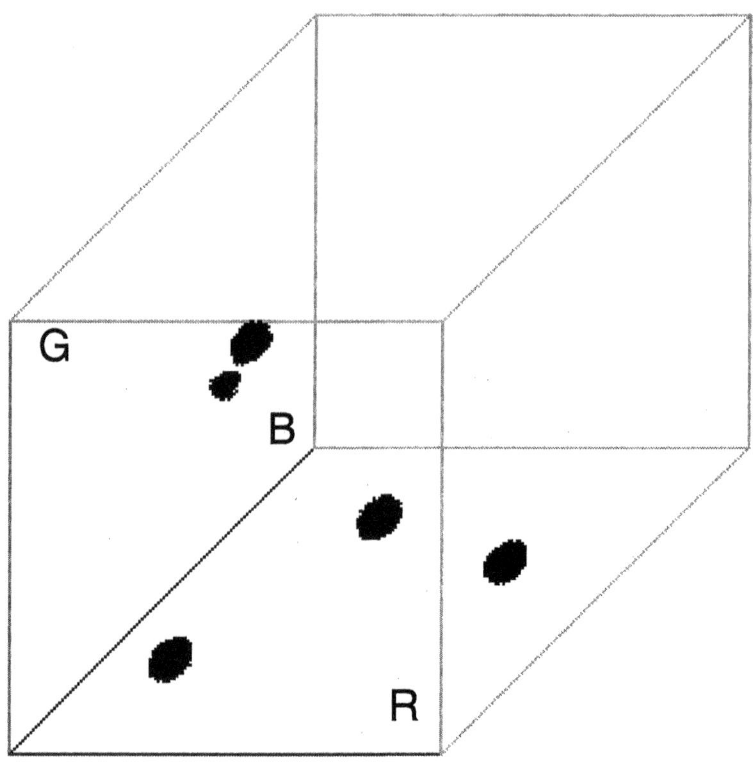

Fig. 20. Modes detected from the image of Fig. 18.
Size of $S_X = 3$ and $TH = 0.5(1 - 0.1) = 0.45$.

References

1. Sharma G., Trussell H.J., *Digital color imaging*, IEEE Transactions on Image Processing, IP7(6), 1997, pp. 901–932.
2. Cheng H.D., Jiang X.H., Sun Y., Wang J.,*Color image segmentation: advances and prospects*, Pattern Recognition, PR-34(6), 2001, pp. 2259-2281.
3. Bloch I., Maître H., *Fuzzy mathematical morphologies : a comparative study*, Pattern Recognition, PR-9(28), 1995, pp. 1341-1387.
4. Botte-Lecocq C., Zhang R.D., Postaire J.G., *Cluster analysis by binary morphology*, IEEE Transactions on Pattern Analysis and Machine Intelligence, PAMI-15(2) , 1993, pp. 170-180.
5. Devijver P.A., KittlerJ., *Pattern recognition : a statistical approach*, Prentice-Hall - Englewood Cliffs, New Jersey, 1982.
6. Gillet A., Botte-Lecocq C., Macaire L., Postaire J.G., *Application of fuzzy mathematical morphology for unsupervised color pixels classification*, Data Analysis, Classification and Related Methods, H.A.L. Kiers and all editor, Springer-Verlag, 2000, pp. 69-75.

Fig. 21. Image of Fig. 18 segmented with the detected modes of Fig. 20.

7. Haralick R.M., Sternberg S.R., Zhuang X., *Image analysis using mathematical morphology*, IEEE Transactions on Pattern Analysis and Machine Intelligence, PAMI-9(4), 1987, pp. 532-550.
8. Huntsberger T.L., Jacobs C.L. , Cannon R.L., *Iterative fuzzy image segmentation*, Pattern Recognition, PR 18(2), 1985, pp. 131-138.
9. Lambert P., Macaire L., *Filtering and segmentation: the specificity of color images*, CGIP'2000 - International Conference on Color in Graphics and Image Processing, Saint-Etienne - France, 2000, pp. 57-71.
10. Lim Y.W., Lee S.U., *On the color image segmentation algorithm based on the thresholding and the fuzzy c-means techniques*, Pattern Recognition, PR(23)9, 1990, pp. 935-952.
11. Ohta Y.I., Kanade T., Sakai T., *Color information for region segmentation*, Computer Graphics and Image Processing,(13), 1980, pp. 222-241.
12. Park S.H., Yun I.D., Lee S.U., *Color image segmentation based on 3-D clustering : morphological approach*, Pattern Recognition, PR-31(8), 1998, pp. 1061-1076.
13. Pham T.D., Yan H., *Color image segmentation using fuzzy integral and mountain clustering*, Fuzzy Sets and Systems, FSS(107), 1999, pp. 121-130.
14. Postaire J.G., Vasseur C., *A convexity testing method for cluster analysis*, IEEE Transactions on Systems, Man and Cybernetics, SMC(10), 1980, pp. 145-149.
15. Schettini R., *Segmentation algorithm for color images*, Pattern Recognition Letters, PRL (14), 1993, pp. 499-506.
16. Serra J., *Image analysis and mathematical morphology*, Academic Press, 1988.
17. Shaffarenko L., Petrou M., Kittler J., *Histogram-based segmentation in a perceptually uniform color space*, IEEE Transactions on Image Processing, IP-7(9), 1998, pp. 1354-1358.
18. Tominaga S., *Color classification of natural color images*, Color Research and Application, CRA(17)4, 1992, pp. 230-239.
19. Turpin-Dhilly S. , Botte-Lecocq C., *Application of fuzzy mathematical morphology for pattern classification*, Advances in Data Science and Classification, IFCS'98, 1998, pp. 125-130.

Chapter 8

Fast and Robust Fuzzy Edge Detection

Hamid R. Tizhoosh

University of Waterloo
Department of Systems Design Engineering
Pattern Recognition and Machine Intelligence Group
200 University Ave. West, Waterloo, ON, Canada
Email: tizhoosh@uwaterloo.ca

Summary. In recent years, fuzzy techniques have been applied to develop new edge detection techniques because they offer a flexible framework for edge extraction with respect to specific requirements. These techniques, however, are usually expensive in computing compared to classical approaches like the Sobel operator. In many practical applications we need fast edge detection. In this chapter, several fast methods are proposed which are suitable for cases where a rough edge estimation is required. On the other side, the result of edge detection techniques in noisy environments is often not satisfactory. In this chapter, also a robust algorithm based on fuzzy if-then rules is proposed that can detect edges and lines in noisy images.

1 Introduction

Edge detection is one of the fundamental tasks in computer vision. It can be regarded as a special type of segmentation. Extracted edges can be used for measurement and/or recognition purposes. A large number of algorithms already exists in literature. In recent years, also fuzzy techniques have been used to develop new edge detectors.

Due to inherent vagueness and ambiguity of edge classification, the fuzzy techniques seem to be a promising approach. Some algorithms already have been introduced [1,2,5,3,4,6,9–11]. These new approaches offer a flexible framework for edge detection, however, they usually need more computation time. Additionally, their development and/or adjustment consumes more time than classical approaches. The need for fast fuzzy edge detectors is still high. On the other side, since the real images are often corrupted by different kinds of noise, robust techniques are also necessary.

In this chapter, several fast fuzzy edge detectors are described for practical cases where a rough edge map is needed in a short time. First, a brief overview of existing fuzzy edge detectors is provided. For a better understanding, some preliminary definitions are also presented. Afterwards, three fuzzy approaches to edge detection are proposed based on heuristic membership functions, simple fuzzy if-then rules, and complement-based edge extraction. Finally, a robust fuzzy edge detector is also presented which extracts edges in noisy images.

2 Existing fuzzy edge detectors

The inherent vagueness and ambiguity of edge detection tasks in image processing has induced many researchers to apply the concept of fuzziness to this problem. In literature, there exist already some fuzzy approaches to edge detection. Following, the most popular ones are briefly described:

FIRE Operators [9,10] FIRE operators (Fuzzy inference ruled by else-action) are a family of nonlinear operators which adopt fuzzy rules to detect edges. For each pixel, a set of neighboring pixels is considered. The FIRE operator processes this neighborhood information by using fuzzy rules in order to estimate the location of edges.

FEDGE [3,4] This approach is based on the idea of forming fuzzy edges and by classifying them using Fuzzy Categorization and Classification Method (FCC). FEDGE extracts edges from a new image by finding the similarity between the input image and the templates in the example set (example in Fig. 1).

Fig. 1. FEDGE Results: original image, results without/with α-cuts [3,4].

Takagi-Sugeno Approach [1,2,11] Using Takagi-Sugeno fuzzy model, this approach provides a means to blend the gradient outputs of the horizontal and vertical Sobel operators as opposed to traditional choices such as city-block or Euclidean norm for blending (example in Fig. 2).

Perception-Based Approach [5,12] Based on human perception, this approach defines primary, secondary, tertiary edges. Adjustable membership functions and different α-cuts are used to extract appropriate edges (example in Fig. 3).

The interested reader may follow the references to find out more details about these approaches. These techniques are flexible and robust but generally very

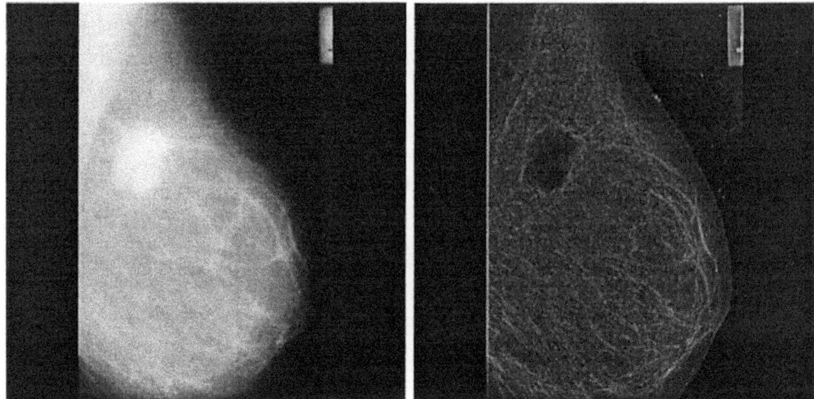

Fig. 2. Takagi-Sugeno approach for Mammography (MIAS) Data Set [11].

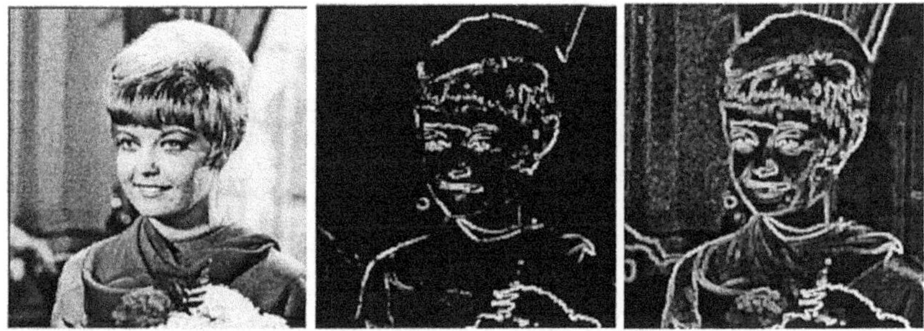

Fig. 3. Perception-based edge detection [5,12].

expensive in computing. Also the design and implementation of these techniques require more time and effort compared with classical methods. Therefore, they may be used in applications if specific requirements exist which can only be met by integration of expert knowledge.

3 Images as fuzzy sets

Let a $M \times N$ image X be the set of all pixels $g_{mn} \in [0, L]$ with maximal intensity L, then X can be regarded as an array of fuzzy singletons $\mu_{mn} \in [0, 1]$ indicating the degree of brightness of each gray level g_{mn} [7,12,13]:

$$X = \bigcup_{m=1}^{M} \bigcup_{n=1}^{N} \frac{\mu_{mn}}{g_{mn}}. \tag{1}$$

The membership function μ_{mn} can be achieved, among other possibilities, by a simple normalization:

$$\mu_{mn} = \frac{g_{mn}}{\max\limits_{i\in[1,M],j\in[1,N]}(g_{ij})}. \tag{2}$$

The image X' containing all edges is:

$$X' = \bigcup_{m=1}^{M}\bigcup_{n=1}^{N}\frac{\hat{\mu}_{mn}}{g_{mn}}, \tag{3}$$

where $\hat{\mu}_{mn}$ indicates the *degree of edginess* for each pixel. The task of edge detection is, therefore, the determination of the membership function $\hat{\mu}_{mn}$ for each pixel. Following, three different ways for fast calculation of edginess will be proposed.

4 Heuristic membership functions

The simplest way for defining a fuzzy edge detector is the determination of a proper membership function $\mu_{mn}^{edginess}$ for each pixel at the position (m, n) with a surrounding $N \times N$ spatial window. The definition is heuristic, and in this sense, not unique. Based on general properties of an edgy neighborhood and based on empirical expert knowledge different membership functions can be introduced. An example for such membership functions is the following formula based on gray level differences in each neighborhood:

$$\mu_{mn}^{edginess} = 1 - \frac{1}{1 - \frac{1}{\Delta}\sum_i\sum_j|g_{ij} - g_{mn}|}, \tag{4}$$

$$= \frac{\sum_i\sum_j|g_{ij} - g_{mn}|}{\Delta + \sum_i\sum_j|g_{ij} - g_{mn}|}, \tag{5}$$

where $\Delta \in [0, L]$ is a proper parameter. Meaningful values are in $[L/2, L]$. The lower Δ, the more edges are detected. An example is illustrated in Fig. 4.

Another example for the heuristic membership functions is the following equation:

$$\mu_{mn}^{edginess} = \left(\max\left(0, \min\left(1, \frac{|MAX + MIN - 2 * g_{mn}|}{\Delta}\right)\right)\right)^{\alpha}, \tag{6}$$

where $\Delta < L/2$ and $\alpha \in [0.5, 2]$. It is based on the assumption that if the center pixel in a spatial window belongs to an edge, then its distance from maximum and minimum gray levels in the neighborhood should be more or less equal. Figure 5 shows an example for this membership function.

Fig. 4. From left to right: Original image, edges detected using Eq. (5) with $\Delta = 128$, edges detected using Eq. (5) with $\Delta = 255$ (3×3 processing).

Fig. 5. From left to right: Original image, edges detected using Eq. (6) with $\Delta = 32$ and $\alpha = 2$, edges detected using Eq. (6) with $\Delta = 64$ and $\alpha = 1$.

The advantage of defining the degree of edginess as a fuzzy membership function is that in this case the whole palette of fuzzy set theory is available for further modifications. The linguistic hedges, for instance, or fuzzy inference scheme can then be applied to integrate expert knowledge and modify/improve the result. This approach is very fast but it has a limited performance due to its simple structure.

5 From membership functions to rules

To create a better and more robust/effective edge detector, we have to design the membership functions more carefully. If we regard a typical edgy neighborhood, we recognize, for instance, that the difference between maximum and minimum gray levels, depended on edge strength, is relatively high. However, this can also be the case in noisy environments or even in cases where a single outlier is present. To exclude these cases, we can additionally state that the gray-level intensity of center pixel in an *optimal* edgy neighborhood is more or less between the minimum and maximum gray-levels. The linguistic formulation of this knowledge is:

If the difference of minimum and maximum gray level is high,

and the value of center pixel is between minimum and maximum, **then** the edginess is high.

If W is a $w \times w$ neighborhood surrounding the center pixel g_{mn}, then the corresponding membership functions can be defined as follows:

$$\hat{\mu}^1_{mn} = \min\left(1, \frac{\max\limits_{i,j\in[1,w]} W(i,j) - \min\limits_{i,j\in[1,w]} W(i,j)}{\Delta_1}\right), \tag{7}$$

$$\hat{\mu}^2_{mn} = 1 - \min\left(1, \frac{g_{mn} - \text{MeanMed}_{i,j\in[1,w]} W(i,j)}{\Delta_2}\right). \tag{8}$$

MeanMed is either mean or median value. $\Delta_1 > L/4$ and $\Delta_2 < L/4$ are meaningful boundaries for the parameters. Using simple t-norms the edge map is then

$$X' = \bigcup_{m=1}^{M} \bigcup_{n=1}^{N} \frac{\min\left(\hat{\mu}^1_{mn}, \hat{\mu}^2_{mn}\right)}{g_{mn}}, \tag{9}$$

or

$$X' = \bigcup_{m=1}^{M} \bigcup_{n=1}^{N} \frac{\hat{\mu}^1_{mn} \times \hat{\mu}^2_{mn}}{g_{mn}}. \tag{10}$$

Examples are illustrated in Fig. 6.

Fig. 6. From left to right: Original image, edges using minimum and product operator,($\Delta_1 = 128$, $\Delta_2 = 32$).

6 Fuzzy complement-cased edge detection

Let $\bar{X}'$ be the complement of image X':

$$\bar{X}' = 1 - X' = \bigcup_{m=1}^{M} \bigcup_{n=1}^{N} \frac{1 - \hat{\mu}_{mn}}{g_{mn}}. \tag{11}$$

The degree of fuzziness γ of an edgy image X' can be calculated as follows:

$$\gamma(X') = \frac{2}{MN} \sum_{m=1}^{M} \sum_{n=1}^{N} \mathrm{T}(\hat{\mu}_{mn}, 1 - \hat{\mu}_{mn}), \tag{12}$$

where T is a suitable t-norm. Since we are regarding partial membership values, the intersection of an *edgy image* with its complementary set *non-edgy image* is a non-empty set:

$$\bar{X}' \cap X' \neq \emptyset. \tag{13}$$

Assuming that the image contains only one object on a background, a membership function μ (*bright image*) can be defined using standard S-function:

$$\mu(g) = \begin{cases} 0 & \text{if } i < g_{min}, \\ 2\left(\frac{g-g_{min}}{g_{max}-g_{min}}\right)^2 & \text{if } g \in [g_{min}, g_{mid}], \\ 1 - 2\left(\frac{g-g_{max}}{g_{max}-g_{min}}\right)^2 & \text{if } g \in [g_{mid}, g_{max}], \\ 1 & \text{if } g > g_{max}, \end{cases} \tag{14}$$

where g_{min} and g_{max} are the minimum and maximum gray levels, and

$$g_{mid} = \frac{g_{min} + g_{max}}{2}. \tag{15}$$

Using look-up tables, an appropriate s-norm S for union and a t-norm T for intersection, a histogram-based edge detector can be formulated as follows:

$$\hat{\mu} = 1 - (S(\mu, \overline{\mu}) - T(\mu, \overline{\mu})), \tag{16}$$

$$= 1 - \max(\mu, 1 - \mu) + \min(\mu, 1 - \mu). \tag{17}$$

For a gradual thresholding the intensification operator INT can be applied, whereas the membership values will be intensified in each iteration n based on their values in previous iteration $n - 1$:

$$\hat{\mu}_n = \begin{cases} 2(\hat{\mu}_{n-1})^2 & \text{if } \hat{\mu}_{n-1} \leq 0.5, \\ 1 - 2(1 - \hat{\mu}_{n-1})^2 & \text{otherwise.} \end{cases} \tag{18}$$

Using algebraic sum for s-norm and algebraic product for t-norm, one can show that Eq. (16) represents basically the calculation of histogram fuzziness:

$$
\begin{aligned}
\hat{\mu} &= \overline{S(\mu,\overline{\mu}) - T(\mu,\overline{\mu})}, \\
&= \overline{\mu + \overline{\mu} - \mu\overline{\mu} - \mu\overline{\mu}}, \\
&= \overline{\mu + \overline{\mu} - 2\mu\overline{\mu}}, \\
&= 1 - (\mu + \overline{\mu} - 2\mu\overline{\mu}), \\
&= 1 - \mu - \overline{\mu} + 2\mu\overline{\mu}, \\
&= 1 - \mu - (1-\mu) + 2\mu(1-\mu), \\
&= 1 - \mu - 1 + \mu + 2\mu(1-\mu), \\
&= 2\mu(1-\mu).
\end{aligned}
\tag{19}
$$

The last term corresponds to the index of fuzziness. Figure 7 shows some results for this approach. If the image contains more than one object or it

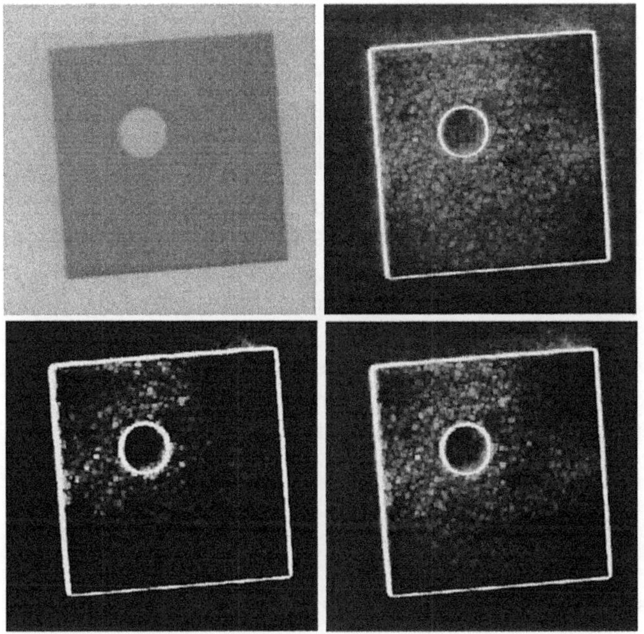

Fig. 7. Histogram-based edge detection. From top left to bottom right: original image, extracted edges with 1, 2 and 3 times application of INT-operator.

represents a complex scene, then we have to modify this approach so that spatial image information can be taken into account.

Assuming an *optimal* transition, the intersection of image and its complement should deliver high values for edgy regions.

Example: Let consider an optimal edge in a 3×3 neighborhood W:

$$W = \begin{bmatrix} 0 & 0 & 0 \\ 50 & 50 & 50 \\ 100 & 100 & 100 \end{bmatrix}.$$

The spatial membership values can be calculate by dividing all pixel by maximum intensity:

$$\mu = \begin{bmatrix} 0 & 0 & 0 \\ 0.5 & 0.5 & 0.5 \\ 1 & 1 & 1 \end{bmatrix}.$$

The fuzzy complement is:

$$1 - \mu = \begin{bmatrix} 1 & 1 & 1 \\ 0.5 & 0.5 & 0.5 \\ 0 & 0 & 0 \end{bmatrix}.$$

The intersection is:

$$\mu \cap (1 - \mu) = \begin{bmatrix} 0 & 0 & 0 \\ 0.5 & 0.5 & 0.5 \\ 0 & 0 & 0 \end{bmatrix}.$$

Interpretation: In optimal cases, the intersection of an edgy image with its complement delivers high values ($\in (0, 0.5]$) for edgy pixels and very low values ($\to 0$) for other pixels. Since edges separate two different regions, they also belong to both of them; the membership value of 0.5 indicates exactly this fact.

Since we have an optimal edge the fuzziness of complement membership matrix is only high for the row/column containing the gray level transition. An edginess measure can be defined in a $w \times w$ neighborhood:

$$\hat{\mu}_{mn} = \min\left(1, \frac{2}{w} \sum_i \sum_j \mathrm{T}(\mu_{ij}, 1 - \mu_{ij})\right), \tag{20}$$

where T is t-norm such as the minimum operator

$$\hat{\mu}_{mn} = \min\left(1, \frac{2}{w} \sum_i \sum_j \min\left(\mu_{ij}, 1 - \mu_{ij}\right)\right), \tag{21}$$

or the product operator

$$\hat{\mu}_{mn} = \min\left(1, \frac{2}{w} \sum_i \sum_j \mu_{ij} \times (1 - \mu_{ij})\right). \tag{22}$$

Since the spatial calculation of membership values is noise sensitive we can extend the function and make it more robust. We have two possibilities: either we correct the membership values after the fuzzy complements are built:

$$\hat{\mu}_{mn} = \min\left(1, \frac{2}{N}\sum_i \sum_j \min\left(\mu_{ij}, 1-\mu_{ij}\right)\right) \times \left(\frac{\max\limits_{spatial}(g_{ij}) - \min\limits_{spatial}(g_{ij})}{\max\limits_{global}(g_{mn})}\right), \quad (23)$$

where μ_{ij} is a simple normalization according to Eq. (2), or we correct the membership values before fuzzy complements are built:

$$\hat{\mu}_{mn} = \min\left(1, \frac{2}{W}\sum_i \sum_j \min\left(\mu_{ij}, 1-\mu_{ij}\right)\right), \quad (24)$$

where μ_{ij} is calculated as follows:

$$\mu_{ij} = \frac{g_{ij} \times \left(\max\limits_{spatial}(g_{ij}) - \min\limits_{spatial}(g_{ij})\right)}{\max\limits_{spatial}(g_{ij}) \times \max\limits_{global}(g_{ij})}. \quad (25)$$

Results for this approach are presented in Fig. 8.

(a) edges using Eq. 23 (b) edges using Eq. 24

Fig. 8. Complement-based edge detection.

6.1 Extension to Sugeno complements

Let μ_{mn} be the membership values of image X. The Sugeno (involutive) complements $\bar{\mu}_{mn}$ are given as follows:

$$\bar{\mu}_{mn} = \frac{1-\mu_{mn}}{1+\lambda\mu_{mn}}, \tag{26}$$

with $\lambda \in (-1, \infty)$ and the property

$$\bar{\bar{\mu}}_{mn} = \bar{\mu}_{mn}\left(\bar{\mu}_{mn}(\mu_{mn})\right) = \mu_{mn}. \tag{27}$$

An edginess measure can be redefined in a $w \times w$ neighborhood:

$$\hat{\mu}_{mn} = \min\left(1, \frac{2}{w}\sum_{i}\sum_{j}\min\left(\mu_{ij}, \frac{1-\mu_{ij}}{1+\lambda\mu_{ij}}\right)\right), \tag{28}$$

or by using the product operator:

$$\hat{\mu}_{mn} = \min\left(1, \frac{2}{w}\sum_{i}\sum_{j}\mu_{ij} \times \frac{1-\mu_{ij}}{1+\lambda\mu_{ij}}\right). \tag{29}$$

Results for this approach using different λ values are illustrated in Fig. 9.

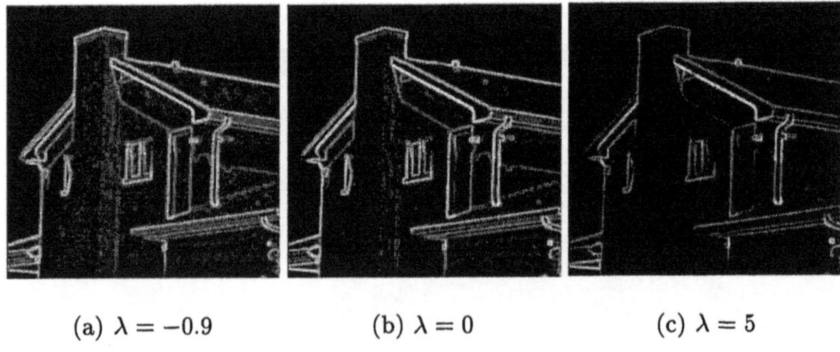

(a) $\lambda = -0.9$ (b) $\lambda = 0$ (c) $\lambda = 5$

Fig. 9. Edge detection using Sugeno complements.

7 Robust estimation of degree of edginess

The aforesaid approaches are not only fast, they can also be implemented easily. But they should mainly serve as edge detectors if we need a rough edge estimation, and if the image noisiness is limited. In this section, a more robust technique will be proposed which requires more computation time but delivers much better results in noisy environments.

The technique is simple in implementation and though fast in computation compared to some existing fuzzy methods. It also can be easily extended to line detection. Defining $N \times N$ neighborhoods in an image (usually 3×3), the gray-level differences between the center pixel g_{mn} and its surrounding pixels g_{ij} can be computed and stored in a matrix E:

$$E = \begin{bmatrix} g_{mn} - g_{i-1,j-1} & g_{mn} - g_{i,j-1} & g_{mn} - g_{i+1,j-1} \\ g_{mn} - g_{i-1,j} & 0 & g_{mn} - g_{i,j} \\ g_{mn} - g_{i-1,j+1} & g_{mn} - g_{i,j+1} & g_{mn} - g_{i+1,j+1} \end{bmatrix}. \quad (30)$$

If the neighborhood belongs to a homogenous region, then E contains values near zero. In the case of edges, the matrix E possesses N values (nearly) equal to zero in vertical, horizontal or oblique direction. These divide E in two halves, one with positive and another with negative difference values (see Fig. 10):

$$E = \begin{bmatrix} \alpha & \alpha & \alpha \\ 0 & 0 & 0 \\ -\alpha & -\alpha & -\alpha \end{bmatrix}, E = \begin{bmatrix} \alpha & 0 & -\alpha \\ \alpha & 0 & -\alpha \\ \alpha & 0 & -\alpha \end{bmatrix}, E = \begin{bmatrix} 0 & -\alpha & -\alpha \\ \alpha & 0 & -\alpha \\ \alpha & \alpha & 0 \end{bmatrix}. \quad (31)$$

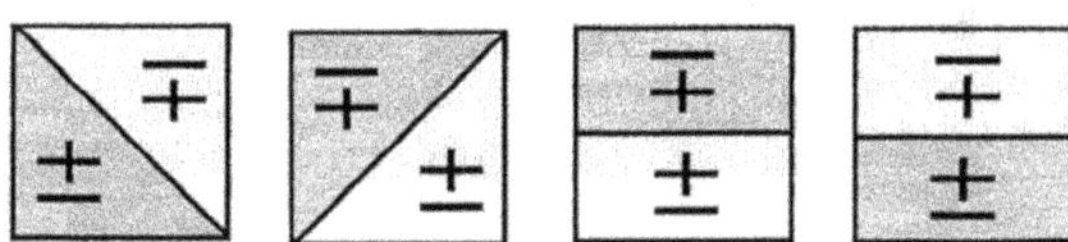

Fig. 10. Possible edge orientation with respect to gray level differences.

The principle is actually very simple: If one half of the neighborhood contains positive/negative differences with center pixel, and if other half contains negative/positive differences with center pixel, then the neighborhood is an edge (Fig. 10). If one interprets this rule strictly, then only ideal, noise-free edges are detected. The auxiliary matrices E^+ and E^- are defined as follows:

$$E^+(i,j) = \begin{cases} 1 & \text{if} \quad d_{ij} \geq 0, \\ 0 & \text{else,} \end{cases} \quad (32)$$

$$E^-(i,j) = \begin{cases} 1 & \text{if} \quad d_{ij} < 0, \\ 0 & \text{else,} \end{cases} \quad (33)$$

where $d_{ij} = g_{mn} - g_{ij}$. Sub-membership values $\hat{\mu}_{ij}^{0^\circ}$ and $\hat{\mu}_{ij}^{90^\circ}$ for horizontal and vertical edges can be calculated as follows:

$$\hat{\mu}_{ij}^{0^\circ} = \frac{2}{N^2 - N} \max \left[\left(\sum_{j=1}^{N/2-1} \sum_{i=1}^{N} E_{ij}^{+} \right) \cdot \left(\sum_{j=N/2+1}^{N} \sum_{i=1}^{N} E_{ij}^{-} \right), \left(\sum_{j=1}^{N/2-1} \sum_{i=1}^{N} E_{ij}^{-} \right) \cdot \left(\sum_{j=N/2+1}^{N} \sum_{i=1}^{N} E_{ij}^{+} \right) \right]^2, \tag{34}$$

$$\hat{\mu}_{ij}^{90^\circ} = \frac{2}{N^2 - N} \max \left[\left(\sum_{j=1}^{N} \sum_{i=1}^{N/2-1} E_{ij}^{+} \right) \cdot \left(\sum_{j=1}^{N} \sum_{i=N/2+1}^{N} E_{ij}^{-} \right), \left(\sum_{j=1}^{N} \sum_{i=1}^{N/2-1} E_{ij}^{-} \right) \cdot \left(\sum_{j=1}^{N} \sum_{i=N/2+1}^{N} E_{ij}^{+} \right) \right]^2. \tag{35}$$

These equations are the numerical implementation of the rules of following form:

If one half of neighborhood possesses positive differences
and the opposite half possesses negative differences,
then the edginess is high.

Analogously, membership values $\hat{\mu}_{ij}^{45^\circ}$ and $\hat{\mu}_{ij}^{135^\circ}$ for oblique edges can be calculated. Finally, we have the following expression for the membership value:

$$\hat{\mu}_{ij}^{edginess} = \max(\hat{\mu}_{ij}^{0^\circ}, \hat{\mu}_{ij}^{45^\circ}, \hat{\mu}_{ij}^{90^\circ}, \hat{\mu}_{ij}^{135^\circ}). \tag{36}$$

This type of edge detection differs from classical procedures like the gradient approach. Edge detectors look usually for the explicit edge location (the zero-crossing of derivatives), this approach detects the entire area where the edge could be (see Fig. 11).

7.1 Extension to lines

The lines as fine structures play an important role in many applications. The line detection, therefore, is always discussed as a special task in image processing. Many techniques already exist for line extraction. To make sure that lines will not be eliminated or manipulated during the preprocessing steps, and to avoid an expensive line detection, one may extend the membership values to lines. Following rules can be used:

R_1 (horizontal bright line):
if the differences of the horizontal middle axis are zero, **and**
the upper half of the neighborhood has positive differences, **and**

the lower half of the neighborhood has positive differences,
then the horizontal middle axis is a line.

R_2 (horizontal dark line):
if the differences of the horizontal middle axis are zero, **and**
the upper half of the neighborhood has negative differences, **and**
the lower half of the neighborhood has negative differences,
then the horizontal middle axis is a line.

Hence, the first rule calculates the line membership *horizontal bright line* $\hat{\mu}_{ij}^{HBL}$ and the second rule the line membership *horizontal dark line* $\hat{\mu}_{ij}^{HDL}$:

$$\hat{\mu}_{ij}^{HBL} = \left(1 - \min\left(1, \frac{\sum\limits_{i,j=m} d_{ij}}{\frac{2}{N}\max\left(1, \sum\limits_{i,j \le N^2/2-1} d_{ij}\right)}\right)\right) \times \left(1 - \min\left(1, \frac{\sum\limits_{i,j=m} d_{ij}}{\frac{2}{N}\max\left(1, \sum\limits_{i,j \ge N^2/2+1} d_{ij}\right)}\right)\right) \times \left(\sum_{j=1}^{N/2-1}\sum_{i=1}^{N} E_{ij}^{+}\right) \times \left(\sum_{j=N/2+1}^{N}\sum_{i=1}^{N} E_{ij}^{+}\right), \quad (37)$$

$$\hat{\mu}_{ij}^{HDL} = \left(1 - \min\left(1, \frac{\sum\limits_{i,j=m} d_{ij}}{\frac{2}{N}\max\left(1, \sum\limits_{i,j \ge N^2/2+1} d_{ij}\right)}\right)\right) \times \left(1 - \min\left(1, \frac{\sum\limits_{i,j=m} d_{ij}}{\frac{2}{N}\max\left(1, \sum\limits_{i,j \le N^2/2-1} d_{ij}\right)}\right)\right) \times \left(\sum_{j=N/2+1}^{N}\sum_{i=1}^{N} E_{ij}^{-}\right) \times \left(\sum_{j=1}^{N/2-1}\sum_{i=1}^{N} E_{ij}^{-}\right). \quad (38)$$

If we define appropriate rules for all other directions, then the total line membership $\hat{\mu}_{ij}^{line}$ can be calculated. Thus, the edge membership can be extended to lines:

$$\hat{\mu}_{ij}^{edge/line} = \max(\hat{\mu}_{ij}^{0^\circ}, \hat{\mu}_{ij}^{45^\circ}, \hat{\mu}_{ij}^{90^\circ}, \hat{\mu}_{ij}^{135^\circ}, \hat{\mu}_{ij}^{line}). \tag{39}$$

The computation time for estimation of the line membership is actually very small compared with some classical approaches (e.g. Hough transformation). Embedded into the procedure of the filtering, the extension of edge detection to lines, however, leads inevitably to the deceleration of the image processing. As a function of the respective applications on the one hand and the selected filters on the other hand it should be therefore considered whether this is really necessary. Because as long as lines possess a strength/width of more than one pixel, they (suitable window size presupposed) are captured by the edge membership. For some images, for example satellite photographs, lines represent significant information. In such cases one must possibly accept the higher expenditure.

Rule-based methodology for edge/line detection has a higher robustness compared with conventional approaches (Fig. 11). Also with regard to computational efficiency this technique can quite compete with other well-known algorithms. Additionally, this method combines also the two, usually in image processing separated tasks of edge and line detection. For the latter, a more expensive computation is usually accepted.

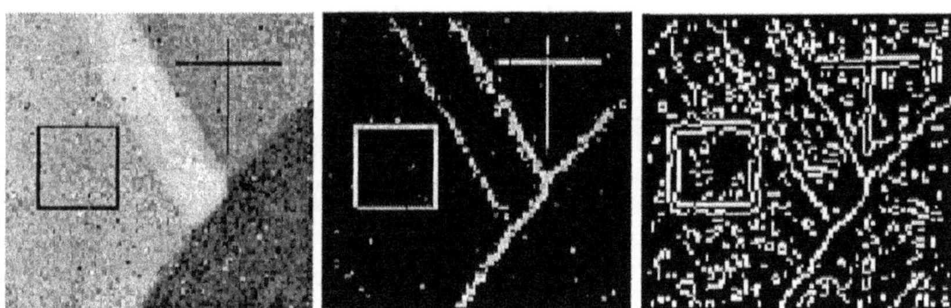

Fig. 11. Comparison between rule-based edge detection and Canny-Operator. From left to right: noisy test image, result of rule-based detector, result of Canny operator.

8 Conclusions

Three fuzzy approaches to edge detection are proposed in this chapter: heuristic membership functions, simple fuzzy rules, and complement-based edge extraction. These methods are simple in implementation and fast in computing. They are suitable for practical cases where a rough edge estimation is necessary in a short time.

Several tests have been accomplished in MatLab environment to compare the computation times of the proposed techniques heuristic membership functions (HMF), simple rule-based approach (SRB) and fuzzy complement-based method (FC) with Sobel operator. Results are presented in Table 1. The MatLab codes were not optimized so that the developed techniques possibly still can be accelerated.

Table 1. Computation times compared to Sobel.

Sobel	HMF	SRB	FC
1	1.08	1.82	1.20

Further, a more robust fuzzy approach to edge and line detection has been described which detects edges/lines in noisy images. It requires more computation time compared to aforesaid techniques but is still faster than some existing methods. A comparison to existing fuzzy and classical techniques with respect to detection performance for all proposed approaches will be subject to future works.

References

1. Bezdek, J.C., Chandrasekhar, R., and Attikiouzel, Y., *A geometric approach to edge detection*, in: IEEE Transactions on Fuzzy Systems, 6 (1), 1998, pp. 52–75.
2. Bezdek J.C., Shirvaikar M., *Edge Detection using the Fuzzy Control Paradigm*, in: Proc. of the 2nd European Congress on Intelligent Techniques and Soft Computing (EUFIT'94), Aachen, Germany, 1994.
3. Ho, K.H.L., *Fuzzy Categorisation and Classification in Pattern Recognition and Computer Vision*, in: Proc. of the 7th Australian Joint Conference on Artiticial Intelligence (AI'94), Armidale N.S.W., Australia, 1994.
4. Ho,K.H.L., *FEDGE - fuzzy edge detection by fuzzy categorization and classification of edges*, in: IJCAI'95 Workshop, Montréal, Canada, 1995, pp. 182–196.
5. Gupta, M.M., Knopf, G.K., Nikiforuk, P.N., *Edge Perception Using Fuzzy Logic*, in: Fuzzy Computing, M.M. Gupta and T.Yamakawa (editors), Elsevier Science Publishers, 1988, pp. 35–51.
6. Law, T., Itoh, H., Seki, H., *Image Filtering, Edge Detection, and Edge Tracing Unsing Fuzzy Reasoning*, in: IEEE Transactions on Pattern Analysis and Machine Intelligence, vol. 18 (5), 1996, pp. 481–491.
7. Pal, S.K., Dutta Majumder, D.K., *Fuzzy Mathematical approach to pattern recognition*, John Wiley and Sons, 1985.
8. Pal, S.K., *A Measure of Edge Ambiguity Using Fuzzy Sets*, in: Patern Recognition Letters 4, 1986, pp. 51–56.
9. Russo F., Ramponi G., *Edge Extraction by FIRE Operators*, in: Third IEEE International Conference On Fuzzy Systems, Orlando, vol. 1, 1994, pp. 249–253.

10. Russo, F., *FIRE operators for image processing*, Fuzzy Sets and Systems, 103 (2), 1999, pp. 265–275.
11. Sutton, M.A., Bezdek, J., *Enhancement and analysis of digital mammograms using fuzzy models*, in: Proceedings of the 26th Applied Imagery and Pattern Recognition (AIPR) Workshop: Exploiting New Image Sources and Sensors (SPIE Vol. 3240). J.M. Selander, ed. Bellingham, WA: SPIE Press, 1998, pp.179–190.
12. Tizhoosh, H.R., *Fuzzy Image Processing* (in German), Springer, Heidelberg, 1997.
13. Tizhoosh, H.R., Haußecker, H., *Fuzzy Image Processing: An Overview*, in: Jähne, B., Haußecker, H., Geißler, P. (editors), Handbook on Computer Vision and Applications, Academic Press, vol. 2, 1999, pp. 683–727.

Chapter 9

Fuzzy Data Fusion for Multiple Cue Image and Video Segmentation

Spyros Ioannou, Yannis Avrithis, Giorgos Stamou and Stefanos Kollias

National Technical University of Athens
Department of Electrical and Computer Engineering
Image, Video and Multimedia Systems Laboratory
9 Iroon Polytechniou St., 157 73 Zographou, Athens, Greece
email: sivann,iavr@image.ntua.gr, gstam@softlab.ntua.gr, stefanos@cs.ntua.gr

Summary. Fusion of multiple cue image partitions is described as an indispensable tool towards the goal of automatic object-based image and video segmentation, interpretation and coding. Since these tasks involve human cognition and knowledge of image semantics, which are absent in most cases, fusion of all available cues is crucial for effective segmentation of generic video sequences. This chapter investigates fuzzy data fusion techniques which are capable of integrating the results of multiple cue segmentation and provide time consistent spatiotemporal image partitions corresponding to moving objects.

1 Introduction

Fusion of multiple cue segmentations has proved to be an indispensable tool towards the goal of automatic object-based video segmentation and coding, mainly in the framework of the MPEG-4 standard [20]. State of the art video analysis/coding systems such as the SESAME system and the European COST 211 Analysis Model [10] achieve content-based spatiotemporal segmentation of video sequences employing fusion of multiple cue image partitions. Individual cues usually include color, motion as well as motion-compensated partitions of previous video frames, mainly for the purpose of object tracking [1,9]. The fusion process itself is accomplished by generation of hierarchical image partition structures, construction of decision trees and heuristic rule-based partition processing. In many cases, current approaches mainly focus on video coding, usually through rate-distortion criteria [17]. Other approaches perform simultaneous fusion and segmentation, making it hard to extend the fusion process to multiple cues [16]. Although such systems offer robust distinction between objects with different motion patterns, such as moving and stationary objects, many limitations exist.

First, it has become clear that additional cues, including depth, texture and shape apart from color and motion, are generally necessary for accurate object detection. Each additional cue can offer enhanced segmentation performance in specific cases where other cues fail. Moreover, each segmentation is given as an image partition, or, in a more general approach, is represented by a set of image partitions at different levels of detail, forming a partition tree derived from a multiscale coarse-to-fine strategy [17]. Unified handling and processing of more complex data representations becomes difficult, especially in the presence of uncertainty.

Moreover, the fusion process should interact with object identification, categorization and recognition; cognitive vision systems require use of a priori knowledge about specific objects for robust segmentation fusion [4]. Strict associations between image partitions make existing systems susceptible to noise, hence fuzziness should be introduced in all levels of partition representation. Adaptation mechanisms are needed for adjustment of the fusion modules behavior in real life dynamic environments. Integration of large cue sets, a priori knowledge, fuzzy representation, intelligence and adaptation mechanisms into fusion algorithms will allow their robust operation in generic cognitive vision systems and in real-life conditions [18].

The above issues give rise to fuzzy rule-based fusion systems [15] that are able to handle complex partition tree representations and accommodate for integration of numerous cues including intensity, color, texture, motion and depth. Fuzzy rule based fusion techniques with built-in complex knowledge representation will enable robust image / video segmentation in contrast to conventional fusion techniques based on rate-distortion criteria or elementary heuristic rules. They may also permit reliable multiple object tracking, effectively dealing with occlusion / disocclusion, appearance variation, articulated motion and distraction problems.

In this chapter, we focus on integration of the results from multiple cue segmentation and on derivation of fuzzy decisions for handling inconsistencies and improving segmentation accuracy. The proposed approach takes into account a priori knowledge related to the expected characteristics of the different cue based segments. For instance, motion information is related to moving parts of objects, depth information defines the main object planes/views, but rather imprecisely, and color information provides precise but oversegmented objects. The target of our approach is to fuse individual single cue partitions, group image regions into objects and derive a representation of candidate objects by projecting and combining the generated cue segments.

In more detail, the generic properties of each candidate cue segmentation is given below, illustrating that segment partitions generated by single cues cannot be directly exploited for semantic object extraction:

- color / intensity segmentation, derived on the basis of spatial homogeneity criteria, allows very accurate definition of region boundaries, but fails to identify objects composed of regions with varying color characteristics

- texture segmentation permits extraction of areas with specific color / intensity patterns but still generates a large number of segments for each object
- motion segmentation, usually based on parametric motion models, produces a more limited number of regions, but produces coarse region boundaries due to matching errors and occlusions, while it segments objects with articulated motion
- depth segmentation, either though stereoscopic analysis or relative motion / occlusion between neighboring regions, achieves reliable approximation of real objects; however it fails to detect distant, stationary objects while depth regions have inaccurate boundaries too
- motion-compensated partitions of previous video frames provide a clear clue of the expected object positions and are necessary for temporal object tracking and detection of newly exposed objects, but need refinement using partitions derived from other cues

In general, image partitions obtained from color, texture, motion and depth segmentation form a hierarchical tree structure, in the sense that a depth region is composed by one or more motion region, motion regions contain several texture regions and so on. Fusion can then be accomplished by means of projection and hierarchical grouping of low-level partitions based on high-level ones. However, this is not always the case, since an object might be decomposed into independent non-overlapping partitions, according to color or motion. For this reason the proposed fusion approach performs intelligent region grouping based on a priori knowledge.

Specifically, based on existing efforts on fusion of color and motion segmentations, an initial set of heuristic rules is constructed, for instance:

1. projection of a color region onto motion regions identifies its degree of membership in each motion region
2. grouping of color regions that belong to the same motion region provides a single region with accurate boundaries
3. projection of such regions onto motion-compensated regions of previous frames permits object matching and tracking
4. discrepancies between motion regions and motion-compensated regions of previous frames identify appearance of new objects or object splitting due to articulated motion.

This set is enriched by additional rules to accommodate for depth segmentation. The system presented in this chapter focuses on color, motion, depth and motion-compensated segments of previous frames; however, the formulation of the proposed fusion approach permits integration of an arbitrary number of cues in a uniform and consistent way. Artificial intelligence techniques, and in particular fuzzy rule based systems are employed for the implementation of the fusion process [14]. Adaptation is possible, allowing addition, removal and updating of rules. Experiments on natural video sequences are presented in this chapter to illustrate the performance of the proposed technique. In particular, the output of the intelligent segmentation fusion and grouping procedure is shown to be of better quality than the one provided by each separate segmentation process.

The remaining of this chapter is organized as follows. In Section 2 the proposed integrated system architecture is described and details on the extraction of color, motion and depth segments are given. Section 3 deals with the formulation of the fusion process and provides a detailed description of the proposed fuzzy segmentation fusion approach. In Section 4, experimental results on natural video sequences are presented and compared to single cue segmentation approaches. Finally, conclusions are drawn and a brief description of further work is given in Section 5.

2 The proposed system architecture

In order to take advantage of depth segmentation in the fusion process, depth estimation is performed on stereoscopic video sequences consisting of two channels. Each stereoscopic sequence is first separated into two sequences, one for the left and one for the right channel. The left-channel sequence is used by the color, depth and motion segmentation modules. The right-channel sequence is only used by the depth segmentation module for the creation of the disparity map. For each frame, when all the modules produce their final segmentation, the segmentation maps are passed to the fuzzy fusion system which produces the final segmentation map. This final segmentation is fed back to the system to be used by the scene detection [11] and the motion estimation / compensation [12] modules. All segmentation modules use the M-RSST algorithm to create the image partitions. The block diagram of the proposed system along with its modules is shown in Fig. 1. A description of the M-RSST algorithm and its application to each module follows.

2.1 The M-RSST algorithm

The Recursive Shortest Spanning Tree (RSST) algorithm [5] is our basis for cue segmentation of each frame in a given video shot. Despite its relative computational complexity, it is considered as one of the most powerful tools for image segmentation, compared to other techniques (including cue clustering, pyramidal region growing and morphological watershed). Initially an image I of size $M \times N$ pixels, is partitioned into $M \times N$ regions (segments) of size 1 pixel and links are generated for all 4-connected region pairs. Each link is assigned a weight equal to the distance between the two respective regions, which for example in the case of color segmentation could be defined as the Euclidean distance between the average color components of the two regions, using a bias for merging small regions. All link weights are then sorted in

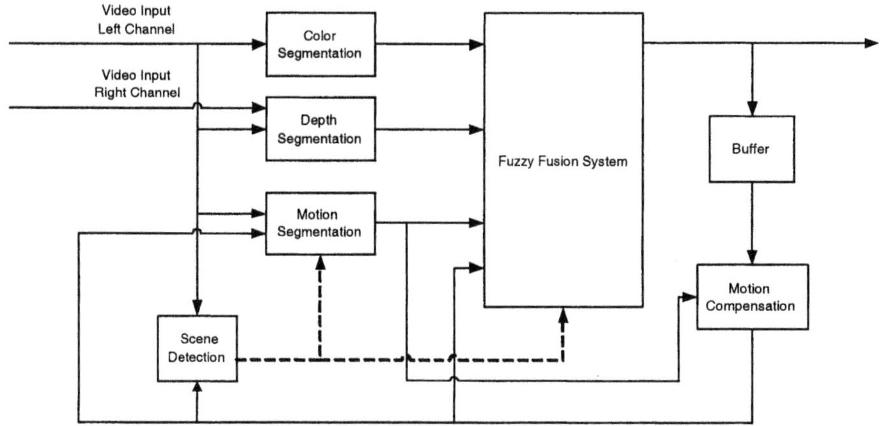

Fig. 1. General System Architecture

ascending order, so that the least weighed link corresponds to the two closest regions. The iteration phase of the RSST is then initiated, where neighboring regions are recursively merged by applying the following actions in each iteration:

1. The two closest regions are merged and the new region cue components and size are calculated.
2. The new region link weights from all neighboring regions are recalculated and sorted.
3. Any duplicated links are removed.

The iteration terminates when either the total number of regions or the minimum link weight (distance) reaches a target value (threshold). A distance threshold is in general preferable since it provides a result that is independent of the image content. The execution time of the RSST is heavily dependent upon the choice of the sorting algorithm, which is certainly a bottleneck of the algorithm. For this reason, a new approach has been proposed, the Multiresolution RSST [6] which recursively applies the RSST algorithm on images of increasing resolution. Initially a multiresolution decomposition of image I is performed with a lowest resolution level of L_0 so that a hierarchy of frames $I(0) = I, I(1), \ldots, I(L_0)$ is constructed, forming a truncated image pyramid, with each layer having a quarter of the pixels of the layer below.

The RSST initialization takes place for the lowest resolution image $I(L_0)$ and then an iteration begins, involving the following steps:

1. Regions are recursively merged using the RSST iteration phase.
2. Each boundary pixel of all resulting regions is split into four new regions, whose cue components are obtained from the image of the next higher resolution level.

3. The new link weights are calculated and sorted.

This 'split-merge' procedure is repeated until the highest resolution image $I(0)$ is reached. It is shown in [6] that effectively only the segment contour shapes are affected at each iteration, since no segments are created or destroyed, hence the multiresolution segmentation approach yields much faster execution compared to RSST. Although its computational complexity is not straightforward to calculate, since it depends on the number, shape and size of segments, it is shown by experiments that it is at least 400 times faster than RSST for a typical image size of 768x576 pixels.

2.2 Color segmentation

The M-RSST algorithm is directly applied on the left-channel image for color segmentation. Using the RGB color space, a distance measure between two adjacent regions (segments) S_1 and S_2 is defined as

$$d^C(S_1, S_2) = \left[(R_{S_1} - R_{S_2})^2 + (G_{S_1} - G_{S_2})^2 + (B_{S_1} - B_{S_2})^2\right]^{1/2} \frac{A_{S_1} A_{S_2}}{A_{S_1} + A_{S_2}} \tag{1}$$

where R_{S_i}, G_{S_i} and B_{S_i} respectively represent the average R, G and B values of all pixels inside region S_i and A_{S_i} is the number of pixels within this region.

2.3 Motion segmentation

In order to solve the motion segmentation problem, numerous different techniques can be employed, such as direct intensity based methods, optical flow based methods, or simultaneous motion estimation and segmentation methods [3,21]. Each method has its own advantages and disadvantages, restricting its use to specific applications. For example, simultaneous motion estimation and segmentation methods are unattractive due to their high computational complexity, while direct intensity based methods cannot handle camera noise and illumination changes. Optical flow methods are quite popular and widely used both for video coding and image analysis and understanding. In this system a block matching algorithm is used to create the motion field. In low-texture areas, the produced motion field can be noisy, thus, a post-processing step for motion field smoothing is indispensable. Median filtering is selected for this purpose due to its speed and its ability to preserve object contours.

Motion field segmentation is performed by dividing each frame into regions of homogeneous motion. The M-RSST algorithm described above is now applied on the motion field with the following distance:

$$d^M(S_1, S_2) = \left[(X_{S_1} - X_{S_2})^2 + (Y_{S_1} - Y_{S_2})^2\right]^{1/2} \frac{A_{S_1} A_{S_2}}{A_{S_1} + A_{S_2}} \tag{2}$$

where X_{S_i}, Y_{S_i} represent the average x and y coordinates of the motion vectors on segment S_i.

2.4 Depth segmentation

The use of three-dimensional (3-D) video, obtained by stereoscopic or multi-view camera systems provides very efficient visual representation. The problem of content-based segmentation is addressed more precisely since video objects are usually composed of regions belonging to the same depth plane. Therefore, 3-D video enables efficient handling and manipulation of video objects by exploiting depth information provided by stereoscopic image analysis [2,7].

Stereoscopic video sequences consist of two channel images obtained by two cameras and provide the perspective projection of 3-D points onto the two 2-D image planes of the cameras. In the framework of this chapter, the technique proposed in [8] is employed to generate a disparity map from the two images. As described in [19], a depth map is then obtained from the disparity map, occluded areas are detected and compensated for, and finally depth segmentation is performed by applying the M-RSST algorithm on the final depth map. The following distance is used for depth segmentation:

$$d^D(S_1, S_2) = \left| D_{S_1} - D_{S_2} \right| \frac{A_{S_1} A_{S_2}}{A_{S_1} + A_{S_2}} \tag{3}$$

where D_{S_i} represents the average depth of segment S_i.

3 Fuzzy segmentation fusion

We have now extracted the color, motion and depth segments, using the M-RSST algorithm. The next step is to fuse the color and motion segments as these segments identify the object boundaries very accurately. Additionally depth segments can provide us with a coarse separation of the semantic objects because usually objects consist of segments that are located at about the same depth. So by using depth segments as a constraining mask we can project color and motion segments onto this mask. Then for every color segment we find in which depth it belongs and finally we fuse the color segments of the same depth into one new segment.

The Fuzzy Segmentation Fusion (FSF) takes as input the color, motion, depth and motion-compensated previous segmentations and gives to the output the final fused segmentation.

Let us first provide the reader with the formal mathematical description of our work. We denote with

$$S^C = \{s_1^C, s_2^C, \cdots, s_N^C\}$$
$$S^M = \{s_1^M, s_2^M, \cdots, s_K^M\}$$
$$S^D = \{s_1^D, s_2^D, \cdots, s_L^D\}$$
$$S^P = \{s_1^P, s_2^P, \cdots, s_Q^P\}$$

the color, motion, depth and motion-compensated previous segmentations (the inputs of the FSF module), while

$$S^F = \{s_1^F, s_2^F, \cdots, s_U^F\}$$

stands for the fused segmentation (the output of the FSF module). For simplicity reasons, we did not involve the time in the above notation. Thus, we assume that time t is implied, unless something else is clearly stated. The above mentioned segmentations are actually *crisp* image partitions.

The FSF module tries to take advantage of the information provided by S^C, S^M, S^D and S^P in order to improve the precision and the semantic framework of the segmentation. Since the color semgnetation provides the most precise boundaries, it is reasonable to use S^C as the basis on which S^F will be constructed. Furthermore, we assume that S^C provides an over-segmented image partition. Thus, FSF actually produces a more "unified" partition S^F in comparison with S^C. This means that:

$$\forall i \in \mathbb{N}_N \quad \text{and} \quad j \in \mathbb{N}_U \quad \text{we have} \quad s_i^C \subseteq s_j^F$$

where

$$N_a = 1, 2, \cdots, a.$$

In order to compute S^F out of S^C, S^M, S^D and S^P we use the M-RSST algorithm. Roughly speaking, we take all the possible color segment pairs provided by S^C and compute the distances between them. Then, the two closest segments (smallest distance) are united (if their distance is less than a threshold) and the whole process is repeated until no segments are united. Obviously, the intelligence of the method lies on the way in which the above distances are computed. Let us now describe this process.

We first introduce some useful operators. For each pair of segments s_1, s_2 the *projection operator* $R(s1, s2)$ is defined by:

$$R(s1, s2) = \frac{|s1 \cap s2|}{|s1|} \tag{4}$$

where $|\cdot|$ denotes the cardinality of the set (in this case the number of the pixels of the segment).

Given a segment s and a set of segments S, the *θ-neighborhood operator* $I_\theta(s, S)$ is defined by:

$$I_\theta(s,S) = \{i \in N_{|S|} : |R(s,si) - \max_{j \in \mathbb{N}_{|S|}} R(s,si)| \leq \theta\}$$

where *theta* is a threshold and $R(\cdot,\cdot)$ the projection operator defined by 4.

The *cue-based distance* of two segments is defined (for any cue like motion, depth, motion-compensated previous) with the aid of the θ-neighborhood operator. For example, the *motion-based distance* $D^M(s_i^C, s_j^C)$ of two color segments $s_i^C, s_j^C, (i,j \in \mathbb{N}_N)$ is defined by:

$$D^M(s_i^C, s_j^C) = a \cdot b \cdot \delta_1 - a \cdot b \cdot \delta_2$$

where

$$\begin{aligned}
a &= \max_{k \in \mathbb{N}_K} R(s_i^C, s_k^M) \\
b &= \max_{k \in \mathbb{N}_K} R(s_j^C, s_k^M) \\
\delta_1 &= \delta(|I_\theta(s_i^C, S^M) \cap I_\theta(s_j^C, S^M)|) \\
\delta_2 &= \delta\Big(\big|[I_\theta(s_i^C, S^M) \cup I_\theta(s_j^C, S^M)] - [I_\theta(s_i^C, S^M) \cap I_\theta(s_j^C, S^M)]\big|\Big) \\
\delta &= \begin{cases} 1, \text{if } x \neq 0 \\ 0, \text{if } x = 0 \end{cases}
\end{aligned}$$

The distance $D(s_i^C, s_j^C)$ between two color segments s_i^C and s_j^C, used in the M-RSST segmentation process of the FSF module, is derived from the

motion-based $(D^M(s_i^C, s_j^C))$
depth-based $(D^D(s_i^C, s_j^C))$
and previous-based $(D^P(s_i^C, s_j^C))$

distances computed with the aid of the above process. This is actually the intelligent part of the FSF module that provides the system with the ability to fuse the various cues using a priori knowledge in the form of fuzzy linguistic rules [13]. Figure 2 shows the structure of the fuzzy inference system that implements the above idea. The system takes three inputs: the motion-based distance, the depth-based distance and the previous-based distance. Obviously, they are fuzzy values (between -1 and 1) representing the degree in which the specific cue "advises" the fusing fuzzy inference system to provide or not a high value of "unity" between the two color segments (its output).

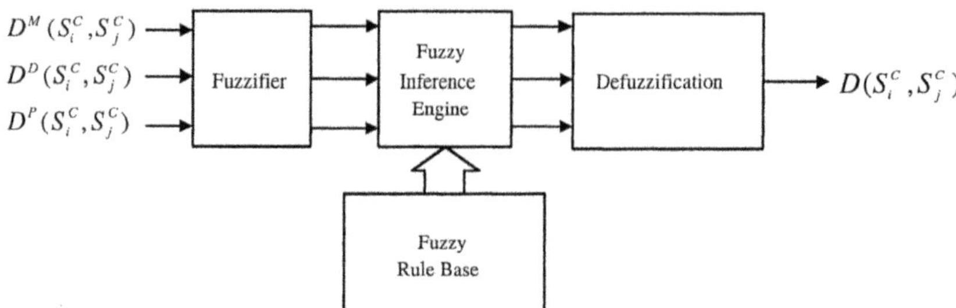

Fig. 2. Fusing Fuzzy Inference System

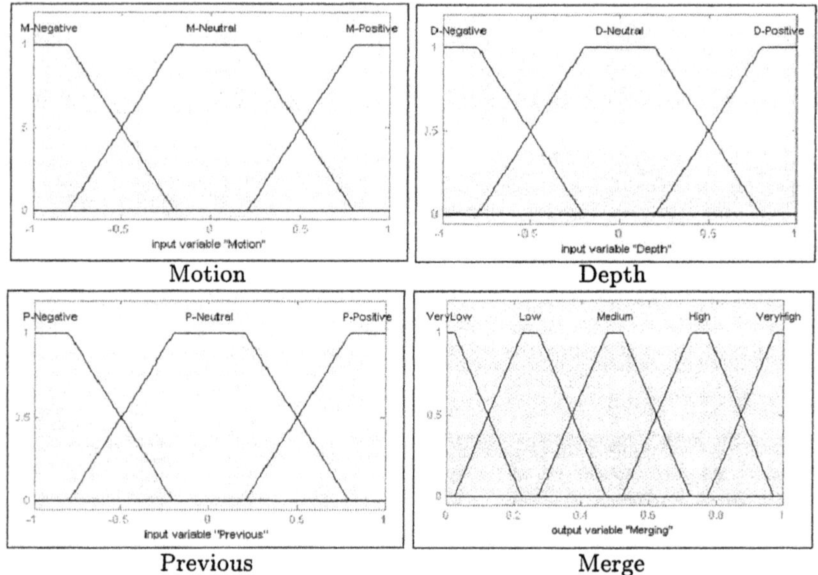

Fig. 3. Definition of the fuzzy partitions

The operation of the fusing fuzzy inference system is based on a set of linguistic rules provided by experts. The rules map the input linguistic variable to the output one, i.e. the fuzzy partitions defined on

$$D^M(s_i^C, s_j^C), D^D(s_i^C, s_j^C) \quad \text{and} \quad D^P(s_i^C, s_j^C)$$

to the fuzzy partition defined on $D(s_i^C, s_j^C)$ (Figure 3). Table 1 summarises this set of rules. Finally, in Figure 4 the surface view of the system input-output mapping is shown.

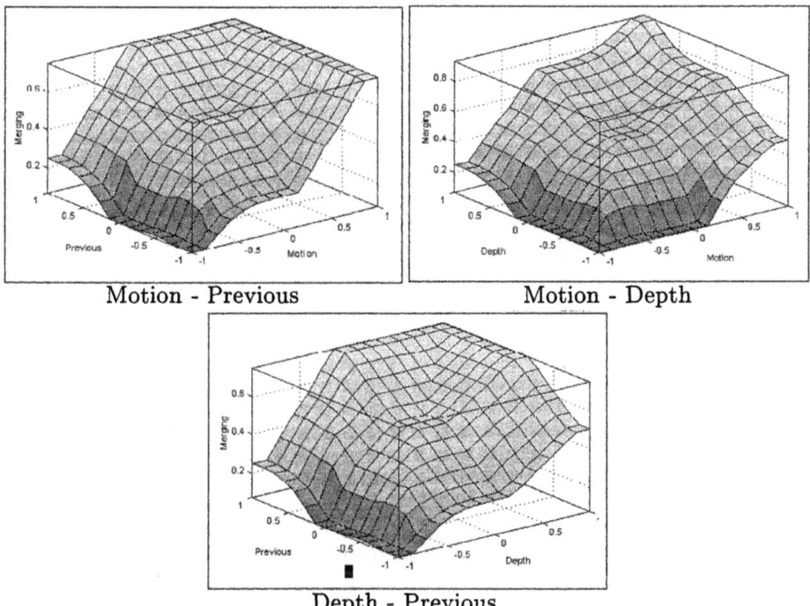

Fig. 4. Schematic representation of the input output associations

4 Simulation results

4.1 Equipment

We tested our algorithm in a real 3-D video sequence. This sequence was obtained with the aid of the "SX2000 Stereo-Optix" lens from NuView connected to a normal MiniDV video camera. The SX2000 lens projects two images on the camera's focus with the aid of a mirror; those two images are not recorded simultaneously but sequentially by taking advantage of the interlaced nature of the CCD video signal. Therefore the even lines of the resultant video sequence represent the left view and the odd lines represent the right view. The focal distance can be adjusted through mechanical means. Apparently this method diminishes the vertical resolution by half (720×288) but it is still satisfactory for our use.

The obtained stereo video sequence is first analyzed and for each pair of stereo frames a depth map is estimated. The application of the M-RSST algorithm on the depth map results on the depth segmentation. The color and motion segmentation cues are then constructed from the left channel image.

If Motion is	and Depth is	and Previous is	then Merging is
M-Negative	D-Negative	P-none	VeryLow
M-Negative	D-Neutral	P-Negative	VeryLow
M-Negative	D-Neutral	P-Neutral	VeryLow
M-Negative	D-Neutral	P-Positive	Low
M-Negative	D-Positive	P-Negative	VeryLow
M-Negative	D-Positive	P-Neutral	Low
M-Negative	D-Positive	P-Positive	Medium
M-Neutral	D-Negative	P-Negative	VeryLow
M-Neutral	D-Negative	P-Neutral	VeryLow
M-Neutral	D-Negative	P-Positive	Low
M-Neutral	D-Neutral	P-Negative	Low
M-Neutral	D-Neutral	P-Neutral	M-Positive
M-Neutral	D-Neutral	P-Positive	High
M-Neutral	D-Positive	P-Negative	Medium
M-Neutral	D-Positive	P-Neutral	High
M-Neutral	D-Positive	P-Positive	High
M-Positive	D-Negative	P-Negative	Low
M-Positive	D-Negative	P-Neutral	Medium
M-Positive	D-Negative	P-Positive	Medium
M-Positive	D-Neutral	P-none	High
M-Positive	D-Positive	P-none	VeryHigh

Table 1. The set of linguistic rules of the fusing fuzzy inference system

4.2 Color

Since the fusion process merges color segments we want those segments to have precise boundaries. Thus, for the color segmentation we use a level 1 resolution (of size 2×2 pixels) on the M-RSST. Since we also want an over-segmented image, we choose a very low distance threshold for the region merging process. For this experiment we used a distance threshold of 0.05% in the RGB space, but any low threshold would produce similar results. An enlarged segmentation result is illustrated in Fig. 5 where the segments are clearly shown. An interesting feature of the RSST algorithm is that it can stop at a predifined number of segments and this would be an alternative way of achieving an oversegmented color image.

4.3 Motion

A block-matching algorithm is used to construct the motion field. The search area of the block matching in our test was set at 10 pixels in both directions and the block size is 3×3 pixels. Despite the small blocksize, the motion field is quite accurate. During the block matching process for the motion field creation, if a block's minimum distance from the test block is less than 20%

greater than the distance of the respective zero-motion block, then we consider the corresponding motion vector as zero. This step reduces the motion noise in low texture areas. A median filter is then applied at the motion field before the segmentation process. For the motion segmentation we apply the M-RSST with a highest resolution level of $L_0 = 3$ (block resolution of 8×8 pixels) to speed up the segmentation process.

4.4 Depth

A disparity map is estimated from the left and right channel images. A depth map is then obtained from the disparity map, occluded areas are detected and compensated for, and finally depth segmentation is performed by applying the M-RSST algorithm on the final depth map. The search area for the disparity estimation is 10 pixels with a block size of 3×3 pixels. The depth segmentation is created with a block resolution of 4×4 pixels.

4.5 Fusion

The fusion subsystem is then started with the color, motion and depth segmentation maps as inputs along with the cue fusion result of the previous frame. In the case of the first frame where no previous fusion results exist, a blank map is used instead. The fusion subsystem continues with the color segment merging by considering now their fuzzy distance.

4.6 Results

An example of the fusion process is illustrated in figure 5, where the color, motion and depth segmentation results are shown along with the fusion result of the previous frame. More examples are given in figures 6–17 for two video sequences. On the first one the camera is still and the person is moving from left to right with his right hand extended. A 6-frame sample of this sequence is shown in figures 6–11. In this sample, the color segmentation produces about 140-150 segments; the motion segmentation produces two segments, one for the background and one for the moving person; the depth segmentation produces three to five segments: one for the background, one for the person's body, one for his hand and additionally some false segments especially in areas with low energy in the high horizontal frequencies.

It can be seen that the person's boundaries in the fused result are more precise than the boundaries of the cue sources. This is not due to the higher resolution of the color segmentation but due to the fusion of color and depth segmentations; a depth segmentation at higher resolution would not have the accuracy of the color segmentaton. The person on the fused image has been seperated succesfully from the background; his hand, having a different depth from his body, is split in a different segment. At frame 59 in figure 10 the hand is separated in two segments due to the vertical expansion of the hand segment of the depth channel.

On the second sequence, both the camera and the person are still but the person is waving his hand. A 6-frame sample is shown in figures 12–17. As it can be seen, up to two motion vectors are created on the palm and arm areas which when combined with the depth and color channels result in the fused images. Note that the area below the the person's right arm (at the bottom left) is considered to be in a different segment than the rest of the background, because the corresponding color segments are not adjacent; they are separated by the arm's color segments. In figure 12 two motion segments have been detected due to the higher linear velocity of the palm. Because of this, the palm is a seperate segment in the fusion result (Fig. 12) and this separation propagates to the following segmentations. If this separation is desirable, the M-RSST algorithm could be adjusted to be more sensitive and produce more segments. If the separation is not desirable, the use of a full parametric motion model would detect the circular motion and thus preserve the arm and palm in one segment.

Because of the inaccurate depth segmentation, one can observe an extra segment in figure 15 near the palm (in black color). This segment was not merged with the rest of the palm because of the previous and motion segmentations, and thus it was not propagated to the following frames.

5 Conclusions

In this chapter, we examined the integration of multiple cue segmentations and the use of fuzzy decisions for handling inconsistencies and improving segmentation accuracy. The resulting segmentation is shown to be superior to the individual cue segmentations.

One way to improve the fusion performance is by augmenting its memory: more than one previous frames could be taken into account thus minimizing the error propagation. Kalman filters could be introduced in combination with the fused-result memory, to implement object tracking and further enhance segmentation reliability.

Fuzziness may also be introduced in the partitions themselves, so that the transition between neighboring regions may be gradual, as well as in the generation of the partition trees, by means of link weights expressing the strength of decomposition of a coarse region into finer ones. Fuzzy partitions can be defined on the partition trees, giving linguistic meaning to low level visual descriptors and allowing the numerical interpretation of inference rules. Moreover the fusion module parameters could be dynamically adapted with the aid of higher level object description information.

Color Segmentation

Motion Segmentation

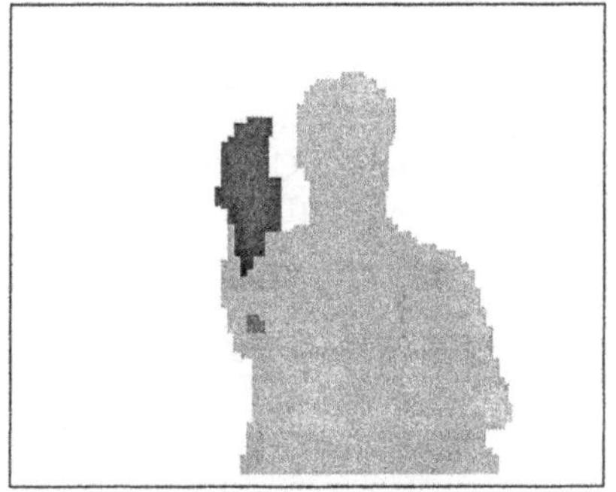

Depth Segmentation

Previous Fusion

Fusion Result

Fig. 5. Sequence 2, Frame 60, Large format

References

1. T. Meier and K. N. Ngan *Automatic Segmentation of Moving Objects for Video Object Plane Generation* IEEE Trans. Circuits and Systems for Video Technology, Vol. 8, No. 5 pp. 525-538 1998
2. M. Pardas and P.Salembier *3-D morphological segmentation and motion estimation for image segmentation* Signal Processing, vol. 38 pp. 31-43 Sept. 1994
3. M. M. Chang, A. M. Tekalp, and M. I. Sezan *Motion field segmentation using an adaptive MAP criterion* Proc. of IEEE Int. Conf. Acoust., Speech, Signal Processing ICASSP'93 Minneapolis, MN, Apr. 1993, vol. V pp.33-36
4. D. Geiger and A. Yuille *A common framework for image segmentation* Int. Journal Comput. Vision, vol 6 pp.227-243 1991
5. O. J. Morris, M. J. Lee and A. G. Constantinides *Graph Theory for Image Analysis: an Approach based on the Shortest Spanning Tree* IEE Proceedings, Vol. 133 pp.146-152 April 1986
6. Y. Avrithis, A. Doulamis, N. Doulamis and S. Kollias *A Stochastic Framework for Optimal Key Frame Extraction from MPEG Video Databases* Computer Vision and Image Understanding July 1999
7. R. C. Gonzalez and R. E. Woods *Digital Image Processing* Addison-Wesley 1992
8. D. Tzovaras, N. Grammalidis and M. G. Strintzis *Disparity Field and Depth Map Coding for Multiview 3D Image Generation* Image Communication, No. 11 pp. 205-230 1998
9. C. Gu and M.-C. Lee *Semiautomatic Segmentation and Tracking of Semantic Video Objects* IEEE Trans. on Circuits and Systems for Video Technology, Vol. 8, No. 5 pp. 572-584 Sept. 1998
10. A.A. Alatan, L. Onural, M. Wollborn, R. Mech, E. Tuncel, and T. Sikora *Image Sequence Analysis for Emereging Interactive Multimedia Services - The European COST 211 Framework* IEEE Trans. on Circuits and Systems for Video Technology, Vol. 8, No. 7 pp. 802-813 Nov. 1998
11. Chong-Wah Ngo, Ting-Chuen Pong, and R. T. Chin *Video Partitioning by Temporal Slice Coherency* IEEE Trans. on Circuits and Systems for Video Technology, Vol. 11, No. 8 pp. 941-953 Aug. 2001
12. A. Smolic, T. Sikora, and J.R. Ohm *Long-Term Global Motion Estimation and Its Application for Sprite Coding, Content Description, and Segmentation* IEEE Trans. on Circuits and Systems for Video Technology, Vol. 9, No. 8 pp. 1227-1242 Dec. 1999
13. G.B. Stamou and S.G. Tzafestas *Fuzzy Relation Equations and Fuzzy Inference Systems: An Inside Approach* IEEE Transactions on Systems, Man and Cybernetics - PART B: Cybernetics, Vol. 29, No. 6 Dec. 1999
14. G.J. Klir and B. Yuan *Fuzzy Sets and Fuzzy Logic: Theory and Applications* Englewood Cliffs, NY: Prentice Hall, 1995
15. L.A. Zadeh *Fuzzy logic=computing with words* IEEE Transactions on Fuzzy Systems pp. 103-111 1996
16. E. Saber and A. Tekalp and G. Bozdagi *Fusion of Color and Edge Information for Improved Segmentation and Edge Linking* Image and Vision Computing vol 15 pp. 769-780 1997
17. P.Salembier, F.Marques, M. Pardas, J.R.Morros, I. Corset, S.Jeannin, L.Bouchard, F.Meyer, and B.Marcotegui *Segmentation-Based Video Coding*

System Allowing the Manipulation of Objects IEEE Trans. on Circuits and Systems for Video Technology, Vol. 7, No. 1 pp. 60-74 Feb. 1997
18. I. Koprinska, S. Carrato *Temporal video segmentation: A survey* Signal Pricessing: Image Communication 16 pp. 477-500 2001
19. N. Doulamis, A. Doulamis, Y. Avrithis, K. Ntalianis, and S. Kollias *Efficient Summarization of Stereoscopic Video Sequences* IEEE Trans. Circuits and Systems for Video Technology, Vol. 10, No. 4 pp. 501- 517 June 2000
20. T. Sikora *The MPEG-4 Video Standard Verification Model* IEEE Trans. Circuits and Systems for Video Technology, Vol. 7, No. 1 pp. 19-31 Feb. 1997
21. A.M. Tekalp *Digital Video Processing* Prentice Hall 1995

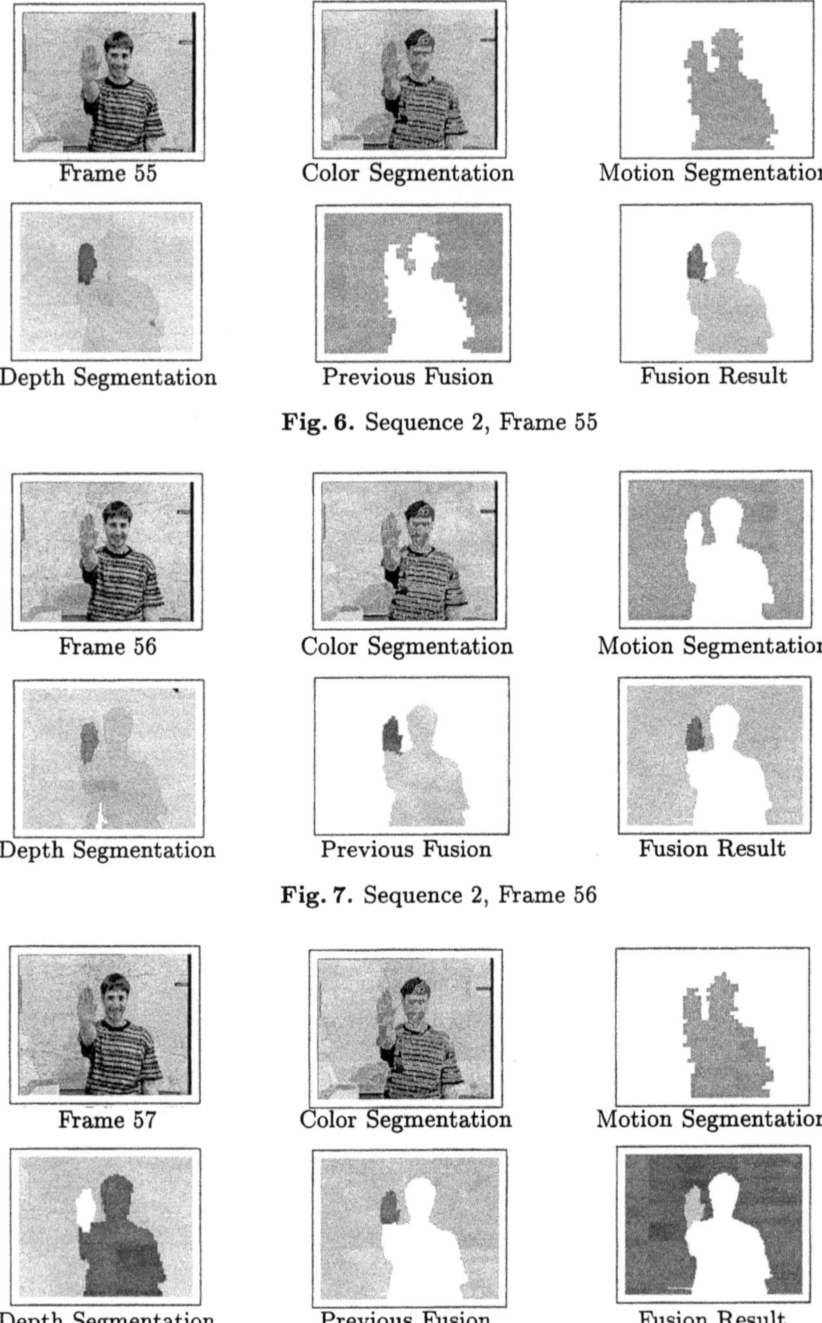

Fig. 6. Sequence 2, Frame 55

Fig. 7. Sequence 2, Frame 56

Fig. 8. Sequence 2, Frame 57

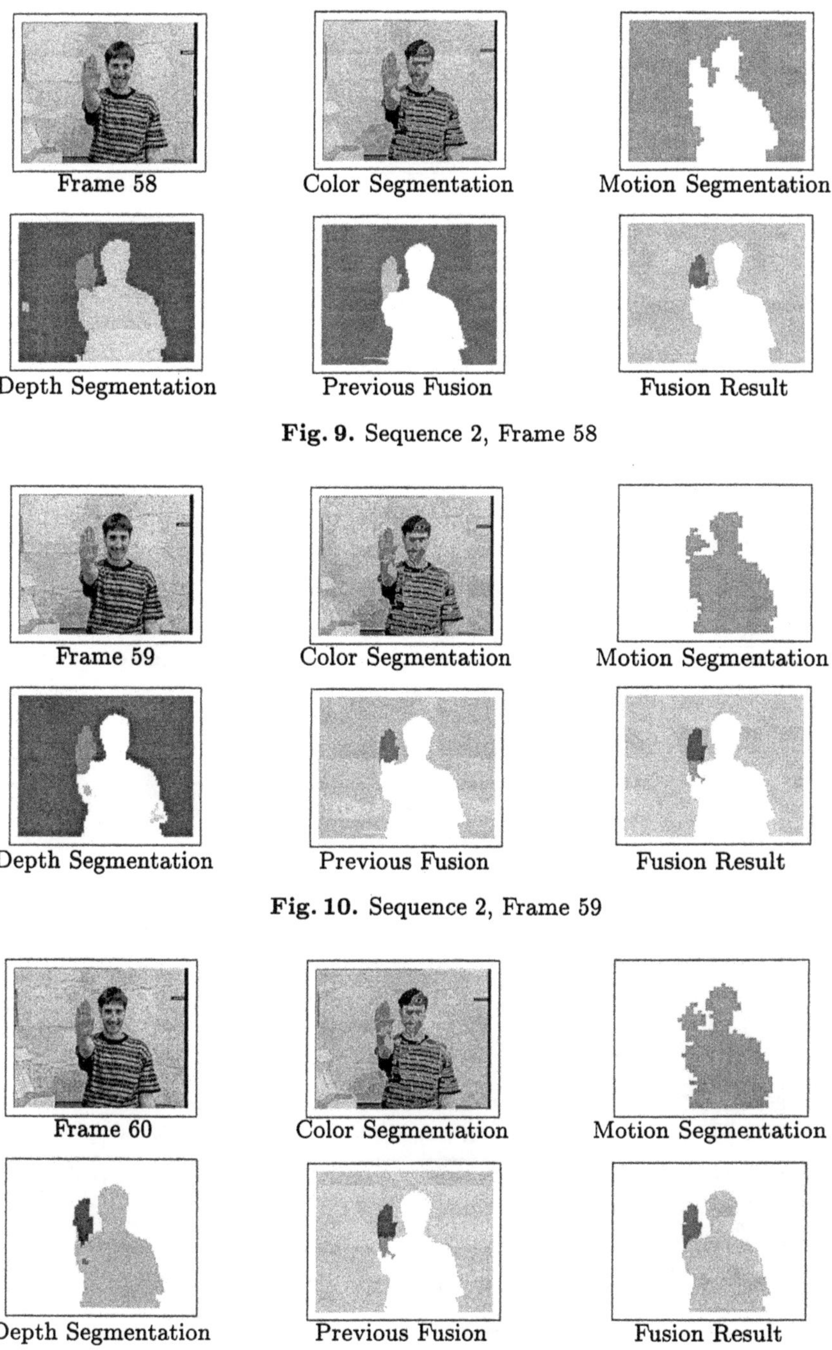

Fig. 9. Sequence 2, Frame 58

Fig. 10. Sequence 2, Frame 59

Fig. 11. Sequence 2, Frame 60

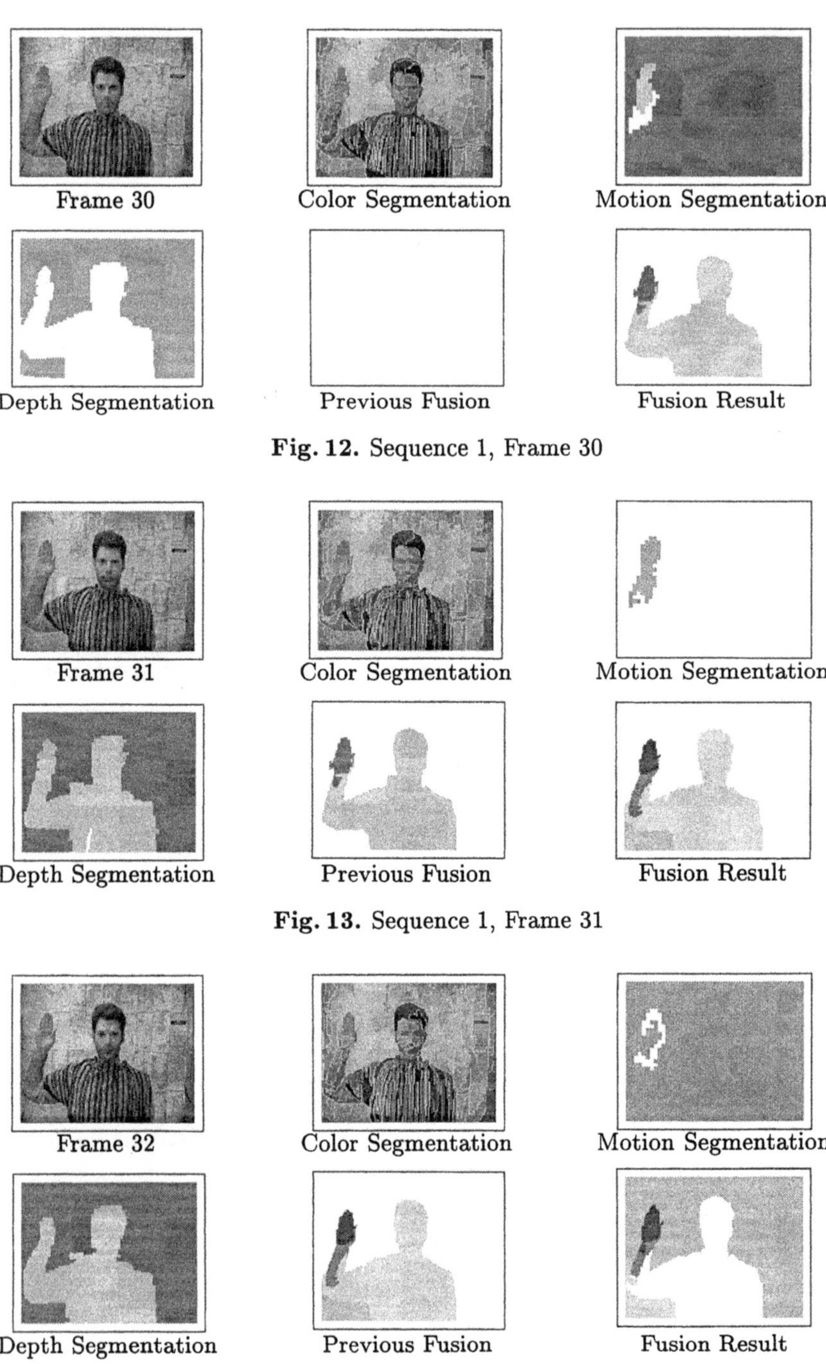

Fig. 12. Sequence 1, Frame 30

Fig. 13. Sequence 1, Frame 31

Fig. 14. Sequence 1, Frame 32

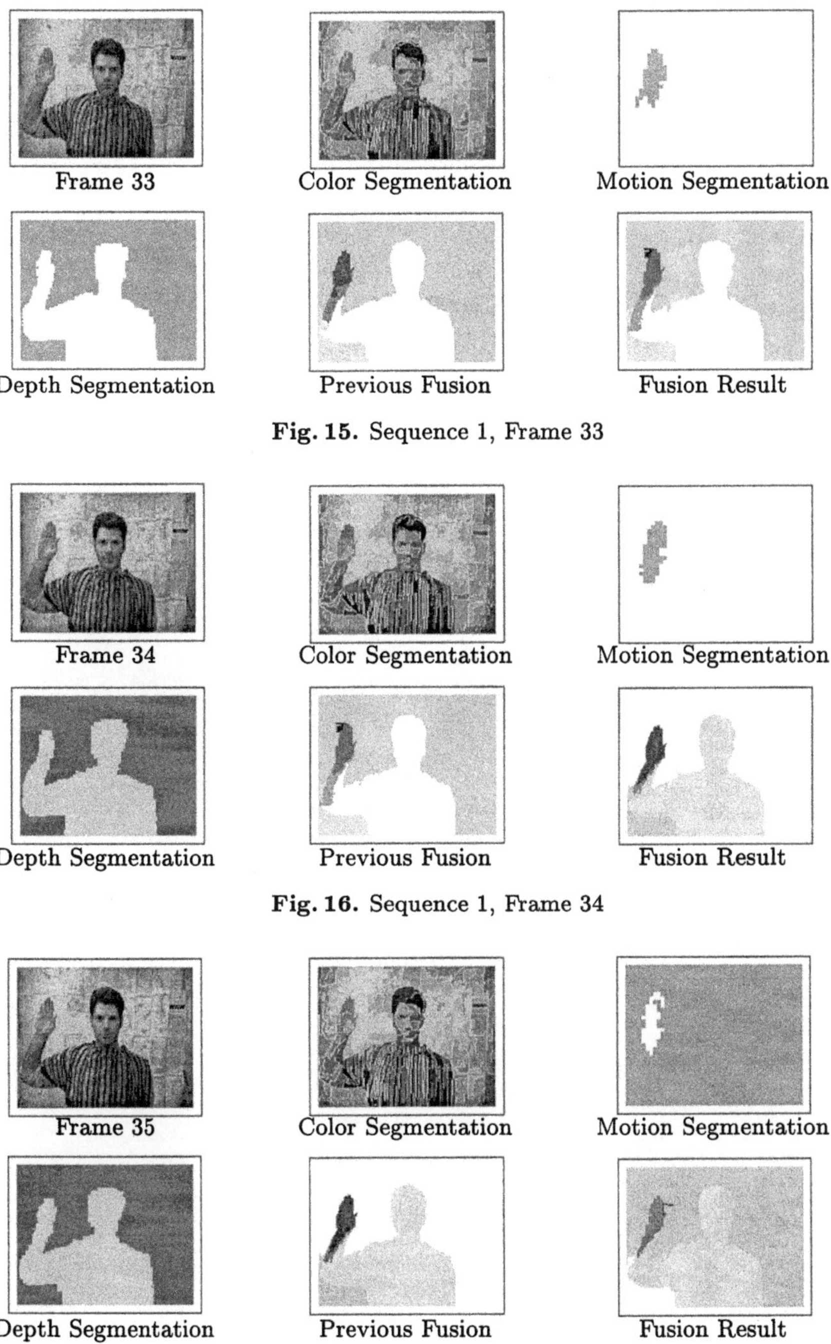

Fig. 15. Sequence 1, Frame 33

Fig. 16. Sequence 1, Frame 34

Fig. 17. Sequence 1, Frame 35

Part III

Fuzzy Filters for Image Enhancement

Chapter 10

Fuzzy Image Enhancement in the Framework of Logarithmic Models

Vasile Pătraşcu, Vasile Buzuloiu and Constantin Vertan

Politehnica University of Bucureşti
Faculty of Electronics and Telecommunications
Applied Electronics and Information Engineering Department
Image Processing and Analysis Laboratory (LAPI)
313, Splaiul Independenţei, Bucureşti, Romania
email: vpatrascu@tarom.ro, vbuzuloiu@alpha.imag.pub.ro,
cvertan@alpha.imag.pub.ro

Summary. The logarithmic model offers new tools for image processing. An efficient method for image enhancement, is to use an affine transformation with the logarithmic operations: addition and scalar multiplication. By adding a fuzzy setting to our model we gain flexibility and better results are possible. We define some criteria for automatically determining the parameters of the processing and this is done via the fuzzy mean and fuzzy variance computed by logarithmic operations.

1 Introduction

Image enhancement is a pre-processing step applied to any input image of an image analysis system. The enhancement methods are specific for every application and the physical nature of pictures as well as the mathematical models for their representation are equally important for choosing the right enhancement method.

For the images obtained by transmitted/reflected light through absorbing/reflecting media, where the effects are naturally of a multiplicative form, Stockham [13] proposed an enhancement method based on the homomorphic theory introduced by Oppenheim [8]. The key to this approach is a homomorphism which transforms the product into a sum (by logarithm) and so allows one to use the classical linear filtering in the presence of additive components. Another logarithmic representation evolving from multiplicative properties of the transmission of light and of the human visual perception was developed by Jourlin and Pinoli [6]. Their model was more elaborate from the mathematical point of view and it inspired the one presented and used in this chapter. Still, both mentioned models do work with semi-bounded sets of values. Perhaps this is the main reason which limits extending them to color spaces. Our logarithmic model [9–12] does work with bounded real sets: the range of values of our images, say $(0, M)$, is linearly applied onto the standard set $(-1, 1)$ and it is this interval that plays the central role in the model: it is endowed with the structure of a linear (moreover: Euclidean) space over the scalar field of real numbers, $\mathbf{R}$. In this framework we can obtain very efficient enhancement algorithms with just one affine transformation (a point operation) if, for instance, the image has a bad exposure [10]. In other, more difficult cases, as e.g. of a variable illumination over the image, such a simple solution is not anymore efficient.

One can improve the results by adding fuzziness [18] to our representation: we adopt a fuzzy partition P with 3 classes for the gray level set (here already the interval $(-1, 1)$), namely A_0 - dark pixels, A_2 - bright pixels and A_1 - middle valued pixels. Further we define three (log-) affine transforms parameterized by fuzzy means and fuzzy variances obtained for each class. The final enhanced image is obtained by merging through weighting the three images obtained by the application of the individual transforms.

The chapter starts with the construction of the logarithmic representation of images with a bounded value set as mentioned previously. Then we use fuzzification procedure sketched above and we apply it to various examples. The algorithm is then extended to color images. The results are presented by comparing them with those of classical (not fuzzy) algorithms. We stress that the whole exercise presented in this chapter refers to point operations while our new framework allows us to rebuild the entire image processing algorithms (neighborhood as well as integral operations).

2 The fundamentals of the logarithmic model

The most usual mathematical model for the gray level images is the real valued function defined on a bounded subset of $\mathbf{R}^2$; and a real vector-valued function for color or multicomponent images. Having in mind that the function value is a point (x, y) representing the luminosity of that pixel $\mathbf{R}^2$ or reflectancy or transparency - it becomes clear that the functions we use are bounded ones (say, they take values in a bounded interval $(0, M)$).

The problems appear when processing an image: the mathematical operations on real valued functions use implicitly the algebra of the real numbers i.e. on the whole real axis and we are faced with results that do not belong anymore to the interval $[0, M]$ – the only ones with physical meaning. Nevertheless this situation is generally accepted by using a very rough "stratagem": truncation of the result. Sometimes a chain of operations is accomplished before truncation, sometimes the truncation is applied to the intermediate results. Always the correction is only done for the values out of the range $[0, M]$.

The framework used in the sequel, from this point of view, is radically different: namely, we want our set $(0, M)$ or, more generally, $(0, M)^n$ in the case of vector-valued images, to be a stable set with respect to the algebraic operations we use – addition and scalar multiplication (with real scalars) [1].

For convenience we prefer to build this structure on the set $(-1,1)$ and linearly move $(0,M)$ onto it by an affine transformation. The key to our construction on $(-1,1)$ is an isomorphism between $\mathbf{R}$ and $(-1,1)$. The generalization for vector valued images is almost straightforward.

2.1 The vector space of gray levels

Let us consider the space of gray levels as the set $\mathbf{E} = (-1,1)$. In the set of gray levels we will define the addition $\oplus$ and the multiplication $\otimes$ by a real scalar and then, defining a scalar product $\langle\cdot,\cdot\rangle$ and a norm $||\cdot||$, we shall obtain an Euclidean space of gray levels.

The addition We shall define the sum of two gray levels, $v_1 \oplus v_2$, by:

$$v_1 \oplus v_2 = \frac{v_1 + v_2}{1 + v_1 v_2}, \ \forall v_1, v_2 \in \mathbf{E}. \tag{1}$$

All the right-hand operations are the usual real scalar operations.

The neutral element for the addition is $\theta = 0$. Each element $v \in \mathbf{E}$ has an opposite w, which verifies $v \oplus w = \theta$. The addition $\oplus$ is stable, associative, and commutative. Thus, it follows that this operation induces on $\mathbf{E}$ a commutative group structure.

We shall define the subtraction operation $\ominus$ by:

$$v_1 \ominus v_2 = \frac{v_1 - v_2}{1 - v_1 v_2}, \ \forall v_1, v_2 \in \mathbf{E}. \tag{2}$$

Using the defined subtraction $\ominus$, we shall denote the opposite of v by $\ominus v$.

The multiplication by a scalar We shall define the multiplication of a gray level v by a scalar λ as:

$$\lambda \otimes v = \frac{(1+v)^\lambda - (1-v)^\lambda}{(1+v)^\lambda + (1-v)^\lambda}, \ \forall v \in \mathbf{E}, \ \forall \lambda \in \mathbf{R}. \tag{3}$$

The operations defined above, the addition $\oplus$ (1) and the scalar multiplication $\otimes$ (3) induce on $\mathbf{E}$ a real vector space structure. In this vector space the usual associativity and distributivity properties hold:

$$\lambda \otimes (v_1 \oplus v_2) = (\lambda \otimes v_1) \oplus (\lambda \otimes v_2). \tag{4}$$

$$\lambda_1 \otimes (\lambda_2 \otimes v) = (\lambda_1 \lambda_2) \otimes v. \tag{5}$$

$$(\lambda_1 + \lambda_2) \otimes v = (\lambda_1 \otimes v) \oplus (\lambda_2 \otimes v). \tag{6}$$

[1] This is needed for the framework here: an Euclidean vector space. One can go further to build an algebra i.e. defining the multiplication between two elements of our set (see Pătraşcu [12]).

The fundamental isomorphism The vector space of gray levels $(\mathbf{E}, \oplus, \otimes)$ is isomorphic to the space of real numbers $(\mathbf{R}, +, \cdot)$ by the function $\varphi : \mathbf{E} \to \mathbf{R}$, defined as:

$$\varphi(v) = \frac{1}{2} \ln \left(\frac{1+v}{1-v} \right) \tag{7}$$

The isomorphism φ verifies:

$$\varphi(v_1 \oplus v_2) = \varphi(v_1) + \varphi(v_2), \ \forall v_1, v_2, \in \mathbf{E}. \tag{8}$$

$$\varphi(\lambda \otimes v) = \lambda \cdot \varphi(v), \ \forall \lambda \in \mathbf{R}, \ \forall v \in \mathbf{E}. \tag{9}$$

The particular nature of this isomorphism induces the logarithmic character of the mathematical model.

The Euclidean space of gray levels The scalar product of two gray levels, $\langle \cdot, \cdot \rangle_{\mathbf{E}} : \mathbf{E} \times \mathbf{E} \to \mathbf{R}$ is defined with respect to the isomorphism from (7) as:

$$\langle v_1, v_2 \rangle_{\mathbf{E}} = \varphi(v_1) \cdot \varphi(v_2), \ \forall v_1, v_2 \in \mathbf{E}. \tag{10}$$

Based on the scalar product $\langle \cdot, \cdot \rangle_{\mathbf{E}}$ the gray level space becomes an Euclidean space. The norm $|| \cdot ||_E : \mathbf{E} \to \mathbf{R}^+$ is defined via the scalar product:

$$||v||_{\mathbf{E}} = \sqrt{\langle v, v \rangle_{\mathbf{E}}} = |\varphi(v)|, \ \forall v \in \mathbf{E}. \tag{11}$$

This norm verifies the following relations:

$$||v_1 \oplus v_2||_{\mathbf{E}} = |\varphi(v_1) + \varphi(v_2)| \tag{12a}$$
$$||v_1 \ominus v_2||_{\mathbf{E}} = |\varphi(v_1) - \varphi(v_2)| \tag{12b}$$

The asymptotical behaviour of the gray level operations In the neighborhood of the zero value (and thus in the range of middle-valued gray levels), the algebraic operations $\oplus$ and $\otimes$ are similar to the classical operations for real numbers. Considering some gray levels u and v, with $|u| \ll 1$ and $|v| \ll 1$, the approximations $|uv| \ll 1$ and $(1 \pm v)^\lambda \simeq 1 \pm \lambda v$ hold. Thus,

$$u \oplus v \simeq u + v.$$
$$\lambda \otimes v \simeq \lambda v.$$

2.2 The vector space of gray level images

A gray level image is a function defined over a 2-dimensional compact set $\mathbf{D} \subset \mathbf{R}^2$ and takes values within the gray level set $\mathbf{E}$. We shall denote by $F(\mathbf{D}, \mathbf{E})$ the set of gray level images defined over the spatial domain $\mathbf{D}$. The previously defined gray level operations can be naturally extended to the set of gray level images, by considering the basic operation at each spatial location (pixel) within the images. Thus the addition of two images becomes:

$$(f_1 \oplus f_2)(x,y) = f_1(x,y) \oplus f_2(x,y), \ \forall f_1, f_2 \in F(\mathbf{D},\mathbf{E}), \ \forall (x,y) \in \mathbf{D}. \quad (14)$$

The neutral element is the identical null function. The addition $\oplus$ is stable, associative, commutative, and provides an opposite for each gray level image. Thus, this operation establishes a commutative group structure on the set $F(\mathbf{D},\mathbf{E})$.

The multiplication of an image f by a scalar λ becomes:

$$(\lambda \otimes f)(x,y) = \lambda \otimes f(x,y), \ \forall f \in F(\mathbf{D},\mathbf{E}), \ \forall \lambda \in \mathbf{R}, \ \forall (x,y) \in \mathbf{D}. \quad (15)$$

The two operations, addition $\oplus$ and scalar multiplication $\otimes$, induce a real vector space structure on $F(\mathbf{D},\mathbf{E})$.

For any two square integrable functions in $F(\mathbf{D},\mathbf{E})$, we shall define their scalar product by:

$$\langle f_1, f_2 \rangle_{L^2(\mathbf{E})} = \int_{\mathbf{D}} \langle f_1(x,y), f_2(x,y) \rangle_{\mathbf{E}} dxdy, \ \forall f_1, f_2 \in F(\mathbf{D},\mathbf{E}). \quad (16)$$

With the scalar product the space $F(\mathbf{D},\mathbf{E})$ becomes a Hilbert space.

The norm is defined based on the scalar product as:

$$\|f\|_{L^2(\mathbf{E})} = \sqrt{\langle f, f \rangle_{L^2(\mathbf{E})}}, \ \forall f \in F(\mathbf{D},\mathbf{E}). \quad (17)$$

2.3 The vector space of colors

We shall consider next the space of colors, defined as the cube $\mathbf{E}^3 = (-1,1)^3$. We will denote by r, g and b the three components of any vector $\mathbf{v} \in \mathbf{E}^3$ (assumed their are the usual red, green and blue color components). In the cube $\mathbf{E}^3$ we must re-define the addition $\oplus$ and the real scalar multiplication $\otimes$. The sum of two colors $\mathbf{v}_1 = (r_1, g_1, b_1)$ and $\mathbf{v}_2 = (r_2, g_2, b_2)$ is defined as:

$$\mathbf{v}_1 \oplus \mathbf{v}_2 = (r_1 \oplus r_2, \ g_1 \oplus g_2, \ b_1 \oplus b_2), \ \forall \mathbf{v}_1, \mathbf{v}_2 \in \mathbf{E}^3. \quad (18)$$

The neutral element for addition is $\theta = (0,0,0)$. Each element $\mathbf{v} = (r, g, b)$ has an opposite element $\mathbf{w} = (\ominus r, \ominus g, \ominus b)$, and obviously $\mathbf{v} \oplus \mathbf{w} = \theta$. The addition $\oplus$ is stable, associative, and commutative. Thus, this operation induces on $\mathbf{E}^3$ a commutative group structure.

We can also define the subtraction operation $\ominus$ by:

$$\mathbf{v}_1 \ominus \mathbf{v}_2 = (r_1 \ominus r_2, \ g_1 \ominus g_2, \ b_1 \ominus b_2). \quad (19)$$

We shall denote the opposite of color $\mathbf{v}$ by $\ominus \mathbf{v}$.

The external multiplication between a real scalar λ and a color vector $\mathbf{v}$ is defined as:

$$\lambda \otimes \mathbf{v} = (\lambda \otimes r,\ \lambda \otimes g,\ \lambda \otimes b)\,,\ \forall \lambda \in \mathbf{R},\ \forall \mathbf{v} \in \mathbf{E}^3. \tag{20}$$

The two operations, addition $\oplus$ (18) and scalar multiplication $\otimes$ (20), induce a real vector space structure onto $\mathbf{E}^3$.

The scalar product of two color vectors, $\langle \cdot, \cdot \rangle_{\mathbf{E}^3} : \mathbf{E}^3 \times \mathbf{E}^3 \to \mathbf{R}$ is defined using the basic isomorphism (7) as:

$$\langle \mathbf{v}_1, \mathbf{v}_2 \rangle_{\mathbf{E}^3} = \varphi(r_1) \cdot \varphi(r_2) + \varphi(g_1) \cdot \varphi(g_2) + \varphi(b_1) \cdot \varphi(b_2),\ \forall v_1, v_2 \in \mathbf{E}^3. \tag{21}$$

The norm $||\cdot||_{\mathbf{E}^3} : \mathbf{E}^3 \to \mathbf{R}^+$ is defined by:

$$||\mathbf{v}||_{\mathbf{E}^3} = \sqrt{\varphi^2(r) + \varphi^2(g) + \varphi^2(b)},\ \forall \mathbf{v} \in \mathbf{E}^3. \tag{22}$$

2.4 The vector space of color images

A color image is a function defined over a 2-dimensional compact set $\mathbf{D} \subset \mathbf{R}^2$ and having values within the color set $\mathbf{E}^3$. We shall denote by $F(\mathbf{D}, \mathbf{E}^3)$ the set of color images defined over the spatial domain $\mathbf{D}$. The previously defined color operations can be naturally extended to the set of color images, by considering the basic operation at each spatial location (pixel) within the images. Thus the addition, scalar multiplication, scalar product and norm for color images are deduced by the same formulae as for the gray level images.

3 Image fuzzification

3.1 The fuzzification of the gray level set

We shall represent the original gray level set $\mathbf{E}$ by three fuzzy sets, that will correspond to the dark, middle-range and respectively bright gray levels. Thus we introduce a partition of the set $\mathbf{E}$ into the fuzzy sets $\{A_0, A_1, A_2\}$. The membership degree of any gray level v from $\mathbf{E}$ within the mentioned fuzzy sets will be given by some functions $\mu_{A0}, \mu_{A1}, \mu_{A2}$. These functions could be any Łukasiewicz functions representing the natural language description that a number v from the set $\mathbf{E}$ is small, medium or large. A particular example for such functions is given in Eqs. (23a)-(23c) below, inspired by the particular nature of the fundamental isomorphism from (7):

$$\mu_{A0}(v) = \left(\frac{1-v}{2}\right)^2,\ \forall v \in \mathbf{E}. \tag{23a}$$

$$\mu_{A1}(v) = 1 - v^2,\ \forall v \in \mathbf{E}. \tag{23b}$$

$$\mu_{A2}(v) = \left(\frac{1+v}{2}\right)^2,\ \forall v \in \mathbf{E}. \tag{23c}$$

The usual model used for such a partitioning is the probabilistic model: the membership degrees of any gray level within the three sets must sum to one. In this case the membership degrees become sharing degrees (or sharings) w_{Ak} of any gray level between the different classes (sets) of gray levels A_k, $k = \{0, 1, 2\}$. Thus, the following relation must hold:

$$w_{A0}(v) + w_{A1}(v) + w_{A2}(v) = 1, \ \forall v \in \mathbf{E}. \tag{24}$$

Since the basic Łukasiewicz functions μ_{Ak} that describe the three fuzzy sets are not probabilistic normalized as required by (24), we will consider their normalization. Aiming at a further flexibility we shall introduce a real, positive tuning parameter γ, and define the family of coupled sharing degrees for the three gray level subsets as:

$$w_{Ak}(v) = \frac{(\mu_{Ak}(v))^\gamma}{(\mu_{A0}(v))^\gamma + (\mu_{A1}(v))^\gamma + (\mu_{A2}(v))^\gamma}, \text{ with } k = \{0, 1, 2\}. \tag{25}$$

Figure 1 shows some particular partitions of the gray level set $\mathbf{E}$.

3.2 Gray level image fuzzification

We shall now consider an entire gray level image f. The three considered gray level fuzzy subsets A_0, A_1, A_2 (defined by their Łukasiewicz functions from (25)) will induce a representation of the original gray level image by three fuzzy images. The fuzzy images f_{A0}, f_{A1}, f_{A2} can be defined as arrays of fuzzy singletons [14,15] (fuzzy sets with only one supporting point), which stand for the membership degree of the gray level of each pixel within the fuzzy gray level subset A_k:

$$f_{Ak} = \bigcup_{(x,y)\in\mathbf{D}} \mu_{Ak}\left(f(x,y)\right), \text{ with } k = \{0, 1, 2\}. \tag{26}$$

In the equation above, $\mu_{Ak}\left(f(x,y)\right)$ stands for the membership degree of the pixel within the image f, located at coordinates $(x, y) \in \mathbf{D}$, with respect to the fuzzy set A_k.

For each fuzzy image f_{Ak} defined in (26) and characterized by the Łukasiewicz function μ_{Ak} of its supporting gray level subset, the fuzzy cardinality is computed as:

$$N(f_{Ak}) = \sum_{(x,y)\in\mathbf{D}} \mu_{Ak}\left(f(x,y)\right), \text{ with } k = \{0, 1, 2\}. \tag{27}$$

Further on, we shall use fuzzy statistics of the gray level image f, defined with respect to each of the fuzzy images (and thus, implicitly, with respect to each of the three gray level subsets). We shall mainly use the fuzzy [gray level] mean and the fuzzy [gray level] variance. Their definitions naturally

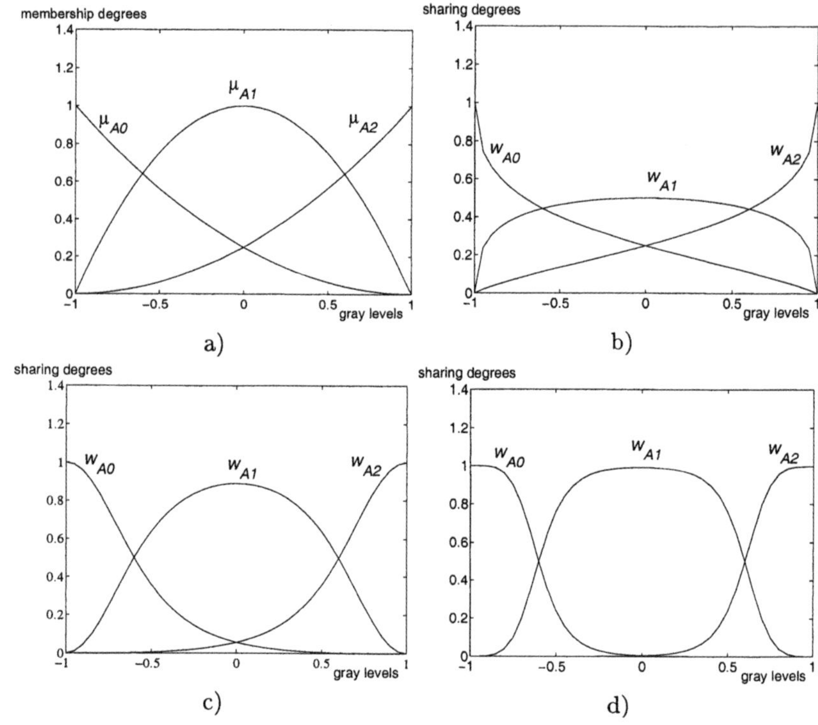

Fig. 1. The Łukasiewicz functions that define the fuzzy sets A_0, A_1 and A_2, based on the particular basis functions (23a)-(23c) and the probabilistic model described in (25), with various values of the tuning parameters γ: a) membership degrees according to the initial μ_{Ak} functions; b) - d) sharing degrees w_{Ak} with b) $\gamma = 0.5$; c) $\gamma = 2$; d) $\gamma = 4$.

extend the definitions of the classical statistical mean and variance. Still, the gray level addition and scalar multiplication will be implemented by the $\oplus$ (1) and $\otimes$ (3) operators, induced on the set of gray levels $\mathbf{E}$ by the logarithmic model.

Thus, the fuzzy [gray level] mean for each fuzzy-image f_{Ak} is a_k, defined as:

$$a_k = \bigoplus_{(x,y)\in\mathbf{D}} \left(\frac{\mu_{Ak}\left(f(x,y)\right)}{N(f_{Ak})} \otimes f(x,y) \right), \text{ with } k = \{0,1,2\}. \qquad (28)$$

The fuzzy [gray level] variance for each fuzzy image f_{Ak} is σ_k^2, defined as:

$$\sigma_k^2 = \sum_{(x,y)\in\mathbf{D}} \frac{\mu_{Ak}\left(f(x,y)\right)}{N(f_{Ak})} \left\| f(x,y) \ominus a_k \right\|_{\mathbf{E}}^2, \text{ with } k = \{0,1,2\}. \qquad (29)$$

Using the property (12b) the relations above become:

$$\sigma_k^2 = \sum_{(x,y)\in\mathbf{D}} \frac{\mu_{Ak}(f(x,y))}{N(f_{Ak})} \left(\varphi(f(x,y)) - \varphi(a_k)\right)^2, \text{ with } k = \{0,1,2\}. \quad (30)$$

3.3 Color image fuzzification

The approach described above can be extended in order to deal with color images. There are two basic ways for processing color (as a particular case of vector-valued information). The most obvious method (but not the most appealing, from a computational or an algorithmic design point of view) is to process the color information as vectors [2,16]. The alternative method is to process the color vectors only by scalar operations; these methods follow the framework of pre-ordering principles, as introduced by Barnett [1,16]. According to this framework, there are four basic means for processing a vector information in a scalar way: marginal, reduced, conditional and partial. The marginal and reduced processing are the most used paradigms and will be further discussed.

The marginal processing implies that every component of the color image is processed independently. Any correlation that naturally exists between the color components is thus ignored and the processing results may not be always as good as expected (compared with their scalar counterparts) [16]. We may define a three-set fuzzy partition $\{A_0, A_1, A_2\}$ for every color component of the image. Each color component (r, g or b) will be thus viewed as a scalar, gray level image, that we will denote by f_p, with $p = \{r, g, b\}$. For each color component image f_p we will define the fuzzy images induced by the fuzzy partition of the pixel values set $\mathbf{E}$ and, subsequently, the fuzzy cardinality, the fuzzy mean, and the fuzzy variance according to the same definitions as in the previous, gray level construction (i.e. Eqs. (27), (28), and (30)):

$$N(f_{p,Ak}) = \sum_{(x,y)\in\mathbf{D}} \mu_k(f_p(x,y)), \quad (31a)$$

$$a_{p,k} = \bigoplus_{(x,y)\in\mathbf{D}} \left(\frac{\mu_k(f_p(x,y))}{N(f_{p,Ak})} \otimes f_p(x,y) \right), \quad (31b)$$

$$\sigma_{p,k}^2 = \sum_{(x,y)\in\mathbf{D}} \frac{\mu_k(f_p(x,y))}{N(f_{p,Ak})} \left(\varphi(f_p(x,y)) - \varphi(a_{p,k})\right)^2, \quad (31c)$$

with $k = \{0,1,2\}$ and $p = \{r, g, b\}$.

The reduced processing implies that a single scalar image is computed from the original color image, and all the processing is based on that new scalar image [1]. Most often, the scalar image is a measure of the image luminosity (or luminance, or light intensity), and is computed according to the usual luminance formula (but using the $\oplus$ and $\otimes$ operations), as:

$$f_Y = 0.3 \otimes f_r \oplus 0.6 \otimes f_g \oplus 0.1 \otimes f_b. \quad (32)$$

For this luminance image the fuzzy cardinality $N_{f_{Y,A_k}}$, fuzzy mean $a_{Y,k}$ and fuzzy variance $\sigma^2_{Y,k}$ are computed according to the formulae introduced for gray level images ((27), (28), and (30) respectively).

We may note that the two proposed fuzzy color models are different from the classically-used fuzzy color models. Usually one associates to each color $\mathbf{c}$ a Łukasiewicz function, $\mu_{\mathbf{c}} : \mathbf{E}^3 \to [0,1]$ that measures the membership degree of any color $\mathbf{c}'$ from $\mathbf{E}^3$ within the class "color $\mathbf{c}$". Thus, $\mu_{\mathbf{c}}(\mathbf{c}')$ is a scalar within [0,1] that expresses how similar is the color $\mathbf{c}'$ with respect to the color $\mathbf{c}$. The analytical definition of the function $\mu_{\mathbf{c}}$ must take into account the natural perception and thus $\mu_{\mathbf{c}}$ must be decreasing with respect to the inter-color distance $d(\mathbf{c},\mathbf{c}')$ (regardless the definition of that distance). A typical model is the one proposed by Haffner [4] (in the framework of comparison of color histograms for image indexing):

$$\mu_{\mathbf{c}}(\mathbf{c}') = \exp\left(-\sigma\left(\frac{d(\mathbf{c},\mathbf{c}')}{d_{\max}}\right)^2\right), \tag{33}$$

with

$$d_{\max} = \max_{\mathbf{c},\mathbf{c}' \in \mathbf{E}^3} \left(d(\mathbf{c},\mathbf{c}')\right). \tag{34}$$

The tuning parameter σ in (33) allows to consider color similarity functions that are more or less localized, and thus to modify the inter-color confusion. We may also notice that the expression in (33) implicitly depends on the maximal dimension of the color gamut (through $d_{\max}$) and thus takes into account the color quantization. Once the color quantization is chosen and thus $d_{\max}$ is determined, the tuning parameter σ can be easily related to the imposed similarity degree $\mu 1$ associated to a color lying at an unitary distance ($d(\mathbf{c},\mathbf{c}') = 1$) from the target color $\mathbf{c}$, by $\sigma = -d^2_{\max} \ln \mu 1$. A similar model (but oriented for the processing within the Lab color space) was proposed in [17] as:

$$\mu_{\mathbf{c}}(\mathbf{c}') = \begin{cases} 1, \text{ if } d(\mathbf{c},\mathbf{c}') \le JND \\ \max\left(0, 1 - \frac{d(\mathbf{c},\mathbf{c}')}{\sigma JND}\right), \text{ if } d(\mathbf{c},\mathbf{c}') > JND \end{cases}. \tag{35}$$

In the equation above, *JND* is the just noticeable color difference, which, for the Lab color space, equals 2.3.

4 Gray level image enhancement by affine transforms

Several authors [2,6,19] have defined image enhancement as a collection of techniques that seek to improve the visual appearance of an image, or to convert the image to a form better suited for analysis by a human or a machine. Image enhancement is one of the pre-processing steps applied to any image in image processing and analysis systems; the particular enhancement methods are very application-specific and user-specific [19]. Moreover, the physical nature of the image and the corresponding acquisition procedure, and their associated mathematical models are equally important for choosing an enhancement procedure.

Most of the image enhancement algorithms are simple point transforms [2,5,3], for which the gray level of the processed pixel depends only on the initial gray level of the same pixel. Thus, these image enhancement transforms are mappings (or functions) of the initial gray level values. Linear (affine) transforms are simple, but general-enough instances of such enhancement procedures. A linear transform ψ of the gray level image f is an image where the value of each pixel is transformed by the function ψ as:

$$\psi(v) = \alpha \cdot v + \beta, \ \alpha, \beta \in \mathbf{R}. \tag{36}$$

The most used enhancement affine transform is the contrast stretching [2,5], which aims to maximize the dynamic range of the image, by spreading the values of the gray levels within the entire allowed range. Still, the gray level transform is followed (or embeds) a thresholding step for the limitation of the values that could lie (due to the use of usual real addition and multiplication in (36)) outside the allowed range.

We will adapt the class of affine image transform to the framework of the logarithmic image model. We shall thus consider the class of affine transforms on $F(\mathbf{D}, \mathbf{E})$, that is, the transforms, $\psi : F(\mathbf{D}, \mathbf{E}) \to F(\mathbf{D}, \mathbf{E})$, of the form:

$$\psi(f) = \alpha \otimes f \oplus \beta, \ \forall f \in F(\mathbf{D}, \mathbf{E}). \tag{37}$$

In the equation above α is a real scalar and β is a constant image. Obviously, the transform from (37) can be also written as:

$$\psi(f) = \alpha \otimes \left(f \oplus (\frac{1}{\alpha} \otimes \beta) \right) = \alpha \otimes (f \oplus \beta') = \alpha \otimes f'. \tag{38}$$

This second form of the affine transform states that we can process the image in two steps: a translation of the gray level values by the constant term $\beta' = \frac{1}{\alpha} \otimes \beta$, followed by a scalar multiplication of the gray levels of the intermediate image by α. The parameters α and β of the affine transformation must be chosen in order to fulfill an optimality criterion (such as the maximization of the dynamic gray level range). The presence of noise often requires the estimation and maximization of an average dynamic range instead of the dynamic range and thus the problem becomes non-linear.

Our approach consists of choosing α and β' such that the resulting image has the same average gray level and gray level variance as a uniformly distributed random variable within the allowed gray level range $\mathbf{E}$. These conditions imply, in fact, that we linearly approximate by an affine transform of type (38) the non-linear histogram equalization [5,3] transform. Indeed, the histogram equalization supposes that the gray levels are particular realizations of some random variable that should be transformed into an uniform random variable. Since we have chosen $\mathbf{E} = (-1, 1)$, the equivalent random variable has a zero mean and a variance $\sigma_u^2 = 1/3$. Thus, we will choose α and β' such that the mean of the processed image is zero and its variance is σ_u^2. For any given image f, with mean a_f and variance σ_f^2, obviously $\beta' = \ominus a_f$ and $\alpha = \sigma_u/\sigma_f$ and thus the affine transform ψ becomes:

$$\psi(f) = \frac{\sigma_u}{\sigma_f} \otimes (f \ominus a_f). \tag{39}$$

The proposed fuzzy setting allows to control in a more precise way the modification of various gray levels. Each of the fuzzy images f_{Ak} will provide a set of control parameters α and β' that reflects the gray level content in the various gray level classes (dark, middle-valued, bright):

$$\begin{aligned} \alpha_k &= \sigma_u/\sigma_k \\ \beta'_k &= \ominus a_k, \text{ with } k = \{0, 1, 2\}. \end{aligned} \tag{40}$$

Thus, the overall image transform ψ will aggregate, according to some aggregation operator $\biguplus$ the transforms defined according to each fuzzy gray level set A_k, as:

$$\psi(f) = \biguplus_k \left(\frac{\sigma_u}{\sigma_k} \otimes (f \ominus a_k) \right). \tag{41}$$

We will consider aggregation operators that take into account the fuzzy partition of the gray level set $\mathbf{E}$; the simplest form of aggregation uses the sharing degrees of the gray levels v with respect to the fuzzy sets A_k, i.e. $w_{Ak}(v)$ (25), as weights of a linear combination. This idea reflects a crude fuzzy processing model [16] and is a particular appealing choice in the image processing framework. Thus, the affine image enhancement transform becomes:

$$\psi(f) = \bigoplus_k w_{Ak}(f) \otimes \left(\frac{\sigma_u}{\sigma_k} \otimes (f \ominus a_k) \right). \tag{42}$$

The Figs. 2 and 3 present the result of the enhancement of some poor illuminated images by the proposed fuzzy-logarithmic approach. In each figure, the results of the application of some classical enhancement techniques (histogram equalization, contrast stretching) on the same test images are also shown. Based on these typical examples, we claim that the proposed enhancement technique provide significant improvements compared to the classical automatic (or adaptive) gray level image enhancement methods.

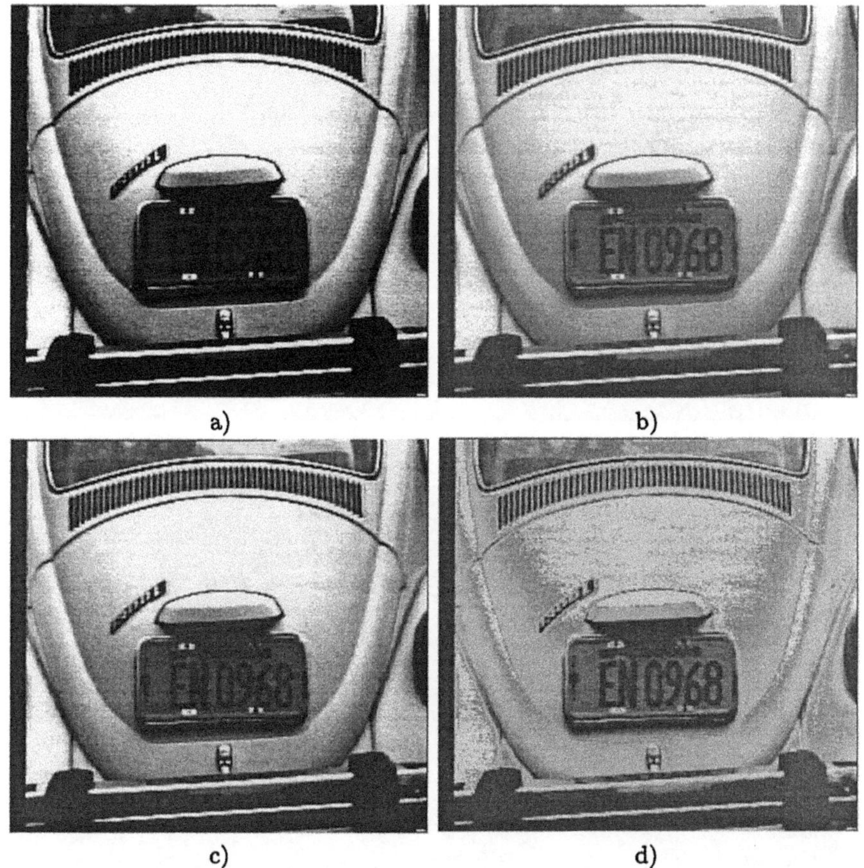

Fig. 2. Gray level image enhancement examples: a) original image (taken form [21]), b) classical histogram equalization, c) contrast streching in the gray level domain d) log-enhanced image according to (42).

5 Color image enhancement by affine transforms

The fuzzy logarithmic image enhancement method described in the previous section will be now extended to accommodate color images. This extension will follow the two main paradigms of scalar processing of vectors: the marginal approach and the reduced approach.

5.1 The marginal approach

Each color component image f_p, with $p = \{r, g, b\}$ will be processed by the same affine transform, deduced according to same principles as in the case of gray level images. Thus, we shall have the general transform, which aggregates, according to some aggregation operator $\uplus$, the transforms defined

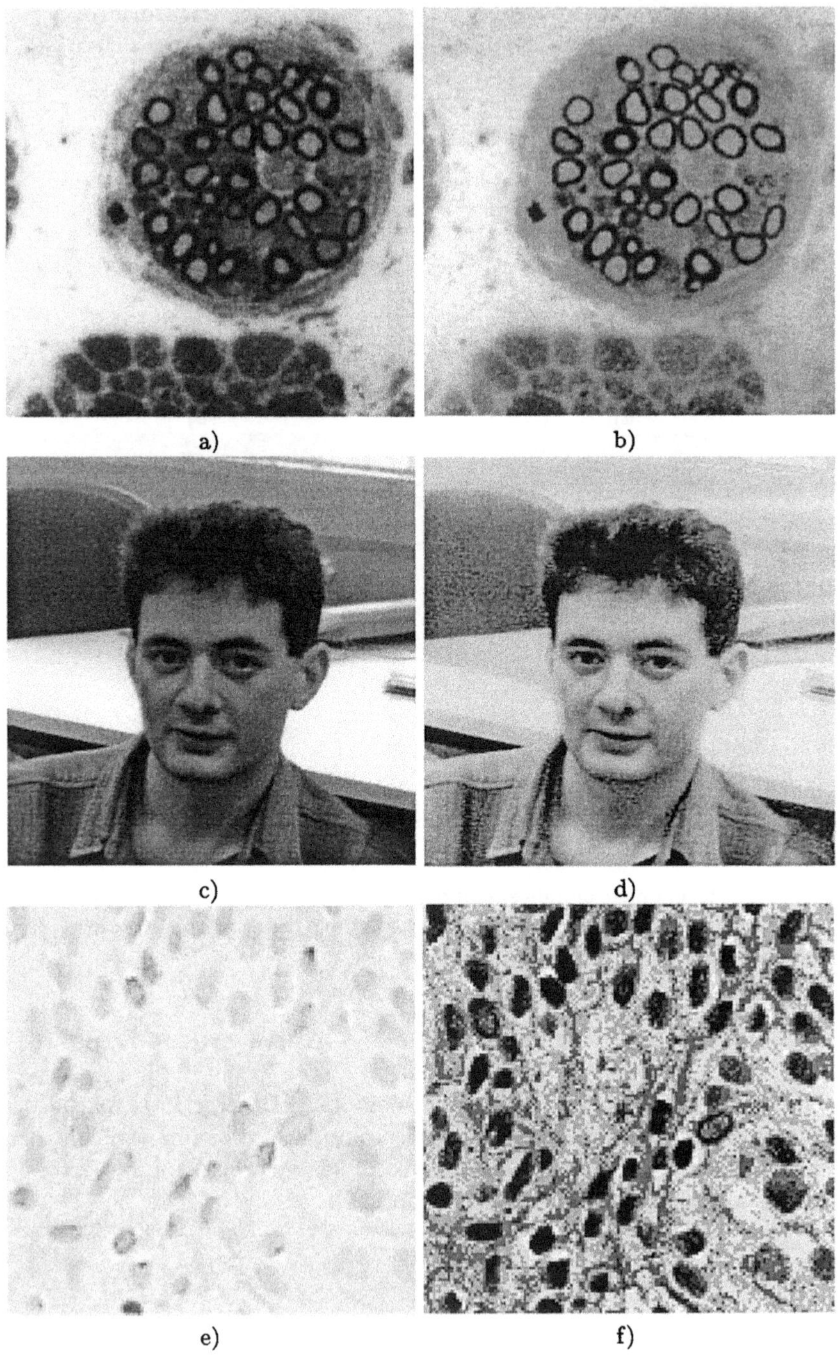

Fig. 3. Gray level image enhancement examples: original images on the left column, log-enhanced images according to (42) on the right column.

according to each fuzzy level set A_k for each color component, as:

$$\psi(f_p) = \biguplus_k \left(\frac{\sigma_u}{\sigma_{p,k}} \otimes (f_p \ominus a_{p,k}) \right), \text{with } k = \{0,1,2\} \text{ and } p = \{r,g,b\}. \quad (43)$$

If the aggregation operator $\biguplus$ is the weighted sum according to the sharing degrees of each color component value with respect to their corresponding image plane, the affine transform becomes:

$$\psi(f_p) = \bigoplus_k w_{Ak}(f_p) \otimes \left(\frac{\sigma_u}{\sigma_{p,k}} \otimes (f_p \ominus a_{p,k}) \right), \quad (44)$$

with $k = \{0,1,2\}$ and $p = \{r,g,b\}$.

5.2 The reduced approach

We shall use the image luminance (32) as the scalar component to be processed instead of the entire color representation. We shall thus deduce the α and β coefficients of the affine transform based on the color luminance only, and apply the resulting transform on each color component image, as:

$$\psi(f_p) = \biguplus_k \left(\frac{\sigma_u}{\sigma_{Y,k}} \otimes (f_p \ominus a_{Y,k}) \right), \text{with } k = \{0,1,2\} \text{ and } p = \{r,g,b\}. \quad (45)$$

If the aggregation operator $\biguplus$ is the weighted sum according to some sharing degrees, these sharing degrees will also be computed based on the color luminance:

$$\psi(f_p) = \bigoplus_k w_{Ak}(f_Y) \otimes \left(\frac{\sigma_u}{\sigma_{Y,k}} \otimes (f_p \ominus a_{Y,k}) \right), \quad (46)$$

with $k = \{0,1,2\}$ and $p = \{r,g,b\}$.

Further on, some practical examples might be seen, as follows: Fig. 4a presents a reddish image, Fig. 4d a greenish one and Fig. 4g a bluish one. They were through different illuminations: thus, the first one under red light, the second one under green light and the last one under blue light. These images with low contrast have been processed by the marginal approach method and the results are shown in Figs. 4c, 4f and 4i. It can be observed that it was realized a color correction in each case, and, moreover, the three images are almost identical. The images processed with the reduced approach method can be seen in Figs. 4b, 4e and 4h. Observe that every image has kept the dominant color, and more, this characteristic has increased. Another experimental result of the reduced approach is in Fig. 5. Thus in Fig. 5a is the original image while Fig. 5d has the log-enhanced one. For comparison two classical methods have been presented: histogram equalization in Fig. 5b and color gamma correction in Fig. 5c.

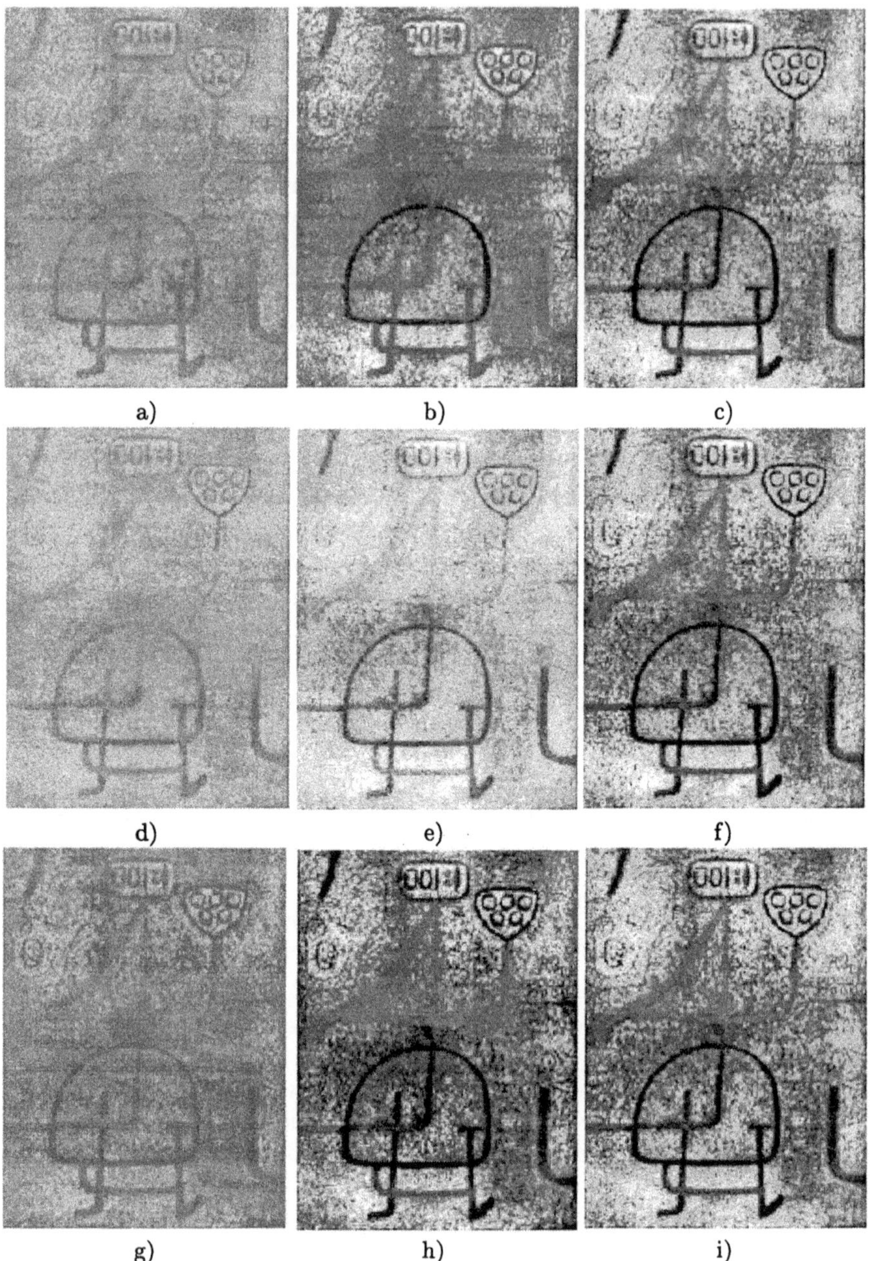

Fig. 4. Color image enhancement examples (all original images are taken form [20]): original images under: a) red, d) green and g) blue illumination, and log-enhancement of the original images by the reduced, luminance-based approach (b), e), h)) and by the marginal approach (c), f), i)).

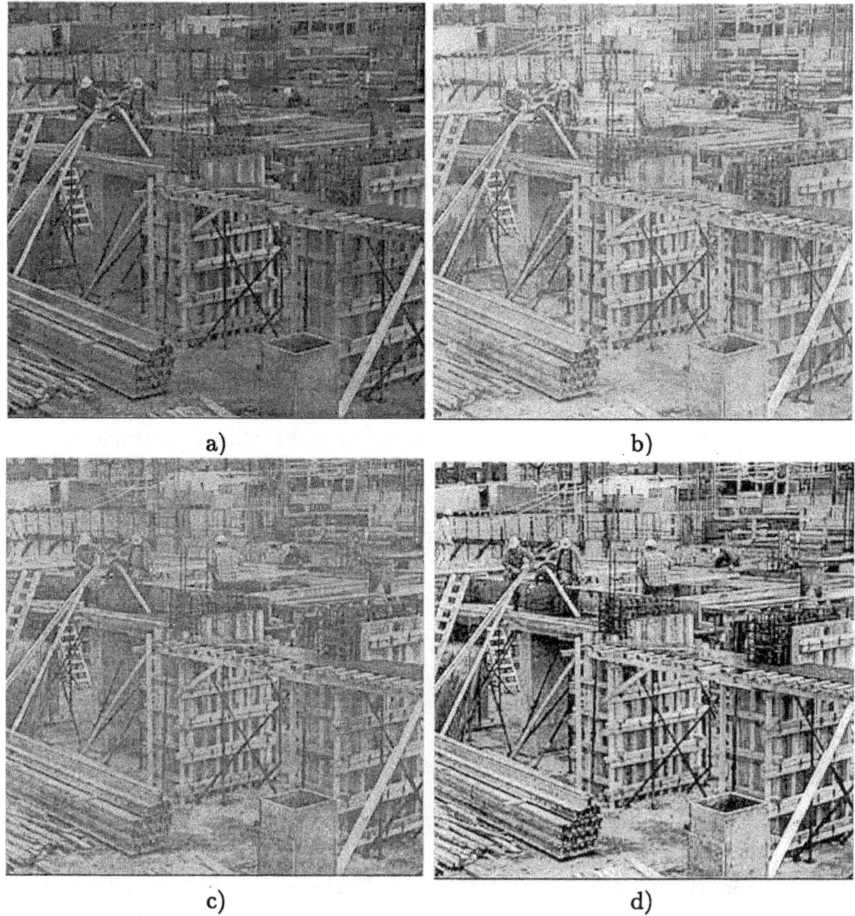

Fig. 5. Color image enhancement examples: a) original, poor illuminated image ("manu1", from [20]), b) classical histogram equalization, c) color gamma correction d) log-enhanced image by the reduced approach.

6 Conclusions

We have presented a mathematical model for the gray level images by defining an algebraic structure on a bounded interval, and by introducing some basic operations (addition, scalar multiplication) and functions (scalar product, norm). This structure, based on a logarithmic model, provides gray level operations which yield results that are always confined to the underlying bounded interval of allowed values, and thus avoiding the need of clipping operations. We propose a fully automatic image enhancement method based on the use of an affine transform. The parameters of the affine transform are computed by the maximization of the dynamic range of the processed image and by approximating the classical histogram equalization technique. The basic, gray level affine transform can be also applied to color images, within the framework of the marginal or reduced (luminance-based) processing models of color.

The tests show that the proposed techniques allow the automatic correction of the illumination problems that occur during the acquisition process (poor illumination or colored illuminants), yielding a better result than some classical image enhancement methods, like the histogram equalization or contrast stretching, for both gray level and color images. The embedding of the fuzzy models into the partition of the value range of the gray levels (or color components) into three fuzzy sets, that represent the naturally low, medium and high values, allows a better control and balance of the overall enhancement. Still, more tests are needed and comparisons are under way, in order to establish the influence of various fuzzy membership functions onto the enhancement performance.

We claim that the proposed method is just another strong argument for the rich potential of the logarithmic image processing models.

References

1. V. Barnett, *The ordering of multivariate data*, in: J. of Royal Stat. Soc. A, vol. 139, part 3, 1976, pp. 318-343.
2. K. R. Castleman, *Digital Image Processing*, Prentice Hall, Englewood Cliffs NJ, 1996.
3. R. C. Gonzales, P. Wintz, *Digital Image Processing*, 2nd Edition, Addison-Wesley, New York, 1987.
4. J. Hafner, H. S. Sawhney, W. Equitz, M. Flickner, W. Niblack, *Efficient color histogram indexing for quadratic form distance functions*, in: IEEE Trans. on Pattern Analysis and Machine Intelligence, vol. 17, no. 7, 1995, pp. 729-736.
5. A. K. Jain, *Fundamentals of Digital Image Processing*, Prentice Hall Intl., Englewood Cliffs NJ , 1989.
6. M. Jourlin, J. C. Pinoli, *Image dynamic range enhancement and stabilization in the context of the logarithmic image processing model*, in: Signal Processing vol. 41, no. 2, 1995, pp. 225-237.
7. M. Jourlin, J. C. Pinoli, *Logarithmic Image Processing. The mathematical and physical framework for the representation and processing of transmitted images*, in: Advances in Imaging and Electron Physics, vol. 115, 2001, pp. 129-196.
8. A. V. Oppenheim, *Supperposition in a class of non-linear system*, in: Technical Report 432, Research Laboratory of Electronics, M.I.T., Cambridge MA, 1965.
9. V. Pătraşcu, I. Voicu, *An Algebraical Model for Gray Level Images*, in: Proc. of the 7th International Conference, Exhibition on Optimization of Electrical and Electronic Equipment, OPTIM 2000, Braşov, Romania, 2000, pp. 809-812.
10. V. Pătraşcu, V. Buzuloiu, *Color Image Enhancement in the Framework of Logarithmic Models*, in: Proc. of the 8th IEEE International Conference on Telecommunications ICT 2001, vol. 1, Bucharest, Romania, 2001, pp. 199-204.

11. V. Pătraşcu, V. Buzuloiu, *A Mathematical Model for Logarithmic Image Processing*, in: Proc. of the 5th World Multi-Conference on Systemics, Cybernetics and Informatics SCI 2001, vol. 13, Orlando, USA, 2001, pp. 117-122.
12. V. Pătraşcu. *A mathematical model for logarithmic image processing.* PhD thesis, "Politehnica" University of Bucureşti, 2001.
13. T. G. Stockham, *Image processing in the context of visual models*, in: Proc. IEEE, Vol. 60, no. 7, 1972, pp. 828-842.
14. H. R. Tizhoosh, *Fuzzy Image Processing*, Springer Verlag, Heidelberg, 1997.
15. H. R. Tizhoosh, H. Haussecker, *Fuzzy Image Processing: An Overview*, in: *Handbook on Computer Vision and Applications*, vol. 2 (B. Jahne, H. Haussecker, P. Geissler eds.), Academic Press, Boston, 1999, pp 683-727.
16. C. Vertan, V. Buzuloiu, *Fuzzy Nonlinear Filtering of Color Images: A Survey*, in: *Fuzzy Techniques in Image Processing*, (E. Kerre, M. Nachtegael eds.), Physica Verlag, 2000, pp. 248-265.
17. C. Vertan, V. Buzuloiu, N. Boujemaa, *A Fuzzy Color Credibility Approach to Color Image Filtering*, in: Proc. of ICIP 2000, Vancouver, Canada, vol. 2, 2000, pp. 808-811.
18. L. A. Zadeh, *Fuzzy sets*, in: Information and Control, vol. 8, 1965, pp. 338-353.
19. P. Zamperoni, *Image Enhancement*, in: Advances in Imaging and Electron Physics, vol. 92, 1995, pp. 1-77.
20. *http://dragon.larc.nasa.gov/retinex.*
21. *http://www.khoros.com.*

Chapter 11

Observer-Dependent Image Enhancement

Hamid R. Tizhoosh

University of Waterloo
Department of Systems Design Engineering
Pattern Recognition and Machine Intelligence Group
200 University Ave. West, Waterloo, ON, Canada
Email: tizhoosh@uwaterloo.ca

Summary. In many image-processing applications the image quality should be improved to support the human visual perception. Image quality evaluation by human observers is, however, heavily subjective in nature. Individual observers judge the image quality differently. In previous works, an observer-dependent system for subjective image enhancement, which is based on fusion of different algorithms, was introduced. In this chapter, more details about the observer-dependent approach are provided. Furthermore, the experimental results for contrast and sharpness/smoothness as interesting image qualities are also presented.

1 Introduction

The image processing gains in the medicine and industry more and more significance. Under all disciplines, which make man-machine communication possible, image processing has a special position because humans mainly perceive their environment visually [7]. Improvement and reconstruction of digital images are important operations in almost all image-processing systems. The captured images are contrast intensified, modified in their brightness distribution, noisy points are eliminated, edges are enhanced and geometrical distortions compensated [1,25]. Such enhancement operations should support visual perception of the expert, who has to make decisions on the basis of image information (e.g. quality control in the industry or diagnosing/monitoring in the medicine). Thus, the interactively or automatically obtained final result depends directly on the success of image enhancement phase. The quality assessment of enhanced images, however, confronts us with various questions, which are unanswered, or at least not satisfyingly clarified.

1.1 The problem

For the evaluation of the image quality mathematical tools are needed, which not only consider the *objective* but also the *subjective* criteria. The objective information is located in nature of data and can be acquired by means of different procedures (e.g. by feature extraction). The subjective information exists generally outside of data. It is the *subjective impression*, the evaluation of the observer, as he perceives and assesses these data. The major problem in pattern recognition often is that one must map the cognitive ability of humans without being able to integrate the human subjectivity into the recognition process by means of mathematical tools. Often, therefore, at the end of the automatic recognition process a subjective evaluation takes place by an expert, who determines whether the result corresponds to his intuitions, expertise and expectations.

Many calculations, classifications and decisions, which are executed/made in the pattern recognition, depend finally on expert opinions. On the basis of suitable features, criteria and measures, algorithms determine to what extent a calculation is precise, a classification valid and a decision correct. These features, criteria and measures are of objective nature, i.e. one can justify and mathematically prove them (e.g. calculation of square errors). Often, however, we have to ask for subjective opinion of an expert in order to judge the results with final certainty (e.g. evaluation of segmented CT-images by a specialist). This subjectivity is usually based on experience and knowledge, which often can not be mathematically formulated. It would be very helpful, if one could move the subjective evaluation of the results to the begin of the processing. Only in this way a fully automatic procedure can be developed. Besides, images could be produced for each observer based on his individual demands.

Usually, however, algorithms are machine-oriented, so that the observer (the expert, who assesses the image content) does not perceive the results as optimal. Which images are called good or bad, it depends certainly on the basic conditions of corresponding system/application. The adjectives *good* and *bad* refer then either to a global impression or to meaningful image features, which vary from image to image and application to application. The problem is originated in the fact that technical criteria for the measurement of image quality often do not match with subjective expectations of the observer. In this regard, it turns out very fast that measures like technical signal-to-noise ratio (SNR) can be easily determined, but they are lastly a bad measure for how the observer perceives the distance of a disturbed image from the original [6, pp. 183]. Mathematical measures such as SNR are, however, popular evaluation tools in image processing. Weeks (1996) examines different criteria like mean square error (MSE) and states:

> The main difficulty with using the MSE or the NMSE as a measure of image quality is that in many instances these values do not match the quality that is perceived by the human visual system [25, pp. 481].

In addition, one knows little about this subjectivity. Therefore, Berbaum et al. propose an *ideal observer*, who is, however, likewise no final solution for the problem:

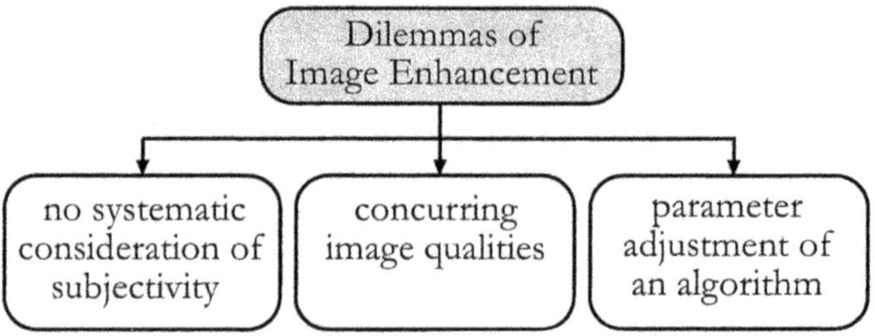

Fig. 1. Problems of Image Enhancement.

> The ideal observer has the following information about the image: (a) the geometry of the image, (b) the distribution of count densities (or magnitudes) in the image; and (c) the probability distribution of the image noise. Also the ideal observer is a poor approximation of the human observer; it is actually a useful approach [7, pp. 254-255].

A further problem results from concurring, mutually exclusive quality criteria such as homogeneity and sharpness. The sequential mode of operations in image processing leads to the fact that one sacrifices an image feature/property at expense of the other. Parallelization would solve the dilemma only if suitable conflict-solving procedures were available, which make an intelligent fusion of individual decisions possible.

Generally one can understand these problems in the image processing also as *parameter adaptation problem.* Various approaches to image enhancement possess a set of parameters, which must be adapted in order to obtain optimal results. This adaptation does not occur in dependency of the subjective analysis of an observer. Thus, the adaptation of the parameters represents likewise a challenge, which is not solved satisfyingly yet.

In summary, one can say that image enhancement is impeded mainly by three problems: First of all it is detached from human perception. The integration of the expert knowledge is, however, not a trivial task. Secondly some image qualities exclude each other, so that one must be sacrificed at expense of the other. Thirdly one could use the adjustment of the parameters in order to solve the two first problems partially. This, however, depends again on how image quality is defined and how expert knowledge is used (Fig. 1).

By a closer look it turns out that these apparently different problems are very strongly interlocked. If knowledge can (in what form ever) be integrated within the process of the image manipulation, then the other problems do not arise any more or will be significantly simplified.

1.2 Existing contributions

Three (non-engineering) disciplines are responsible for research of visual perception: psychophysics, neurophysiology and psychology. Among them, the psychological approach have hardly played a central role, mainly because the psychology saw itself for a long time not responsible for answering questions of image understanding[1].

The discrepancy between objective and subjective quality criteria is also due to the fact that the study of the visual perception is in certain respects still in the starts. Theoretically as empirically, the information extraction during image observation is still to a large extent unexplored [26, pp.10–11]. The knowledge about elementary image understanding processes, however, have been mainly remained unused.

On the other side, in image processing community efforts toward development of more intelligent and more flexible image processing have mainly gone in two completely separated directions: The one direction, which has delivered the majority of results, concentrates on development of new automatic algorithms to solve those dilemmas of the image enhancement (e.g. conflictive objectives such as noise suppression versus edge enhancement) [5,25]. Generally these works can be subsumed under the category *adaptive filtering*. The approach has analytic nature and generally neglects the observer question. Images are enhanced for machines, whereby methods are used, which offer no or only few possibilities for the integration of the expert opinion[2].

The other direction consists of various approaches to integration of human subjectivity, which takes place, however, separately from the first group. The results from this area are rather meager because on the one side they do not rely on the newest psychological models, and on the other side they do not use a proper mathematical framework for capturing and modelling of the knowledge or the subjectivity[3].

2 Communication model of image understanding

Weidenmann[4] regards images as an independent communication medium. He criticizes the past research work, it has neglected the role of the images with the knowledge acquisition and treated images always as *servants of the text.* He states that empirical research in image processing has quite well

[1] David Marr's book *Vision* [11], which has always been quoted in this regard as a psychological response, does not actually contain a psychological approach to image understanding, what Marr himself also stresses [11, pp.325].

[2] The way to design new filters is always analytic in the image processing. One makes assumptions about the nature of the noise, sets up equations, simplifies them by making further assumptions, until a finished formula remains. The observer question is hardly considered (see [25, pp.208 – 210] how by such methodology the MMSE formula is deduced).

[3] In ROC analysis [7](*receiver operating characteristics*), for instance, the subjectivity of the expert is important. This occurs, however, usually completely separated from efforts for automatic image enhancement.

[4] Bernd Weidenmann is professor for educational psychology at the Bundeswehr University of Munich, Germany.

progressed, but also that we are presently still far away from a theory of image understanding [28, p.20].

Weidenmann's model distinguishes between ecological and indicator-based image understanding [26]–[29]. While *the ecological image understanding* runs automatically and without large mental expenditure in a very short time interval (pre-attentive phase), *the indicator-based image understanding* requires more observer effort in order to understand the visual argument (attentive phase). The *visual argument* of an image is its message, its central statement or its most important feature (e.g. a tumor or its position can be the visual argument in a medical image). In the two understanding levels the *normalization need* plays an important role. It is the comparison process of perceived samples with already available knowledge in order to eliminate ambiguities. The communication model defines three classes of images: artistic, entertaining and informative images. The informative images harbor a visual argument and comprise all images used in image processing.

The *recipient* (the observer) perceives an image and processes it whereby complex psychic processes run. To describe these processes Weidenmann puts the term *image understanding* in center of his communicative model and states that understanding means to catch sense data with existing knowledge structures. We have then the impression, the new is familiar to us [29, S.29]. Weidenmann also regards the normalization process as understanding. It aims at elimination of ambiguities. Thus, understanding is a subjective comparison of perceived data with patterns of own knowledge. Weidenmann agrees with Marr that image understanding represents information processing. Fig. 2 illustrates the process of image understanding in Weidenmann's communication model.

3 Necessary requirements

Observer-dependent techniques for image enhancement should fulfil following requirements:

- the techniques have to be able to represent and process *vagueness*. Particularly during the representation of patterns (called *schemes* in psychology) the models should not be precise but extremely flexible.
- the techniques should be able to eliminate *ambiguities*. This is obviously the condition that image understanding process continues and not be aborted in the pre-attentive phase. In addition, in attentive phase ambiguities are effaced (or *normalized* as psychologists say). Particularly, the knowledge activation by informative images depends strongly on handling ambiguous evidences or observations.
- the techniques have to represent *knowledge* in a suitable way (Fig. 3). Only so the ascent to indicator-based image understanding becomes possible. This condition is in close relationship with the modelling or minimization of vagueness and ambiguity.

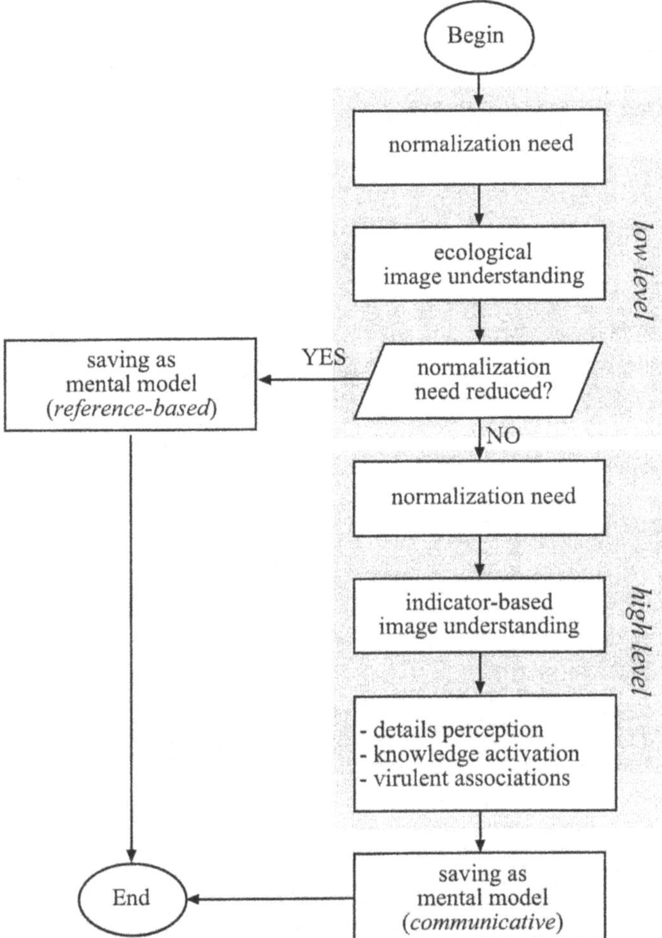

Fig. 2. Image understanding according to commnication model.

- the techniques have to be able to execute *inferences.* This plays particularly a crucial role in the attentive phase, where mental models are used to understand *visual arguments.*
- the techniques have to be able to adapt the knowledge base continuously and acquire new knowledge.

These requirements predestine fuzzy approaches as potential solution. The fuzzy sets and techniques are particularly suitable for handling of vagueness. The fuzzy measures are suitable tools in order to make certain decisions in conflict situations. Finally, fuzzy if-then rules provide a flexible framework for the knowledge representation and processing.

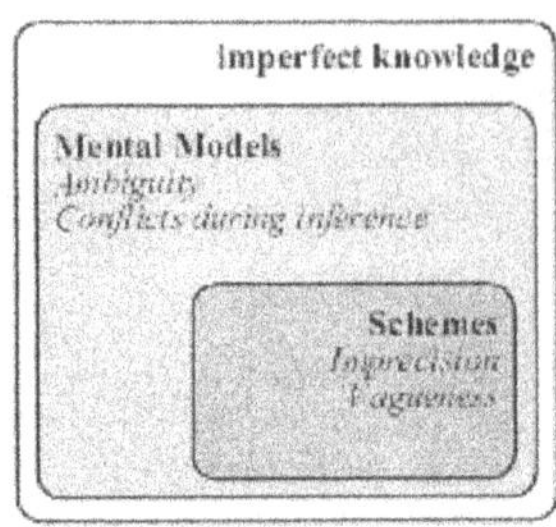

Fig. 3. During image understanding imperfect knowledge will be processed.

Based on model of Weidenmann, two new terms *ecological* and *indicator-based* image quality are introduced here. They serve to classify existing and future algorithms better.

Ecological image quality: The ecological image quality is an image quality, which can be ecologically understood in the pre-attentive phase. The observer does not need a large mental expenditure in order to build his judgement about image quality. Ecological image qualities are, for instance, contrast, brightness, sharpness, homogeneity and noiselessness. The observer is able to evaluate the image quality *at the first sight.* No explicit knowledge is necessary.

Indicator-based image quality: If the observer tries to understand the visual argument of the image beyond the pre-attentive phase, then a high indicator-based image quality can surely reduce the necessary mental expenditure. In contrast to the ecological image quality the listing of the examples for indicator-based image qualities is relatively difficult. One must know much more about the image, observers and the situative context in order to describe such qualities in details. However one can indicate that here the image segments and their geometrical relations play a central role.

It has to be noticed that in many practical cases, the increase of ecological image quality influence or facilitate the following indicator-based image understanding directly. A visual argument can not be completely perceived by the observer due to poor ecological quality. The following approach to observer-dependent image enhancement is, therefore, primarily designed for ecological image qualities.

4 Structure of an observer-dependent system

In order to fulfil the mentioned requirements, a multi-stage aggregation system has been developed (Fig. 4 and 5) [19,22,23]. The system is trained first off line (Fig. 4). The input image is enhanced thereby by means of different algorithms or an algorithm with different parameters. The improvement refers in each case to the interesting image quality (contrast, sharpness, homogeneity etc.). Subsequently, objective and subjective quality measures are calculated, which evaluate the image quality numerically/linguistically. In an aggregation phase these measures are combined to a degree of the compatibility and a degree of compromise. In addition, fuzzy integrals and the Dempster rule (Fig. 4) will be used. Finally, in an inference phase an aggregation matrix Υ is generated, which shows the quality of individual images in consideration of the inter-individual differences among all observers.

The on-line version of the system (Fig. 5) uses the same algorithms, and the generated aggregation matrix Υ in order to execute the fusion of the sub-results for a certain observer. The image fusion in Fig. 5 by different aggregation operators can be determined, however, in dependency of the respective image quality. Here, the image fusion for contrast and sharpness will be described (see Sects. 5.1 and 5.2). Generally valid fusion rules cannot be indicated because of the complexity of the subject. Thus, the fusion scheme should be redefined for other image qualities.

In the following, all phases of the observer-dependent system will be described in details. Afterwards, for verification of the system performance contrast adjustment and edge enhancement will be discussed as examples.

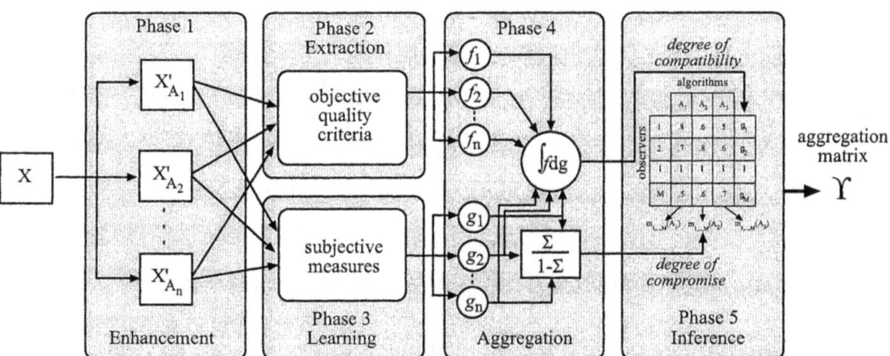

Fig. 4. The off-line structure of the observer-dependent system. At the output the aggregation matrix Υ is available, whose rows contain the relevance of individual algorithms for individual observers.

4.1 Phase 1: image enhancement

In the first phase the input image is enhanced regarding the interesting quality (contrast adjustment, noise suppression, edge enhancement etc.). Either

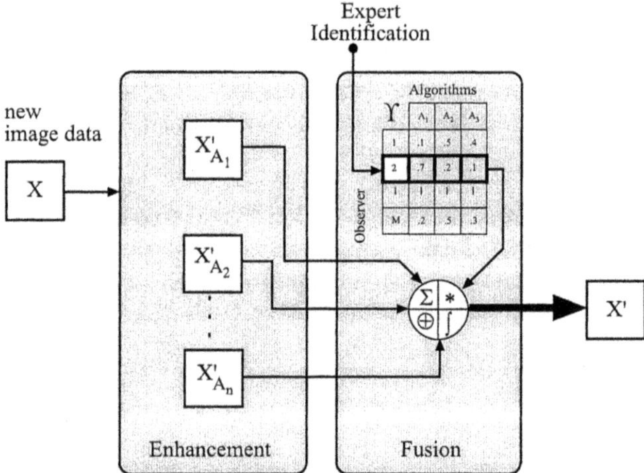

Fig. 5. The on-line structure of the observer-dependent system. Differently enhanced images are combined to a final result in consideration of subjective feeling of the observer. In addition, algorithms are observer-dependently weighted with the help of the aggregation matrix Υ. The concrete image fusion is accomplished depending on selected image quality either globally or with respect to spatial information.

different algorithms or only one algorithm with different parameters can be used to generate different results. Although in the first phase we concentrate here on *enhancement*, which is the main objective of this chapter, one can extend the system with some modifications also to segmenting techniques. The system can likewise be used for subjective image compression, as far as the quality of compressed images should be judged by observers.

4.2 Phase 2: extraction of objective quality criteria

The obtained image quality should be numerically evaluated in the extraction phase. Measures for contrast, brightness, sharpness etc. are calculated in order to quantify the image quality. As example in this chapter new measures for contrast (Eq. 19) and sharpness (Eq. 28) are introduced.

If the result X'_{A_i} of an algorithm A_i for image X is evaluated by means of an objective function $h(X'_{A_i})$, then this evaluation differs usually from the subjective evaluation by an observer. This is to be expected but should not be an unsolvable problem. The evaluation function $h(X'_{A_i})$, however, should tends to agree with intuition. This does not apply to many existing quality measures in image processing. Therefore, if necessary, a new evaluation function must be introduced as a function of the corresponding quality (see Sects. 5.1 and 5.2).

Although the system should be observer-dependent, the integration of objective measure has quite its reasons. First of all, consideration of objective measures counteracts to fluctuations in subjective evaluations, which possibly represent false assessments[5]. Secondly, the system has to keep its general character. In many applications it will be necessary to consider objective and observer-*in*dependent knowledge (e.g. anatomical specifications in a medical application do not depend on subjective impressions). The use of objective measures does not weaken the observer-dependent character of the system, since the subjectivity of the observer is multiply included in calculations of the aggregation phase.

4.3 Phase 3: learning subjective measures

In this phase, subjective impression of observer regarding the image quality should be captured numerically or linguistically. Differently enhanced images are therefore presented to the observer who evaluates their quality subjectively. In order to ensure a systematic and uniform methodology, the subjective evaluation test by observers were adapted after recommendations of International Telecommunication Union ITU-BT 500[6].

Here several test persons evaluate the quality of the images X'_{i,A_k} (generated by the k-th algorithm) using the evaluation scale *excellent* ($\equiv 1$), *good* ($\equiv 2$), *fair* ($\equiv 3$), *poor* ($\equiv 4$) and *bad* ($\equiv 5$)[7]. Subsequently, for M evaluations of the observer b the MOS (*mean opinion score*) is calculated by averaging individual evaluations $p \in \{1, \cdots, 5\}$:

$$\mathrm{MOS}(X'_{i,A_k}) \mid_b = \frac{1}{M} \sum_{i=1}^{M} p_i(X'_{i,A_k}). \tag{1}$$

If the observer b evaluates M results X'_{i,A_k} of the k-th algorithm A_k ($i = 1, \cdots, M$), then MOS provides a measure for how this algorithm is generally evaluated by this observer. A low MOS (≈ 1) indicates, thus, that the observer perceives the results of this algorithm as very good, against what a high MOS (≈ 5) indicates that the algorithm is a bad alternative for the observer.

Using linguistic hedges ([31]), it can be guaranteed that linguistic evaluations, which are not defined MOS scales can, nevertheless, be numerically captured/represented.

[5] Should an observer judges the quality of many images in only one session, then wrong assessments are quite possible due to overfatigue.

[6] the recommendations of the ITU are particularly used in the television engineering [BT.500-9 (11/98) Methodology for the subjective assessment OF the quality OF television of pictures 43pp ITU-R Recommendations, 1997 - BT Series - supplement 2 to of volume 1997]. The ITU-BT 800 prescribes similar guidelines for the subjective evaluation of acoustic signals.

[7] For some test series an additional score *very bd* ($\equiv 6$) was used.

4.4 Phase 4: aggregation of measures

If objective and subjective quality measures are available, then the question about their aggregation to a total measure arises. Two main conflicts are present, which must be solved: On the one hand the evaluations differ over a certain algorithm, i.e. once results of an algorithms are evaluated as *excellent* and once as *not particularly good.* On the other hand different algorithms compete among themselves, if we consider an individual evaluation of a certain observer. If the objective or subjective evaluations of the algorithms are represented in two matrices (Fig. 6), then this becomes clear. A matrix row reflects the competition of the algorithms among each other, independently whether the evaluation takes place objectively or subjectively. A column of the evaluation matrix indicates the additional diverging evaluations of the same observer.

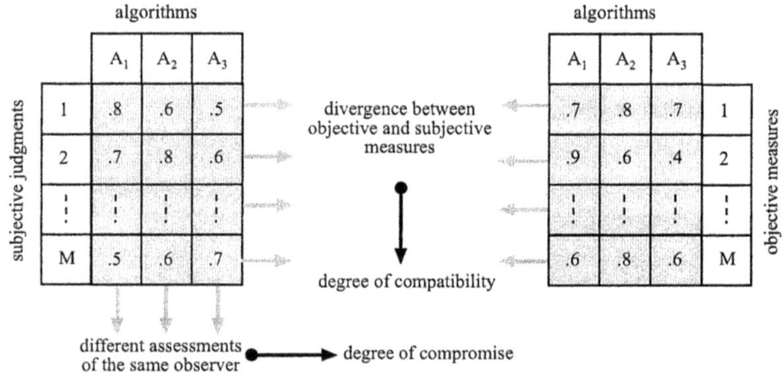

Fig. 6. Resulting problems from evaluation matrix.

In order to solve these problems it must be determined how compatible individual quality criteria are. Additionally, a compromise between different evaluations must be obtained.

The *degree of compatibility* is a measure for the conformity between objective and subjective quality criteria. The more similar objective and subjective evaluations are, the simpler becomes the decision to assess the results of an algorithm as good or bad. The compatibility degree indicates how ambiguous (or how unique) the respective evaluations are.

The *degree of compromise* is a measure for how the result of an algorithm is evaluated in all, if one considers all subjective evaluations of an observer at the same time. Generally, the evaluation of the algorithm varies from reference image to reference image. Sometimes the result is perceived as *very good* and sometimes as *quite fair.* The degree of compromise should, therefore, indicate a reliable total value for each algorithm considering all different evaluations of an observer.

The motivation for the introduction of these measures is the following: If different measures indicate a sub-result as good (high compatibility), and if all or even many evaluations regard the results of a certain algorithm as good (high compromise degree), then the respective sub-result should clearly contribute a greater portion to the final result.

Calculation of compatibility degree with fuzzy integrals Fuzzy integrals have already been successfully used in some image-processing systems [24,16,20]. The Sugeno measure (or λ-fuzzy measure) is a suitable framework for representation of subjective evaluations of image quality by human observers ([14,15]). The motivation for using fuzzy measures is based on the fact that the problem of image quality evaluation is not of additive nature. The observer does not distribute his scores after a mathematical regularity. Consequently MOS values possess always a non-additive character (i.e. super-additive).

The construction of the Sugeno measure g_λ is relatively simple because the subjective evaluations of the observer (the MOS values normalized on the interval $[0,1]$) can be treated as fuzzy density values g^i. For the sub-results X'_{A_1} and X'_{A_2} it follows:

$$g_\lambda(X'_{A_1} \cup X'_{A_2}) = g^1 + g^2 + \lambda \cdot g^1 \cdot g^2. \tag{2}$$

Thus, the individual density values g^i represent the quality of individual results. The parameter λ can be calculated from the condition $g_\lambda(X) = 1$:

$$\lambda + 1 = \prod_{i=1}^{n} (1 + \lambda \cdot g^i). \tag{3}$$

The Sugeno measure for n algorithms can be completely designed thereby on the basis of following recursion:

$$\begin{aligned} g_\lambda(X'_{A_i}) &= g^i \quad \text{if} \quad i = 1, \\ g_\lambda(X'_{A_i}) &= g^i + g_\lambda(X'_{A_{i-1}}) + \lambda g^i g_\lambda(X'_{A_{i-1}}), \quad \text{otherwise.} \end{aligned} \tag{4}$$

The compatibility degree γ_i between objective evaluation values $h(X'_{A_i})$ and subjective measures $g_\lambda(X'_{A_i})$ is calculated by means of an fuzzy integral (also called Sugeno integral):

$$\gamma_i = \int h \circ g_\lambda = \bigvee_{i=1}^{n} [h(X'_{A_i}) \wedge g_\lambda(\{X'_{A_1}, \cdots, X'_{A_i}\})]. \tag{5}$$

The operator $\circ$ is implemented by the operators $\vee$ and $\wedge$ (in the simplest case, in order to save computing time, through maximum or minimum operator). For the calculation of fuzzy integral, the objective values $h(X'_{A_1}), \cdots, h(X'_{A_n})$ are sorted in descending order. The Sugeno measure is constructed afterwards

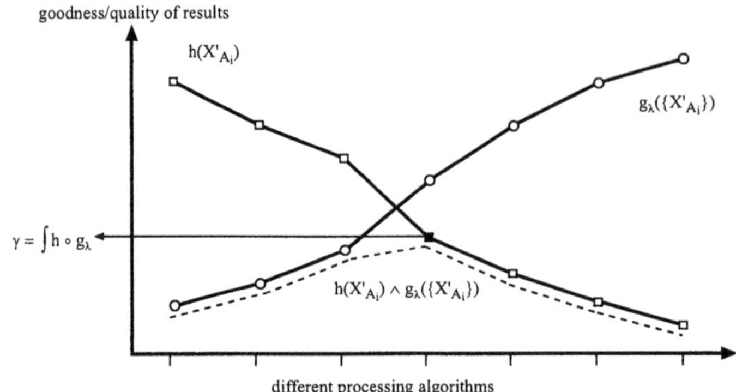

Fig. 7. Calculation of compatibility degree using Sugeno integral.

using Eq. (4). The Sugeno integral aggregates finally the two measures (Fig. 7).

The Sugeno integral is rather suitable for cases in which the fuzzy density values quantify a ranking problem. If the fuzzy density values have a certain meaning (e.g. opinion of the observer), then it is advisable to use the Choquet integral because the order of the data does not play a role any more(see [2, pp. 58]):

$$\gamma_{Choquet} = \sum_{i=1}^{n} (h(X'_{A_i}) - h(X'_{A_{i-1}})) \cdot g_\lambda(X'_{A_i}) \qquad \text{with} \quad h(X'_{A_0}) = 0,$$
$$= \sum_{i=1}^{n} h(X'_{A_i}) \cdot [g_\lambda(X'_{A_i}) - g_\lambda(X'_{A_{i+1}})] \qquad \text{with} \quad X'_{A_{n+1}} = \emptyset. \quad (6)$$

A reason for using fuzzy integrals is that many other aggregation operators can be regarded as special cases of the fuzzy integrals. The objective evaluation function h depends on application and will be calculated from case to case (see Sects. 5.1 and 5.2).

Calculation of compromise degree As previously mentioned, the compromise degree has to solve the conflict between different evaluations and to enable a certain assessment of algorithms. If the different evaluations of the results of an algorithm are regarded as independent evidences (evaluations), then the Dempster rule([3,11,24]) can be applied to calculate the compromise degree [8]. The advantage of the Dempster rule compared to other operators

[8] Evaluations of individual observers regarding the quality of different algorithms can be regarded as independent evidences if the observer is not provided with the information which results originate from which sources. Beyond that it must be ensured that the results are presented to the observer in random order.

is that it calculates a global quality for an algorithm, whereby evaluations of other algorithms are simultaneously considered. Other aggregation possibilities, e.g. averaging, generally calculate expectation values for each individual algorithm without consideration of other algorithms[9].

The evaluations of the observer are normalized for the interval $[0, 1]$ and interpreted as basic probability assignments m_i. A basic probability $m_i(A)$ shows thus the i-th (uncertain) evaluation of an observer of algorithm A. If two basic probabilities m_1 and m_2 are present, then these can be combined by means of the Dempster aggregation rule:

$$(m_1 \oplus m_2)(C) = \frac{\sum\limits_{A_i \cap B_j = C} m_1(A_i) \cdot m_2(B_j)}{1 - \sum\limits_{A_i \cap B_j = \emptyset} m_1(A_i) \cdot m_2(B_j)}. \quad (7)$$

For the calculation of the Dempster rule lookup tables were created (Fig. 8). Since the calculation of the compromise degree is carried out for only one algorithm, the intersection sets with a cardinality> 1 do not play a role (black fields in Fig. 8).

A problem which hinders the use of Dempster rule is its relatively large *computational expenditure* (the cost of computation rises exponentially with number of algorithms). If the number of inputs exceeds a certain limit, then it will practically be impossible to calculate the result in a justifiable time. For the concerns of this work is this, however, not relevant because a small number of algorithms is always used, and the calculation takes place off-line, so that the computing time does not play a central role.

A generalized version of the Dempster rule can also be used (see paragraph 5.2, [30]):

$$(m_1 \oplus m_2)(C) = \frac{\sum\limits_{A \cap B = C} \max\limits_x \mu_{A \cap B}(x) \cdot m_1(A) \cdot m_2(B)}{1 - \sum\limits_{A,B} (1 - \max\limits_x \mu_{A \cap B}(x) \cdot m_1(A) \cdot m_2(B))}. \quad (8)$$

4.5 Phase 5: inferring the aggregation matrix

If results of N algorithms are judged through M evaluations, then the aggregation phase delivers two vectors Γ (compatibility degree) and Φ (compromise degree):

$$\Gamma = (\gamma_1 \quad \gamma_2 \quad \cdots \quad \gamma_M), \quad (9)$$

$$\Phi = (m^*(A_1) \quad m^*(A_2) \quad \cdots \quad m^*(A_N)), \quad (10)$$

[9] If one wants to calculate a global value for an algorithm at the same time, then it is meaningful to take a look at subjective evaluations of other algorithms by the same observer. Only in this way one can find out in test series for an observer, which algorithm he prefers and assigns a higher quality in comparison.

m_2 \ m_1 ∩	{A}	{B}	{C}	{A,B}	{A,C}	{B,C}	{A,B,C}
{A}	{A}	∅	∅	{A}	{A}	∅	{A}
{B}	∅	{B}	∅	{B}	∅	{B}	{B}
{C}	∅	∅	{C}	∅	{C}	{C}	{C}
{A,B}	{A}	{B}	∅		{A}	{B}	
{A,C}	{A}	∅	{C}	{A}		{C}	
{B,C}	∅	{B}	{C}	{B}	{C}		
{A,B,C}	{A}	{B}	{C}				

Fig. 8. Lookup table for the calculation of the Dempster rule.

whereby m^* is in each case the compromise degree calculated with the Dempster rule:

$$m^*(A_i) = ((m_1 \oplus m_2) \oplus \cdots \oplus m_M)(A_i). \tag{11}$$

The motivation for the calculation of these measures was the following consideration: If objective and subjective quality measures concerning the image quality are to a large extent compatible (high γ values) and the individual evaluations of an observer regarding the quality of an algorithm can be summarize to a total value (high m^* values), then the respective method should accordingly contribute to fusion of a final result. However, γ and Φ vectors will have naturally different values. In order to draw reliable conclusions from these values about the quality of an algorithm, the two vectors are used as inputs for an inference approach which generates an aggregation matrix Υ quantifying the ecological quality of individual sub-results. Firstly, the total compatibility γ is calculated by means of the fuzzy expected value FEV [9]:

$$\gamma = \underset{i}{\mathrm{FEV}}\{\gamma_i\}, \tag{12}$$

$$= \max_i\{\min[T(\gamma_i), \mu(\gamma_i)]\}, \tag{13}$$

where (h =frequency)

$$T(\gamma_i) = T(\gamma_{i-1}) - h(\gamma_{i-1}) \text{ with } i = 2, \cdots, n \text{ and } T(\gamma_1) = N, \tag{14}$$

$$\mu(\gamma_i) = \frac{\gamma_i}{\max_i \gamma_i}. \tag{15}$$

Based on the calculated compatibility and compromise degrees, heuristic rules are set up as follows:

if the total compatibility γ is *high* **and**

the compromise degree m^* is likewise *high*,
then the sub-result possesses *high quality*.

If all three linguistic variables compatibility, compromise and quality are quantized (fuzzified) with three subsets *low*, *medium* and *high* (Fig. 9), then a rule base with nine rules can be created (Table 1).

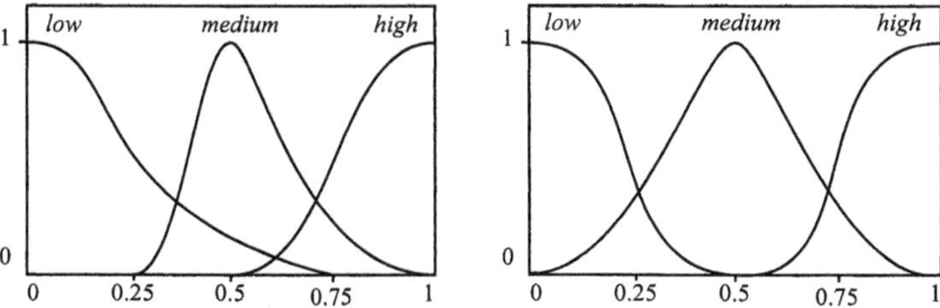

Fig. 9. Membership functions for input (left) and output (right) of the inference phase.

Table 1. Rules for Aggregation of degrees of compatibility and compromise.

	IF	AND	THEN
Rule	γ	m^*	ecological quality
R_1	*low*	*low*	*low*
R_2	*low*	*medium*	*low*
R_3	*low*	*high*	*low*
R_4	*medium*	*low*	*low*
R_5	*medium*	*medium*	*medium*
R_6	*medium*	*high*	*medium*
R_7	*high*	*low*	*medium*
R_8	*high*	*medium*	*high*
R_9	*high*	*high*	*high*

Hence, the inference generates for each observer a row in an aggregation matrix Υ, whose columns represent the applied algorithms. For the observer b and the k-th algorithm A_k the value $\Upsilon(b,k) \in [0,1]$ indicates how strongly this algorithm contributes to fusion of all sub-results to final result. The aggregation matrix shows thus the relevance of individual algorithms for individual observers.

How the final result can be calculated based on information stored in the aggregation matrix depends on which image quality is the target of enhancement. No generally valid rules for the fusion of the algorithms can be provided because global or local criteria must be considered depending on image quality. For the observer b and the contrast as ecological quality, for example, the fusion of the sub-results X'_{A_k} to a final result X' can be done as a global, convex combination (see Sect. 5.1, starting from pp. 255):

$$X' \mid_b = \sum_{k=1}^{N} \Upsilon(b, k) \cdot X'_{A_k}. \tag{16}$$

For other qualities a suitable fusion regulation must be designed in each case. In Sect. 5.2 this will be demonstrated for image sharpness.

After the generation of the aggregation matrix the observer-dependent system is simplified as represented in Fig. 10. Beside new image data, the system is also provided with the information for which observer the results are to be generated (*expert identification*). The differently enhanced images X'_{A_k} are then combined with respect to the weights stored in the aggregation matrix.

It is quite possible, and in many cases even necessary, to produce even for an observer different entries in the aggregation matrix. A particularly typical example for this case are medical applications. Depending on image content the intensity of the qualities required by the observer can vary. In the radiation therapy, for example, physicians operate mainly with three image groups *head*, *thorax* and *pelvis*. Depending on image group (and naturally depending on observer) different enhancement effects are desired. In this case three entries per observer must be available in the aggregation matrix.

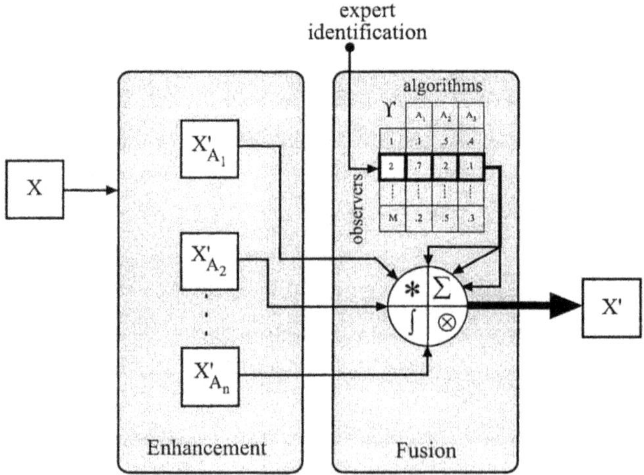

Fig. 10. On-line execution of the system. The image fusion (here represented by the sigma sign) can be indicated only in concrete cases. It can be implemented in the case of the contrast as (global) weighted total (see 5.1), and in the case of the sharpness as (local version) weighting of the algorithms (see 5.2).

5 Experimental verification of system performance

It is usual in image processing to compare the results of individual enhancement algorithms. In addition, for each result a measure is calculated indicating to what extent the result deviates from the original (uncorrupted, optimal) image. The smaller this deviation, the better is the result. Usual measures for this are RMS (*root mean square error*), SNR (*signal to noise ration*), PSNR (*peak to peak SNR*) etc.. This methodology has however several disadvantages:

1. In practice, a noise-free optimal original image usually does not exist. Therefore, there is also no possibility for calculating the deviation from it.
2. Reducing the quality of a complex image into a single number which is not subjectively supported does not seem to be reasonable. Particularly, since these numbers (RMS, SNR, PSNR...) do not give any information about it where and which pixels deviate how strongly from the *ideal status.* It is, therefore, not astonishing that there are many cases in practice, where results with low RMS are perceived by the observer as bad, or images with high RMS perceived by the observer as quite good.
3. The calculation of the deviation between original image and enhanced image is also due to the fact that one operates with synthetic images which are artificially/manually noise-corrupted.

For these reasons one should try to take another approach to image quality evaluation. No technical system is able to judge complex images regarding their quality like humans. Therefore, it seems to be more meaningful to consider the subjective opinion of the expert. As a systematical methodology, the MOS procedure according ITU can be applied (see Sect. 4.3). In order to test the efficiency of the system, two examples, i.e. contrast adjustment and edge enhancement, are briefly discussed in the following.

5.1 Observer-dependent contrast adjustment

The contrast evaluation plays an important role during image enhancement because frequently visual arguments (significant features, central message) either can not be perceived due to bad contrast[10]. In image processing the contrast C is mostly calculated according to the following equation [4,25]:

$$C = \frac{g_{max} - g_{min}}{g_{max} + g_{min}}. \tag{17}$$

Here are g_{max} and g_{min} the maximum/minimum gray levels of the image. This equation harbors, however, some weaknesses. It does not correspond to

[10] For example, *portal images* used in the radiotherapy are very low-contrasted. This leads to the fact that the visual argument (*position of the tumor*) can be hardly perceived even by experienced eyes.

our intuitive expectation since it delivers different values for dark images (underexposure) and bright images (overexposure) (Table 2). For all histogram equalized or stretched images is C identical, since spatial image information is completely ignored.

Table 2. The contrast values determined by Eq. (17) do not correspond to our intuition: Dark images are indicated as very high-contrasted.

Image	Values
	$\begin{bmatrix} g_{min} = 0 \\ g_{max} = 165 \end{bmatrix} \rightarrow C = 1.00$
	$\begin{bmatrix} g_{min} = 84 \\ g_{max} = 177 \end{bmatrix} \rightarrow C = 0.36$
	$\begin{bmatrix} g_{min} = 102 \\ g_{max} = 255 \end{bmatrix} \rightarrow C = 0.43$

An objective contrast measure The contrast should be calculated, therefore, with respect to spatial image information. One will receive a number of individual (spatial) contrast values $\{c_1, c_2, \cdots, c_n\}$. The total contrast C_{total} can be calculated afterwards by means of the fuzzy expected value FEV [9] as a more representative (most typical) value among all local contrasts:

$$C_{total} = \underset{i}{\text{FEV}}\{c_i\}, \tag{18}$$

$$= \max_i\{\min[T(c_i), \mu(c_i)]\}, \tag{19}$$

where (h =frequency)

$$T(c_i) = T(c_{i-1}) - h(c_{i-1}) \text{ mit } i = 2, \cdots, n \text{ and } T(c_1) = n, \tag{20}$$

$$\mu(c_i) = \frac{c_i}{\max_i c_i}. \tag{21}$$

We distinguish between homogeneous and inhomogeneous regions. Darker or brighter neighborhoods are regarded as highly contrasted. For this purpose the contrast c_i in each neighborhood is calculated as follows:

$$c_i = \alpha \cdot c_i^1 + (1 - \alpha) \cdot c_i^2 \quad \text{with} \quad c_i^2 = 1 - c_i^1, \tag{22}$$

whereby in consideration of the local parameters $g_{max,local}$ and $g_{min,local}$, and number of gray levels L we have

$$c_i^1 = \frac{g_{max,local} - g_{min,local}}{L - 1}. \tag{23}$$

c_i^1 represents the contrast of edgy neighborhoods against what its complement c_1^2 (weighted with $1 - \alpha$) is used for dark or bright regions. The aggregation parameter α is calculated as a function of relation between global and local extreme values:

$$\alpha = \frac{g_{max,local} - g_{min,local}}{g_{max,global} - g_{min,global}}. \tag{24}$$

For the proposed local contrast calculation the number of windows N must be determined which usually influences the result strongly. For images with large homogeneous regions the whole image is considered ($N \to 1$), and for highly detailed images as much as sub-images are necessary ($N \gg 1$).

Table 3 shows contrast values for different number of windows. For comparison the global value is indicated in each case. It is remarkable that the number of windows affects the total contrast of bad images (second to fourth rows) only little; the total contrast for 4 or 100 fields remains approximately constant, which speaks for the reliability of the method.

Table 3. Global versus local contrast calculation (N = number of sub-images).

	C_{global}	C_{local} ($N = 4$)	C_{local} ($N = 9$)	C_{local} ($N = 25$)	C_{local} ($N = 100$)	$\mid C_{N=100} - C_{N=4} \mid$
	0.82	0.75	0.68	0.66	0.54	0.21
	1.00	0.57	0.55	0.54	0.51	0.06
	0.36	0.36	0.37	0.37	0.44	0.08
	0.43	0.59	0.57	0.56	0.54	0.05

Beside of invariance regarding number of windows, such measures correspond also to intuitive expectation of the observer or, at least, not to be contradictory to it. In order to verify this, several experiments were accomplished. Test persons were requested to evaluate results of the algorithms for local-adaptive contrast adjustment ([17,18,21]). The results are represented in Table 4. Deviations between objective and subjective viewpoints remain (e.g. the last series). An image which was regarded by observers rather as moderate has a very high contrast value (obviously too much contrast is sensed as disturbing).

Table 4. Objective versus subjective evaluation of image contrast. In a test series observers (P1-P7) were requested to judge the results of different enhancement procedures regarding image contrast. The observers scored the results with $1 \equiv$ as *very well* and with $6 \equiv$ as *very bad*. The normalization by $C_{P_i} = (6 - \text{SCORE}_{P_i})/5$.

image	$C_{local,N=9}$	C_{P_1}	C_{P_2}	C_{P_3}	C_{P_4}	C_{P_5}	C_{P_6}	C_{P_7}	Average
	0.22	–	–	–	–	–	–	–	–
	0.82	0.60	0.80	0.60	1.00	1.00	0.90	0.72	0.80
	0.89	0.86	0.60	0.60	0.80	0.80	1.00	0.84	0.78
	0.91	0.94	0.80	0.94	0.60	0.60	0.80	1.00	0.81
	0.97	0.66	1.00	0.70	0.70	0.40	0.60	0.60	0.66

Results verification For result verification 30 test images were selected, which did not contain a visual argument[11]. Five observers evaluated first 20 images which were enhanced by four algorithms. Here the tests were based on recommendations of the International Telecommunication Union. The four assigned algorithms were implemented locally adaptive and are described in the previous works ([17,18,21]). The remaining 10 images were used in order to test the efficiency of the system. The system output image was presented to the observers in random order together with four other alternatives. The results are represented in Table 5.

Thus, the system result was perceived in 62% of the cases as the best alternative (average among all images and all observers). If one takes into account the evaluations, in which the result of the system was not the best alternative but still good for the observers, then the developed system gained a success rate of 82% (Table 5, σ = standard deviation). The system achieved in this test an MOS of 1.56:

$$\text{MOS} = 0.62(1) + 0.2(2) + 0.18(3) = 1.56. \tag{25}$$

[11] If an image does not contain a visual argument, it means it does not possess a central statement or message. There are no special objects or features that observers have to detect/understand.

Table 5. Success rate of the system in a test series for contrast as ecological image quality. In the ideal case all results of the system should be perceived by the observers as the best alternative (MOS = 1). Test images were enhanced differently (see images in Table 4). The best algorithm achieved a MOS of 2.13 ($\sigma = 0.42$).

Observer	best alternative	good	medium	poor	bad
1	60%	30%	10%	0%	0%
2	50%	30%	20%	0%	0%
3	70%	10%	20%	0%	0%
4	70%	20%	10%	0%	0%
5	60%	10%	30%	0%	0%
total	62%	20%	18%	0%	0%
MOS			1.56		
σ			±0.25		

5.2 Observer-dependent edge enhancement

As an example we treated contrast as interesting image quality. This had not only the advantage that the objective calculation of a contrast measure could take place easily, but also that the aggregation of single results could be executed by a global convex combination. To prove the system performance for a more complex image quality as *sharpness*, we must initially find a way to quantify the image sharpness (extraction phase). Beyond that, another solution must be found for the aggregation of the sub-results because the individual results cannot be fused using a global additive scheme.

An objective sharpness measure If we regard an edge segment which begins at the position $x = a$ and ends at the position $x = b$, then the sharpness S in consideration of the gray level intensity $f(x)$ can be calculated as follows [8]:

$$S = \frac{1}{f(b) - f(a)} \int_a^b \left(\frac{df}{dx}\right)^2 dx. \tag{26}$$

If image is optically sharp, then the *logical* fuzziness is low (the higher the gray-level differences of an edge segment, the lower its logical fuzziness). This fact can be used to redefine the measure S for the optical sharpness. First, local sharpness measures s_{mn} are calculated ($i, j \in [-(n-1)/2, (n-$

$1)/2], \Delta_1 = L-1, \Delta_2 = g_{max})$:

$$
\begin{aligned}
s_{mn} &= [(\text{edginess}) \times (\text{logical sharpness})]^2 , \\
&= \left(\frac{\sum_i \sum_j \mid g_{xy} - g_{mn} \mid}{\Delta_1 + \sum_i \sum_j \mid g_{xy} - g_{mn} \mid} \cdot \left(1 - \frac{\sum_i \sum_j \min \left(g_{xy}, \Delta_2 - g_{xy} \right)}{N^2 \cdot \Delta_2} \right) \right)^2 , \\
&= \left(\frac{\sum_i \sum_j \mid g_{xy} - g_{mn} \mid \cdot \left(N^2 \cdot \Delta_2 - \sum_j \min \left(g_{xy}, \Delta_2 - g_{xy} \right) \right)}{N^2 \cdot \Delta_2 \cdot \left(\Delta_1 + \sum_i \sum_j \mid g_{xy} - g_{mn} \mid \right)} \right)^2 , \quad (27)
\end{aligned}
$$

where $x = m + i$ and $y = n + j$. The first term of this equation guarantees that sharpness is partially weighted (homogeneous areas are not considered). The linguistic modifier *concentration* (the exponent) represents the logical quantor *very (sharp and non-fuzzy)*. The total sharpness S is calculated as follows afterwards for the entire image:

$$
S = \underset{m,n,s_{mn} > \epsilon}{\text{median}} \{s_{mn}\}, \tag{28}
$$

where ϵ represents a threshold which marks the smallest, still relevant edge. A good approximation is[12]

$$
\epsilon = 0.75 \times \max_{m,n} \left(\frac{\sum_i \sum_j \mid g_{m+i,n+j} - g_{mn} \mid}{(L-1) + \sum_i \sum_j \mid g_{m+i,n+j} - g_{mn} \mid} \right). \tag{29}
$$

The threshold ϵ guarantees that the median of the local sharpness values is only looked up for *strong edges*. Figure 11 shows examples of the calculation for total sharpness. It becomes evident that the proposed sharpness measure S meets our intuitive expectations. Although deviations can quite occur in relation to subjective estimations, the sharpness measure follows the tendency of subjective evaluations.

Aggregation of images with different sharpness Although the observer evaluates always images according a global impression, the image sharpness can not be globally aggregated as an ecological image quality. Local image information must be considered. In the aggregation matrix Υ the global values for a certain observer are available. Thus, the aggregation can take place

[12] The indicated approximation is an empirical value. For an observer-dependent determination of this threshold which depends on image content, psychophysical tests can be executed in context of respective applications (see [13]). The smaller ϵ, the more steeply the sharpness drops.

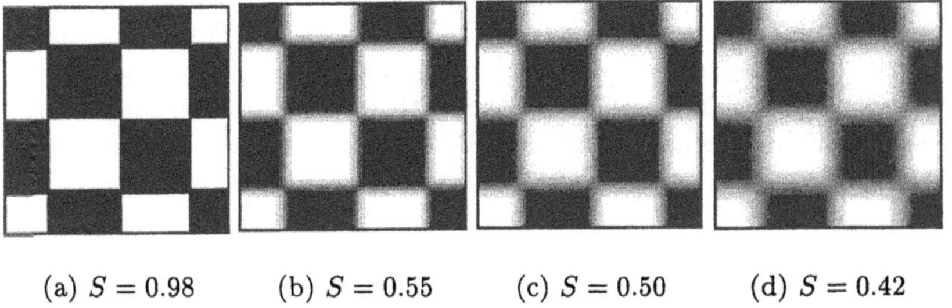

(a) $S = 0.98$ (b) $S = 0.55$ (c) $S = 0.50$ (d) $S = 0.42$

Fig. 11. Calculation of optical sharpness for test images according to Eq. (28).

by means of the following (*spatial*) rule:

if the neighborhood is *edgy*,
then use the aggregation matrix,
otherwise no filtering.

The subjectivity of the observer is considered only if the algorithm detects an edge. The implementation for the observer b is given by the following equation:

$$X'(i,j)\mid_b = \mu_{edge} \cdot \left(\sum_k^n \Upsilon(b,i) \cdot X'_{A_k}(i,j) \right) + (1 - \mu_{edge}) \cdot g_{ij}. \quad (30)$$

Here is A_k the k-th algorithm among n available alternatives. The edginess μ_{edge} can be calculated by suitable membership functions. Besides, conventional edge detectors may be applied likewise.

If we do not use different algorithms $A_1, \cdots, A_n$, but only one algorithm A with n variations and k parameters $a_1^j, \cdots, a_k^j$ ($j = 1, \cdots, n$), then the suitable parameter set for the observer b can be determined as follows:

$$a'^b_1 = \sum_{j=1}^n \Upsilon(b,1) \cdot a_1^j, \quad (31)$$

$$a'^b_2 = \sum_{j=1}^n \Upsilon(b,2) \cdot a_2^j, \quad (32)$$

$$\vdots$$

$$a'^b_k = \sum_{j=1}^n \Upsilon(b,n) \cdot a_k^j. \quad (33)$$

The local aggregation in this case will be as follows:

$$X'(i,j)\mid_b = \mu_{edge} \cdot \underset{(a'^b_1, \cdots, a'^b_k)}{X'_A(i,j)} + (1 - \mu_{edge}) \cdot g_{ij}. \quad (34)$$

Results verification Similar to investigations for contrast, several tests were done for sharpness. Altogether 11 observers evaluated the quality of 80 images (40 for learning, 40 for testing). The system result was presented to observers in random order together with four other alternatives. For the modification of the image sharpness the following convolution kernels were used:

$$\text{Laplace filter (strong)} \quad K_{A_1} = \begin{bmatrix} -1 & -1 & -1 \\ -1 & 9 & -1 \\ -1 & -1 & -1 \end{bmatrix}, \tag{35}$$

$$\text{Unsharp Masking} \quad K_{A_2} = \begin{bmatrix} -.5 & 0 & -.5 \\ 0 & 3 & 0 \\ -.5 & 0 & -.5 \end{bmatrix}, \tag{36}$$

$$\text{Averaging} \quad K_{A_3} = \begin{bmatrix} .11 & .11 & .11 \\ .11 & .11 & .11 \\ .11 & .11 & .11 \end{bmatrix}, \tag{37}$$

$$\text{Laplace filter} \quad K_{A_4} = \begin{bmatrix} 0 & -1 & 0 \\ -1 & 5 & -1 \\ 0 & -1 & 0 \end{bmatrix}. \tag{38}$$

Instead of the Sugeno integral, the Choquet integral was used for this test (Eq. (6)). Further, the Dempster rule (Eq. (7)) was replaced by the generalized version according to Eq. (8).

In Fig. 12 the outputs of the observer-dependent system for two different observers are illustrated. The relevancy of the algorithms A_1 to A_4 using the corresponding kernels in the Eq. (38) was given as follows:

First observer: [0.26 0.29 0.16 0.29]

Second observer: [0.10 0.27 0.35 0.28]

Fig. 12. Inter-individual differences in the system output: Result for two observers.

More examples for enhancement are illustrated in Figs. 13, 14 and 15. The system results are generated for a certain observer in each case. If images are noisy or if we process them using large windows ($N > 3$), then the system result was clearly better than other alternatives. This is due to the spatial rule applied (Eq. (30)). To test the system performance, therefore, noise-free images were enhanced by means of 3 × 3-windows. The difference of the five images, which the observer had to evaluate, was partly insignificant. This was intended so that the observers were forced to go beyond the pre-attentive phase of image understanding and focus on image details. Tables 6 to 16 show the individual evaluations. Table 17 shows the total result for all executed tests. As follows from Table 17, the system was with $\text{MOS}_{total} = 1.75$ clearly the best alternative in this test. The Unsharp Masking (K_{A_2} in Table 17) as the second best alternative is over 15% worse than the system. Beyond that, the system has the lowest variance ($\sigma = \pm 0,19$) which indicates a stable behavior.

Table 6. Assessment by 1. observer.

Algorithm	1	2	3	4	5	MOS_{total}
K_{A_1}	0%	0%	50%	50%	0%	3.5
K_{A_2}	10%	70%	10%	10%	0%	2.2
K_{A_3}	0%	0%	0%	30%	70%	4.7
K_{A_4}	0%	30%	70%	0%	0%	2.7
System	50%	50%	0%	0%	0%	1.5

Table 7. Assessment by 2. observer.

Algorithm	1	2	3	4	5	MOS_{total}
K_{A_1}	0%	0%	20%	40%	40%	4.2
K_{A_2}	30%	50%	20%	0%	0%	1.9
K_{A_3}	0%	0%	10%	50%	40%	4.3
K_{A_4}	10%	30%	50%	10%	0%	2.6
System	70%	30%	0%	0%	0%	1.3

Table 8. Assessment by 3. observer.

Algorithm	1	2	3	4	5	MOS_{total}
K_{A_1}	0%	0%	0%	30%	70%	4.7
K_{A_2}	10%	30%	30%	30%	0%	2.8
K_{A_3}	0%	60%	10%	30%	0%	2.7
K_{A_4}	10%	20%	30%	40%	0%	3.0
System	50%	40%	10%	0%	0%	1.6

Fig. 13. Samples for testing.

Table 9. Assessment by 4. observer.

Algorithm	1	2	3	4	5	MOS_{total}
K_{A_1}	0%	0%	40%	50%	10%	3.7
K_{A_2}	30%	40%	30%	0%	0%	2.0
K_{A_3}	0%	0%	30%	50%	20%	3.9
K_{A_4}	10%	60%	20%	10%	0%	2.3
System	60%	30%	10%	0%	0%	1.5

Table 10. Assessment by 5. observer.

Algorithm	1	2	3	4	5	MOS_{total}
K_{A_1}	0%	0%	0%	90%	10%	4.1
K_{A_2}	0%	50%	30%	20%	0%	2.7
K_{A_3}	10%	20%	10%	50%	10%	3.3
K_{A_4}	0%	30%	60%	0%	10%	2.9
System	20%	50%	20%	10%	0%	2.2

Table 11. Assessment by 6. observer.

Algorithm	1	2	3	4	5	MOS_{total}
K_{A_1}	0%	0%	30%	50%	20%	3.9
K_{A_2}	10%	50%	30%	10%	0%	2.4
K_{A_3}	0%	0%	40%	40%	20%	3.8
K_{A_4}	0%	40%	50%	10%	0%	2.7
System	20%	60%	20%	0%	0%	2.0

(a) K_{A_1} (b) K_{A_2}

(c) K_{A_3} (d) K_{A_4}

(e) System's result (f) System's result

Fig. 14. Image modification with different approaches. The result of the system is generated for one *certain observer* in each case.

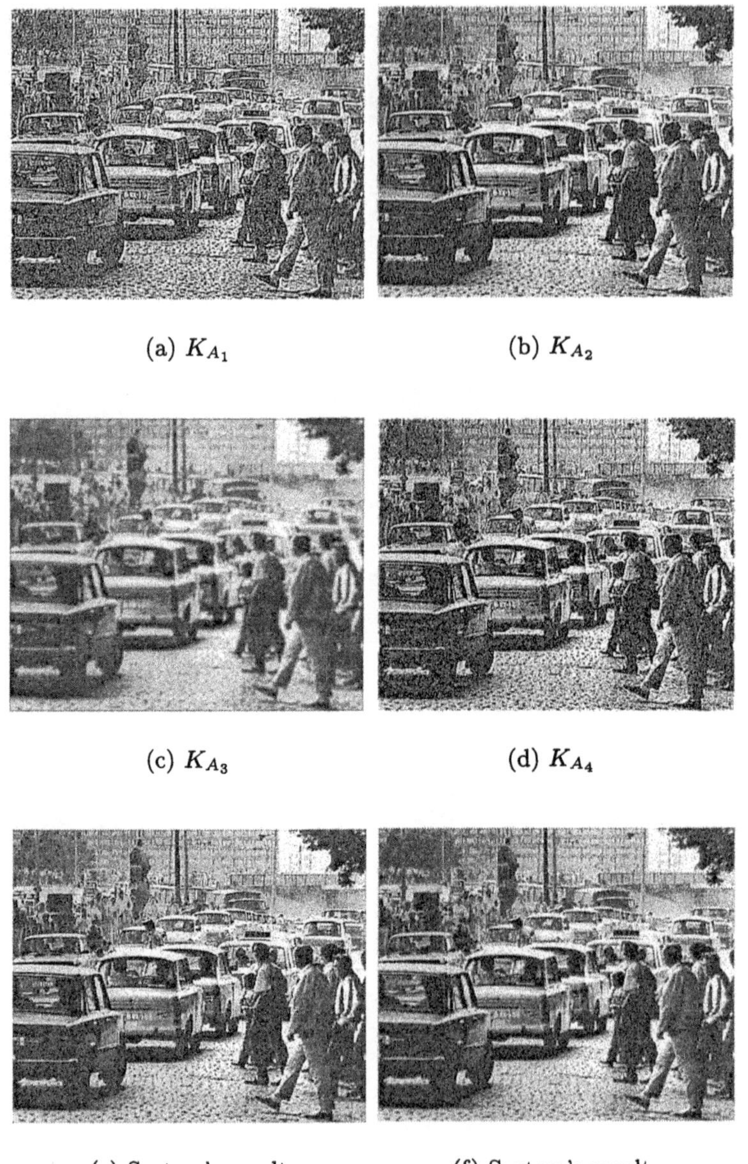

(a) K_{A_1} (b) K_{A_2}

(c) K_{A_3} (d) K_{A_4}

(e) System's result (f) System's result

Fig. 15. Image modification with different approaches. The result of the system is generated for one *certain observer*.

Table 12. Assessment by 7. observer.

Algorithm	1	2	3	4	5	MOS_{total}
K_{A_1}	0%	0%	0%	30%	70%	4.7
K_{A_2}	0%	10%	40%	50%	0%	3.4
K_{A_3}	0%	60%	30%	10%	0%	3.5
K_{A_4}	0%	10%	10%	70%	10%	3.8
System	0%	50%	50%	0%	0%	2.5

Table 13. Assessment by 8. observer.

Algorithm	1	2	3	4	5	MOS_{total}
K_{A_1}	0%	0%	0%	80%	20%	4.2
K_{A_2}	0%	10%	80%	10%	0%	3.0
K_{A_3}	0%	40%	60%	0%	0%	2.6
K_{A_4}	0%	20%	20%	50%	10%	3.5
System	30%	60%	10%	0%	0%	2.4

Table 14. Assessment by 9. observer.

Algorithm	1	2	3	4	5	MOS_{total}
K_{A_1}	0%	10%	50%	40%	0%	3.3
K_{A_2}	40%	40%	20%	0%	0%	1.8
K_{A_3}	0%	20%	40%	40%	0%	3.0
K_{A_4}	30%	30%	40%	0%	0%	3.1
System	60%	20%	20%	0%	0%	1.6

Table 15. Assessment by 10. observer.

Algorithm	1	2	3	4	5	MOS_{total}
K_{A_1}	10%	20%	40%	30%	0%	2.9
K_{A_2}	40%	50%	10%	0%	0%	1.7
K_{A_3}	0%	10%	30%	40%	20%	3.7
K_{A_4}	30%	50%	20%	0%	0%	1.9
System	70%	30%	0%	0%	0%	1.3

Table 16. Assessment by 11. observer.

Algorithm	1	2	3	4	5	MOS_{total}
K_{A_1}	0%	0%	40%	50%	10%	3.7
K_{A_2}	30%	40%	30%	0%	0%	2.0
K_{A_3}	20%	40%	30%	10%	0%	2.3
K_{A_4}	30%	50%	20%	0%	0%	1.9
System	60%	40%	0%	0%	0%	1.4

Table 17. Success rate of the overall system in a test series for sharpness as ecological image quality (σ = standard deviation). In the ideal case all results of the system should be perceived as the best alternative (MOS = 1).

Algorithm	MOS_{total}	σ
K_{A_1}	3.90	±0.31
K_{A_2}	2.35	±0.30
K_{A_3}	3.44	±0.55
K_{A_4}	2.76	±0.35
System	1.75	±0.19

6 Possibilities for performance increase

The goal of the developed system is to obtain a very high success rate with respect to the observers' expectations (MOS $\approx$ 1). In order to achieve this one must initially think about the selected algorithms in the first phase. Depending on application, therefore, appropriate algorithms should be selected. Beyond that alternative techniques can be investigated for a better aggregation of evidences.

The Dempster rule applied to the contrast example has the disadvantage to supply results in strong conflict situations which we do not intuitively expect [13]. In literature, different improvement suggestions have been made [10,24]. In the second test series for image sharpness, therefore, an extended version (Eq. (8)) was used.

No empirical or theoretical investigations are available which provide information about how many images in the learning phase must be used in order to learn the subjectivity of the observer as good as possible. To overcome this problem, the online system can be extended by a observer feedback (Fig. 16). Besides, the aggregation matrix, however, specifications concerning the applied algorithms must also be stored. A suitable correction schemes must be determined likewise.

Probably the best way to increase the performance of the system is a global, psychologically supported interview with observer or the use of more suitable, also psychologically supported ways for testing. Only through co-operation with psychologists we will succeed to obtain the highest possible success rate.

A disadvantage of the system remains to be able to process only *one quality* at each time. However, in many applications only one certain image characteristic is in focus.

[13] A particularly strong conflict situation occurs, for example, if $MOS_{A,1} = 1$ and $MOS_{A,2} = 5$. In the executed test series such cases did not occur, they, however, cannot be excluded.

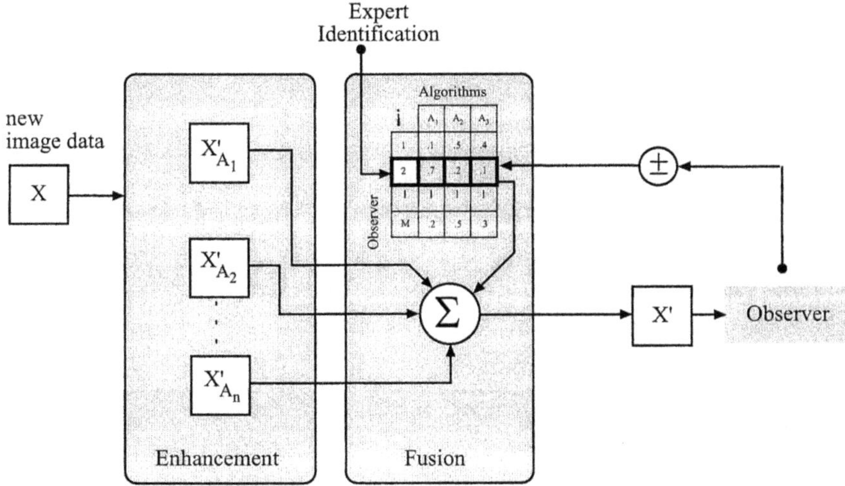

Fig. 16. Extension of observer-dependent system by a user feedback.

References

1. Ballard, D. H., Brown, C. M., *Computer Vision*, Prentice-Hall, New Jersey, 1982.
2. Bouchon-Meunier, B. (editor), *Aggregation and Fusion of Imperfect Information*, Physica-Verlag, Heidelberg, New York, 1998.
3. Dempster, A. P., *Upper and lower probabilities induced by a multivalued mapping*, in: Ann. Math. Statistics, vol. 38, 1967, pp. 325–339.
4. Gonzalez, R. C., Woods, R. E., *Digital Image Processing*, Addison-Wesley, New York, 1993.
5. Haralick, R.M., Shapiro, L.G., *Computer and Robot Vision*, Addison Wesely, 1.Band, 1992.
6. Hauske, G., *Systemtheorie der visuellen Wahrnehmung*, Teubner-Verlag, Stuttgart, 1994.
7. Hendee, W.R., Wells, P.N.T., *The Perception of Visual Information*, Springer-Verlag, Berlin, 1997.
8. Higgins, G.C., Jones, L.A., *The nature and evaluation of the sharpness of photographic images*, in: Journal of Society of Motion Picture and Television Engineers, vol. 58, 1952, pp. 277–290.
9. Kandel, A., Friedman, M., Schneider, M., *The use of weighted fuzzy expected value (WFEV) in fuzzy expert systems*, in: Fuzzy Sets and Systems 31, 1989, pp. 37–45.
10. Lee, E. S., Zhu, Q., *Fuzzy and Evidence Reasoning*, Physica-Verlag, Heidelberg, 1995.
11. Marr, D., *Vision*, Freeman and Co., New York, 1996.
12. Newell, A., *Production systems: Models of control structure*, in: Chase,W.G. (editor), Visual information processing, New York, Academic Press, 1973.

13. Olabarriaga, S.D., Rangayyan, R.M., *Subjective and objective evaluation of image sharpness - Behavior of the region-based image edge profile actuance measure*, in: Proc. SPIE, vol. 2717, 1996, pp. 154–162.
14. Sugeno, M., *Theory of Fuzzy Integrals and Its Applications*, Dissertation, Tokyo, Institute of Technology, Japan, 1974.
15. Sugeno, M., *Fuzzy measures and fuzzy integrals: a survey*, in: Fuzzy Automata and Decision Processes, North-Holland, Amsterdam, 1977, pp. 89–102.
16. Tizhoosh, H.R., *Fuzzy Image Processing* (in German), Springer, Heidelberg, 1997.
17. Tizhoosh, H.R., Krell, G., Michaelis, B., *Locally Adaptive Fuzzy Image Enhancement*, in B. Reusch (editor), Computational Intelligence, Theory and Applications, Proc. Of 5th Fuzzy Days'97, Dortmund, Germany, Springer, 1997, pp. 272–276.
18. Tizhoosh, H.R., Krell, G., Michaelis, B., *On Fuzzy Image Enhancement of Megavoltage Images in Radiation Therapy*, in: FUZZ-IEEE'97, Barcelona, Spanien, 1997, pp. 1399–1404.
19. Tizhoosh, H.R., Michaelis, B., *Improvement of Image Quality Based on Subjective Evaluation and Fuzzy Aggregation techniques*, EUFIT'98, Aachen, Germany, vol. 2, 1998, pp. 1325–1329.
20. Tizhoosh, H.R., Haußecker, H., *Fuzzy Image Processing: An Overview*, in: Jähne, B., Haußecker, H., Geißler, P. (editors), Handbook on Computer Vision and Applications, Academic Press, Boston, 1999, pp. 683–727.
21. Tizhoosh, H.R., Krell, G., Michaelis, B., *Enhancement of Megavoltage Images in Radiation Therapy Using Fuzzy and Neural Image Processing Techniques*, in: P.S.Szczepaniak, P.J.G.Lisboa, S.Tsumoto (editors): Fuzzy Systems in Medicine, Studies in Fuzziness and Soft Computing, Physica-Verlag, 1999.
22. Tizhoosh, H.R., Michaelis, B., *Image Enhancement Based on Fuzzy Aggregation Techniques*, in: IMTC'99, Venedig, Italien, vol. 3, 1999, pp. 1813–1817.
23. Tizhoosh, H.R., Michaelis, B., *Subjectivity, Psychology and Fuzzy Techniques: A New Approach to Image Enhancement*, in: NAFIPS'99, New York, USA, 1999, pp. 522–526.
24. Wang, Z., Klir, G. J., *Fuzzy Measure Theory*, Plenum Press, New York, 1992.
25. Weeks, R. W., *Fundamentals of Electronic Image Processing*, IEEE Press, New York, 1996.
26. Weidenmann, B., *Wissenserwerb mit Bildern: Instruktionale Bilder in Printmedien, Film/Video und Computerprogrammen*, Huber, Göttingen, 1994, pp.9–58.
27. Weidenmann, B., *Psychologie des Lernens mit Medien*, in: Weidenmann, B. & Krapp, A. (editors), Pädagogische Psychologie, PVU, Urban & Schwarzenberg, München, Weinheim, 1986, pp. 495–553.
28. Weidenmann, B., *Psychische Prozesse beim Verstehen von Bildern*, Verlag Hans Huber, Bern, Stuttgart, Toronto, 1988.
29. Weidenmann, B., *Lernen mit Bildmedien: Psychologische und didaktische Grundlagen*, Beltz Verlag, Basel, Weinheim, 1994.
30. Yen, J., *Generalizing the Dempster-Shafer theory to fuzzy sets*, in: IEEE Trans. Systems, Man and Cybernetics, 20(3), 1990, pp. 559–570.
31. Zadeh, L.A., *A fuzzy-set-theoretic interpretation of linguistic hedges*, in: Journal of Cybernetics 2, 1972, pp. 4–34.

Part IV

Specific Applications of Fuzzy Filters

Chapter 12

Fuzzy Techniques in Digital Image Processing and Shape Analysis

Vassilios Chatzis and Ioannis Pitas

Aristotle University of Thessaloniki
Department of Informatics
Thessaloniki, Greece
email: pitas@zeus.csd.auth.gr

Summary. In this chapter, an overview of the applications of several fuzzy operators in image processing and analysis is presented. First, the fuzzy location and scale estimators based on the extension principle, are presented. The definitions of the fuzzy nonlinear means, the fuzzy location and scale estimators based on fuzzy order statistics and other fuzzy scale estimators, e.g., fuzzy sample standard deviation, are also given. Equivalent relations that can be used to calculate the fuzzy estimators using classical arithmetic are derived. The Fuzzy Vector Median is defined as an extension of the classical Vector Median, based on a distance definition between fuzzy vectors. An application of Fuzzy Vector Median for filtering images corrupted by mixed Gaussian and impulsive noise is also given.

Next, the Fuzzy Cell Hough Transform is presented. It is an extension of the classical Hough transform that uses a fuzzy split parameter space in order to detect curves in binary images. The Hough transform parameter space is split into fuzzy cells and the transform is implemented by a fuzzy voting process which is explicitly described. Simulations for the detection of straight lines and circles in artificially generated noisy images are also presented.

Finally, the Generalized Fuzzy Mathematical Morphology is presented. It is based on a Fuzzy Inclusion Indicator, which is a fuzzy set, defined as a measure of the inclusion of one fuzzy set into another one. It is shown that the classical binary and grayscale mathematical morphologies can be considered as special cases of the Generalized Fuzzy Mathematical Morphology. An application on robust 2D and 3D object representation using morphological skeletonization, is also presented.

1 Fuzzy estimators and fuzzy filters

1.1 The extension principle

The *fuzzy set theory* was first introduced by Zadeh [1]. He used the word *fuzzy* to generalize the mathematical concept of the *set* to the *fuzzy set.* Supposing that the available information has uncertain value that can be located inside a closed interval $I \subset \mathbb{R}$, called *interval of confidence*, a *membership function* is defined and maps each element of the interval of confidence to a value in the interval $[0,1]$.

The concept of a fuzzy set X is presented either by its membership function $X = \{(x, \mu_X(x)), x \in I_X\}$ or by the union of its $\alpha - cuts$ $X = \bigcup_\alpha \alpha \cdot X^{(\alpha)} = \bigcup_\alpha \alpha \cdot \left[x_l^{(\alpha)}, x_r^{(\alpha)}\right]$, where $\alpha \in [0,1]$. A fuzzy set is called *normal* if $\exists x : \mu(x) = 1$ or $X^{(1)} \neq \emptyset$. It is called *convex* if $\forall \alpha_1, \alpha_2 \in [0,1], \alpha_1 > \alpha_2 \Leftrightarrow X^{(\alpha_1)} \subseteq X^{(\alpha_2)}$. A normal and convex fuzzy set is called *fuzzy number* [2],[3].

In order to extend mathematical notions and operations of crisp numbers in fuzzy numbers we can use the *extension principle* [4,5]. The extension principle provides the theoretical warranty that fuzzifying the parameters or arguments of a function results in computable fuzzy sets. Let $X_1, X_2, \ldots, X_n$ be n fuzzy sets and $y = f(x_1, x_2, \ldots, x_n)$ be a crisp function. Then the extension principle transfers the fuzziness of $X_1, X_2, \ldots, X_n$ into a fuzzy set Y:

$$Y = \{(y, \mu_Y(y)) \;/\; y = f(x_1, x_2, \ldots, x_n)\}, \tag{1}$$

where:

$$\mu_Y(y) = \begin{cases} 0 & \text{if } f^{-1}(y) = \emptyset \\ \sup_{(x_1, x_2, \ldots, x_n) \in f^{-1}(y)} \min\{\mu_{X_1}, \mu_{X_2}, \ldots, \mu_{X_n}\} & \text{if } f^{-1}(y) \neq \emptyset. \end{cases} \tag{2}$$

The extension principle is used to fuzzify basic operations such as addition and multiplication of fuzzy numbers, multiplication by a crisp number, maximum and minimum of fuzzy numbers. The extensions of the basic operations are presented in Table 1.

Table 1. Basic operations of fuzzy numbers, extended through the extension principle (X_1, X_2: fuzzy numbers).

Operations on Fuzzy Numbers
$X_1 + X_2 = \bigcup_\alpha \alpha \cdot \left[x_{1_l}^{(\alpha)} + x_{2_l}^{(\alpha)}, x_{1_r}^{(\alpha)} + x_{2_r}^{(\alpha)}\right], \quad I_{X_1}, I_{X_2} \subset \mathbb{R}$
$X_1 - X_2 = \bigcup_\alpha \alpha \cdot \left[x_{1_l}^{(\alpha)} - x_{2_r}^{(\alpha)}, x_{1_r}^{(\alpha)} - x_{2_l}^{(\alpha)}\right], \quad I_{X_1}, I_{X_2} \subset \mathbb{R}$
$X_1 \cdot X_2 = \bigcup_\alpha \alpha \cdot \left[\min\{x_{1_l}^{(\alpha)} \cdot x_{2_l}^{(\alpha)}, x_{1_l}^{(\alpha)} \cdot x_{2_r}^{(\alpha)}, x_{1_r}^{(\alpha)} \cdot x_{2_l}^{(\alpha)}, x_{1_r}^{(\alpha)} \cdot x_{2_r}^{(\alpha)}\},\right.$ $\left.\max\{x_{1_l}^{(\alpha)} \cdot x_{2_l}^{(\alpha)}, x_{1_l}^{(\alpha)} \cdot x_{2_r}^{(\alpha)}, x_{1_r}^{(\alpha)} \cdot x_{2_l}^{(\alpha)}, x_{1_r}^{(\alpha)} \cdot x_{2_r}^{(\alpha)}\}\right], \quad I_{X_1}, I_{X_2} \subset \mathbb{R}$
$X^{-1} = \bigcup_\alpha \alpha \cdot \left[\frac{1}{x_r^{(\alpha)}}, \frac{1}{x_l^{(\alpha)}}\right], \quad I_X \subset \mathbb{R}^+$
$\mathrm{MAX}\{X_1, X_2\} = \bigcup_\alpha \alpha \cdot \left[\max\{x_{1_l}^{(\alpha)}, x_{2_l}^{(\alpha)}\}, \max\{x_{1_r}^{(\alpha)}, x_{2_r}^{(\alpha)}\}\right], \quad I_{X_1}, I_{X_2} \subset \mathbb{R}$
$\mathrm{MIN}\{X_1, X_2\} = \bigcup_\alpha \alpha \cdot \left[\min\{x_{1_l}^{(\alpha)}, x_{2_l}^{(\alpha)}\}, \min\{x_{1_r}^{(\alpha)}, x_{2_r}^{(\alpha)}\}\right], \quad I_{X_1}, I_{X_2} \subset \mathbb{R}$

In the following, fuzziness concepts in statistical estimation theory will be addressed. The observation data will be considered to have uncertain values modelled by fuzzy numbers. For example, fuzzy numbers can describe knowledge regarding the conditions of the observation, by changing fuzziness when observation conditions change. Fuzzy numbers can also be considered as the output of fuzzy inference mechanisms. When fuzzy numbers have to be combined in a nonlinear way, fuzzy nonlinear estimators (e.g., fuzzy median) have to be defined. The extension principle will be used to fuzzify location and scale estimators useful in estimation theory as well as in applications on digital signal and image processing.

1.2 Fuzzy location estimators

The nonlinear means have been extensively used as location estimators of convolved and nonlinearly related data. Their definitions on crisp numbers can easily be extended to fuzzy numbers by using the extensions of the basic operations on fuzzy numbers, as they are presented in Table 1. The definitions of fuzzy arithmetic mean, fuzzy harmonic mean, fuzzy geometric mean and fuzzy L_p mean, as well as equivalent relations that can be used to calculate the corresponding fuzzy means by using crisp arithmetic, are given in Table 2 [6].

Table 2. Fuzzy nonlinear means definitions and equivalent relations for their calculation by crisp arithmetic, ($X_i, i = 1, \dots, n$: fuzzy numbers, w_i: crisp weights).

Estimator	Fuzzy Definition	Equivalent Relation
fuzzy arithmetic mean ($\overline{X}$)	$\frac{1}{n}\sum_{i=1}^{n} X_i$	$\bigcup_\alpha \alpha \cdot \left[\frac{1}{n}\sum_{i=1}^{n} x_{i_l}^{(\alpha)}, \frac{1}{n}\sum_{i=1}^{n} x_{i_r}^{(\alpha)}\right]$
fuzzy harmonic mean (Y_H)	$\frac{\sum_{i=1}^{n} w_i}{\sum_{i=1}^{n} \frac{w_i}{X_i}}$	$\bigcup_\alpha \alpha \cdot \left[\frac{\sum_{i=1}^{n} w_i}{\sum_{i=1}^{n} \frac{w_i}{x_{i_l}^{(\alpha)}}}, \frac{\sum_{i=1}^{n} w_i}{\sum_{i=1}^{n} \frac{w_i}{x_{i_r}^{(\alpha)}}}\right]$
fuzzy geometric mean (Y_G)	$\log^{-1}\left(\frac{\sum_{i=1}^{n} w_i \log X_i}{\sum_{i=1}^{n} w_i}\right)$	$\bigcup_\alpha \alpha \cdot \left[\log^{-1}\left(\frac{\sum_{i=1}^{n} w_i \log x_{i_l}^{(\alpha)}}{\sum_{i=1}^{n} w_i}\right), \log^{-1}\left(\frac{\sum_{i=1}^{n} w_i \log x_{i_r}^{(\alpha)}}{\sum_{i=1}^{n} w_i}\right)\right]$
fuzzy L_p mean (Y_{L_p})	$\left(\frac{\sum_{i=1}^{n} w_i X_i^p}{\sum_{i=1}^{n} w_i}\right)^{\frac{1}{p}}$	$\bigcup_\alpha \alpha \cdot \left[\left(\frac{\sum_{i=1}^{n} w_i (x_{i_l}^{(\alpha)})^p}{\sum_{i=1}^{n} w_i}\right)^{\frac{1}{p}}, \left(\frac{\sum_{i=1}^{n} w_i (x_{i_r}^{(\alpha)})^p}{\sum_{i=1}^{n} w_i}\right)^{\frac{1}{p}}\right]$

It is well known that the crisp arithmetic mean minimizes the L_2 error norm. This is sometimes used as an alternative definition of the crisp arithmetic mean. This property is still valid for the fuzzy arithmetic mean since it minimizes the error [7]:

$$E_2 = \sum_{i=1}^{N} d_2 [X_i, M_2], \tag{3}$$

where the norm $d_2[\cdot,\cdot]$ corresponds to the distance of two fuzzy numbers X and Y defined as [2]:

$$d_2 [X, Y] = \int_{\alpha=0}^{1} \left(\left(x_l^{(\alpha)} - y_l^{(\alpha)} \right)^2 + \left(x_r^{(\alpha)} - y_r^{(\alpha)} \right)^2 \right) d\alpha. \tag{4}$$

The definition of the *fuzzy order statistics* is needed to extend location estimators based on order statistics. Let us assume that $X_1, X_2, \ldots, X_n$ are n fuzzy numbers. In order to rank them, we must successively use the fuzzy maximum (MAX) and minimum (MIN) operations presented in Table 1. It can be easily shown that successive fuzzy MAX and MIN operations can be calculated by performing the corresponding successive crisp max, min operations on the crisp limits of the $\alpha - cuts$ of the fuzzy numbers. If we symbolize the fuzzy order statistics as $X_{(1)}, X_{(2)}, \ldots, X_{(n)}$, where $X_{(1)}$ is the fuzzy minimum and $X_{(n)}$ the fuzzy maximum, they can be calculated by the following equations:

$$X_{(i)} = \bigcup_{\alpha} \alpha \cdot \left[x_{(i)_l}^{(\alpha)}, x_{(i)_r}^{(\alpha)} \right], \qquad i = 1, 2, \ldots, n \tag{5}$$

where $x_{(i)_l}^{(\alpha)}$ are the order statistics of $x_{i_l}^{(\alpha)}$ and $x_{(i)_r}^{(\alpha)}$ are the order statistics of $x_{i_r}^{(\alpha)}$, having the following ordering:

$$x_{(1)_l} \leq x_{(2)_l} \leq \ldots \leq x_{(n)_l}, \tag{6}$$

$$x_{(1)_r} \leq x_{(2)_r} \leq \ldots \leq x_{(n)_r}. \tag{7}$$

The fuzzy order statistics do not generally correspond to the input fuzzy numbers, as is the case in crisp arithmetic. For example, let X_1, X_2, X_3, X_4, X_5 be five fuzzy numbers illustrated in Fig. 1a. Fuzzy order statistics $X_{(1)}$, $X_{(2)}$, $X_{(3)}$, $X_{(4)}$, $X_{(5)}$ are illustrated in Fig. 1b. In this example, only $X_{(5)}$ of the fuzzy order statistics is equal to X_3 of the input fuzzy numbers.

The L estimators of location have found extensive applications in digital signal and image processing, since they are explicitly defined and easily calculated. Their definitions [8], which are based on crisp order statistics, can easily be extended to define fuzzy L location estimators. Based on the fuzzy order statistics $X_{(1)}, X_{(2)}, \ldots, X_{(n)}$ of n fuzzy numbers that was previously presented, the fuzzy L location estimators are defined as it is shown in Table 3, where w_i are crisp weights. Special cases of L estimators are the median, the midpoint and the a-trimmed mean, which are all based on order

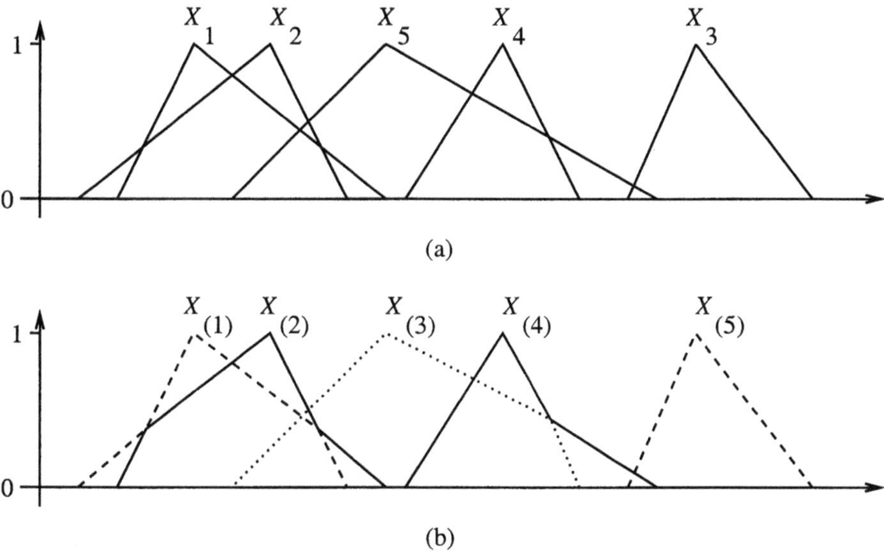

Fig. 1. (a) Five fuzzy numbers X_1, X_2, X_3, X_4, X_5. (b) The fuzzy order statistics $X_{(1)}$, $X_{(2)}$, $X_{(3)}$, $X_{(4)}$, $X_{(5)}$.

Table 3. Fuzzy L location estimator definitions and equivalent relations for their calculation by crisp arithmetic, ($X_i, i = 1, \ldots, n$: fuzzy data, $X_{(i)}, i = 1, \ldots, n$: fuzzy order statistics, w_i: crisp weights).

Estimator	Fuzzy Definition	Equivalent Relation
L estimator (T_n)	$\sum_{i=1}^{n} w_i X_{(i)}$	$\bigcup_{\alpha} \alpha \left[\sum_{i=1}^{n} \min\{w_i x^{(\alpha)}_{(i)_l}, w_i x^{(\alpha)}_{(i)_r}\}, \sum_{i=1}^{n} \max\{w_i x^{(\alpha)}_{(i)_l}, w_i x^{(\alpha)}_{(i)_r}\} \right]$
a-trimmed mean ($\overline{X}_a$)	$\frac{1}{n(1-2a)} \sum_{i=an+1}^{n-an} X_{(i)}$ where $a < 0.5$	$\bigcup_{\alpha} \alpha \cdot \left[\frac{1}{n(1-2a)} \sum_{i=an+1}^{n-an} x^{(\alpha)}_{(i)_l}, \frac{1}{n(1-2a)} \sum_{i=an+1}^{n-an} x^{(\alpha)}_{(i)_r} \right]$
Midpoint ($Midp\{X_i\}$)	$\frac{1}{2}(X_{(1)} + X_{(n)})$	$\bigcup_{\alpha} \alpha \cdot \left[\frac{1}{2}(x^{(\alpha)}_{(1)_l} + x^{(\alpha)}_{(n)_l}), \frac{1}{2}(x^{(\alpha)}_{(1)_r} + x^{(\alpha)}_{(n)_r}) \right]$
Median ($Med\{X_i\}$)	$X_{(\nu+1)}$ if n odd, $n = 2\nu+1$ $\frac{1}{2}(X_{(\nu)} + X_{(\nu+1)})$ if n even, $n = 2\nu$	$\bigcup_{\alpha} \alpha \cdot \left[x^{(\alpha)}_{(\nu+1)_l}, x^{(\alpha)}_{(\nu+1)_r} \right]$ if n odd, $n = 2\nu + 1$ $\bigcup_{\alpha} \alpha \cdot \left[\frac{1}{2}(x^{(\alpha)}_{(\nu)_l} + x^{(\alpha)}_{(\nu+1)_l}), \frac{1}{2}(x^{(\alpha)}_{(\nu)_r} + x^{(\alpha)}_{(\nu+1)_r}) \right]$ if n even, $n = 2\nu$

statistics. Their definitions on fuzzy numbers and equivalent relations that can be used to calculate the corresponding fuzzy estimators by using crisp arithmetic, are also given in Table 3. It is well known that the crisp median minimizes the L_1 error norm, a fact that is also an alternative definition of the crisp median. This property is still valid for the fuzzy median, since it minimizes the error [9]:

$$E_1 = \sum_{i=1}^{n} d_1 [X_i, M_1], \tag{8}$$

where the norm $d_1[\cdot,\cdot]$ corresponds to the distance of two fuzzy numbers X and Y defined as [2]:

$$d_1 [X,Y] = \int_{\alpha=0}^{1} \left(\left| x_l^{(\alpha)} - y_l^{(\alpha)} \right| + \left| x_r^{(\alpha)} - y_r^{(\alpha)} \right| \right) d\alpha. \tag{9}$$

1.3 Fuzzy scale estimators

In the following, the crisp L scale estimators [8] will be generalized to define fuzzy L scale estimators. Let the observation data be n fuzzy numbers $X_1, X_2, \ldots, X_n$. The fuzzy order statistics $X_{(1)}, X_{(2)}, \ldots, X_{(n)}$ of the n fuzzy numbers can be easily computed, as was shown in the previous section. Then, fuzzy L scale estimators are defined as:

$$T_n = \sum_{i=1}^{n} w_i X_{(i)}, \tag{10}$$

where $X_{(i)}$ is the i-th fuzzy order statistic of the fuzzy data and w_i are appropriately chosen crisp weights. If $w_i = -1$ for $i = tn$, $w_i = 1$ for $i = (1-t)n$ and $w_i = 0$ otherwise, then the *fuzzy t-quantile range* is defined. Special cases of fuzzy t-quantile range are the *fuzzy interquantile range* for $t = 1/4$, and the *fuzzy quasi-range* ($W_{(i)}$) for $t \neq 0$ given by:

$$W_{(i)} = X_{(n+1-i)} - X_{(i)}, \qquad 2 \le i \le \left[\frac{n}{2}\right]. \tag{11}$$

A special case of the fuzzy quasi-range for $i = 1$ is the *fuzzy range* (W) given by:

$$W = X_{(n)} - X_{(1)}. \tag{12}$$

Another scale estimator is the *fuzzy thickened range* given by:

$$J_i = W + W_{(2)} + \ldots + W_{(i)}. \tag{13}$$

The most widely known and used scale estimator is the sample standard deviation that is defined by:

$$s = \left[\frac{1}{n-1} \sum_{i=1}^{n} (x_i - \overline{x})^2 \right]^{\frac{1}{2}}, \tag{14}$$

where x_i are the observation data and $\overline{x}$ is the arithmetic mean of x_i. In order to extend the crisp sample standard deviation s to *fuzzy sample standard deviation* S, the crisp functions of the square and the square root must be first extended to fuzzy numbers.

The square X^2 of a fuzzy number X, having its interval of confidence in R, can be defined as the product of two fuzzy numbers X_1, X_2 where $X_1 = X_2 = X$. Then, by using the extension of multiplication given in Table 1, the square of a fuzzy number can be written as:

$$X^2 = \bigcup_{\alpha} \alpha \cdot \left[\min\left\{\left(x_l^{(\alpha)}\right)^2, x_l^{(\alpha)} \cdot x_r^{(\alpha)}, \left(x_r^{(\alpha)}\right)^2\right\}, \max\left\{\left(x_l^{(\alpha)}\right)^2, \left(x_r^{(\alpha)}\right)^2\right\}\right]. \tag{15}$$

Let X be a fuzzy number with its interval of confidence in R. We define the *positive fuzzy numbers* as the subset of fuzzy numbers with positive upper limits $x_r^{(\alpha)} \geq 0$ of the corresponding $\alpha - cuts$ for every $\alpha \in [0,1]$. It is obvious from Eq. (15) that the square of a fuzzy number is always a positive fuzzy number. However, the lower limits of the corresponding $\alpha - cuts$ are not always positive.

The square root can be defined only for positive fuzzy numbers. Let $X^{\frac{1}{2}}$ be the square root of a fuzzy number X. It can be defined as the fuzzy number Y such as $Y^2 = X$. Then, by using Eq. (15), the square root of a fuzzy number is written as:

$$Y = X^{\frac{1}{2}} = \bigcup_{\alpha} \alpha \cdot \left[y_l^{(\alpha)}, y_r^{(\alpha)}\right], \tag{16}$$

where for every $\alpha \in [0,1]$:

$$y_r^{(\alpha)} = \left(x_r^{(\alpha)}\right)^{\frac{1}{2}}, \tag{17}$$

$$y_l^{(\alpha)} = \begin{cases} \left(x_l^{(\alpha)}\right)^{\frac{1}{2}} & \text{if } x_l^{(\alpha)} \geq 0 \\ \frac{x_l^{(\alpha)}}{y_r^{(\alpha)}} & \text{if } x_l^{(\alpha)} < 0. \end{cases} \tag{18}$$

It is supposed that the square roots of crisp numbers used are the positive ones. Then, it is obvious from Eq. (17) that the square root of a fuzzy number is a positive fuzzy number.

Let $X_1, X_2, \ldots, X_n$ be the fuzzy data and $\overline{X}$ be the fuzzy arithmetic mean of X_i, $i = 1, \ldots, n$ given in Table 2. Then, the fuzzy sample standard deviation S is defined as:

$$S = \left[\frac{1}{n-1}\sum_{i=1}^{n}\left(X_i - \overline{X}\right)^2\right]^{\frac{1}{2}}. \tag{19}$$

The square root in Eq. (19) is always defined since the sum of positive fuzzy numbers is always a positive fuzzy number. By using Table 1 and Eqs. (15-18), fuzzy sample standard deviation can be calculated by:

$$S = \bigcup_{\alpha} \alpha \cdot \left[s_l^{(\alpha)}, s_r^{(\alpha)} \right], \tag{20}$$

where for every $\alpha \in [0, 1]$:

$$s_r^{(\alpha)} = \left[\frac{1}{n-1} \sum_{i=1}^{n} \max \left\{ \left(x_{i_l}^{(\alpha)} - \overline{x}_r^{(}(\alpha) \right)^2, \left(x_{i_r}^{(\alpha)} - \overline{x}_l^{(\alpha)} \right)^2 \right\} \right]^{\frac{1}{2}}, \tag{21}$$

$$s_l^{(\alpha)} = \begin{cases} \left[\frac{1}{n-1} z \right]^{\frac{1}{2}} & \text{if } z \geq 0 \\ \frac{1}{n-1} \frac{z}{s_r^{(\alpha)}} & \text{if } z < 0, \end{cases} \tag{22}$$

where:

$$z = \sum_{i=1}^{n} \min \left\{ \left(x_{i_l}^{(\alpha)} - \overline{x}_r^{(\alpha)} \right)^2, \left(x_{i_l}^{(\alpha)} - \overline{x}_r^{(\alpha)} \right) \cdot \left(x_{i_r}^{(\alpha)} - \overline{x}_l^{(\alpha)} \right), \left(x_{i_r}^{(\alpha)} - \overline{x}_l^{(\alpha)} \right)^2 \right\}. \tag{23}$$

It is concluded from Eq. (21) that the fuzzy sample standard deviation is always a positive fuzzy number.

1.4 Fuzzy vectors

In the following, the term *fuzzy vector* will be used to describe the extension of an n-dimensional crisp set $\mathbf{C}$ to an n-dimensional fuzzy set $\mathbf{X}$ defined in an $(n+1)$-dimensional hyperspace, by using a membership function $\mu : \mathbf{C} \to [0, 1]$ as in [10]. In order to define a distance measure between fuzzy vectors our analysis will be restricted in a subset of fuzzy vectors, the *angle decomposed fuzzy vectors* (ADFV) [9]. The use of ADFV gives us the ability to establish a one to one correspondence between the points of two fuzzy vectors that limit their α-cuts on a certain direction.

The one-dimensional α-cuts can easily be extended to describe multidimensional fuzzy sets. The α-cuts of an n-dimensional fuzzy set will be the classical n-dimensional sets $\mathbf{X}^{(\alpha)}$, where $\mathbf{x} \in \mathbf{X}^{(\alpha)} \Leftrightarrow \mu(\mathbf{x}) \geq \alpha$ and μ is a membership function of n variables. An n-dimensional fuzzy set is called *normal*, if $\exists \mathbf{x} : \mu(\mathbf{x}) = 1$ or $\mathbf{X}^{(1)} \neq \emptyset$. It is called *convex*, if $\forall \alpha_1, \alpha_2 \in [0, 1], \alpha_1 > \alpha_2 \Leftrightarrow \mathbf{X}^{(\alpha_1)} \subseteq \mathbf{X}^{(\alpha_2)}$. A normal and convex fuzzy set is called a *fuzzy number* [2],[3]. An n-dimensional fuzzy number is called *convex fuzzy number*, when all its α-cuts, which are classical n-dimensional sets, are convex sets.

Let $\mathbf{X}$ be an n-dimensional fuzzy set, $\mu_{\mathbf{X}}(\mathbf{x})$ be its membership function and $\mathbf{X}^{(\alpha)}$ the corresponding α-cuts. Consider also that there is only one vector $\mathbf{x}_c$ where $\mu_{\mathbf{X}}(\mathbf{x}_c) = 1$. The vector $\mathbf{x}_c$ will be called the center of the fuzzy set. Consider also $n-1$ angles $\theta = (\theta_i, i = 1, 2, \ldots, n-1)$, $\theta_i \in [0, \pi)$. The center of the fuzzy set $\mathbf{x}_c$ and each angle θ_i determine a hyperplane. The union of $n-1$ hyperplanes is a straight line (direction) in the n-dimensional hyperspace, where a function μ_1 can be defined as $\mu_1(x, \theta) = \mu_X(x_1(x, \theta), x_2(x, \theta), \ldots, x_{n-1}(x, \theta), x)$. This function μ_1 can be considered as a membership function of an one-dimensional fuzzy set X^θ. An n-dimensional fuzzy set $\mathbf{X}$ is an Angle Decomposed Fuzzy Vector (ADFV), if, for each vector of angles $\theta = (\theta_1, \theta_2, \ldots, \theta_{n-1})$, the 1-d fuzzy set $X^\theta = \{x, \mu_1(x, \theta)\}$ is a convex fuzzy number.

The distance definition of one dimensional fuzzy numbers can be extended to angle decomposed fuzzy vectors as:

$$D_n[\mathbf{X}, \mathbf{Y}] = \frac{1}{k} \int_{\theta_1=0}^{\pi} \cdots \int_{\theta_{n-1}=0}^{\pi} \int_{\alpha=0}^{1} \left(||\mathbf{x}_l^{\theta\alpha}, \mathbf{y}_l^{\theta\alpha}|| + ||\mathbf{x}_r^{\theta\alpha}, \mathbf{y}_r^{\theta\alpha}|| \right) d\alpha d\theta_{n-1} \ldots d\theta_1, \tag{24}$$

where $\mathbf{x}_l^{\theta\alpha}$, $\mathbf{y}_l^{\theta\alpha}$ and $\mathbf{x}_r^{\theta\alpha}$, $\mathbf{y}_r^{\theta\alpha}$ are the lower and upper points that limit the α-cuts of the corresponding 1-d X^θ fuzzy numbers, $k = 2(n-1)\pi$ and $||\cdot, \cdot||$ denotes a distance norm between classical vectors. The use of ADFVs guarantees that every point that belongs to the line segment from $\mathbf{x}_l^{\theta\alpha}$ to $\mathbf{x}_r^{\theta\alpha}$ belongs also to the α-cut of X^θ fuzzy number.

If the Euclidean norm is chosen in Eq. (24) then the Euclidean fuzzy distance is defined. It can be proven that the Euclidean fuzzy distance between two ADFVs X, Y is given by [9]:

$$D_{e_n}[\mathbf{X}, \mathbf{Y}] = d_{xy}^2 + d_{f_{xy}}^2, \tag{25}$$

where:

$$d_{f_{xy}}^2 = \frac{1}{k} \int_{\theta_1=0}^{\pi} \cdots \int_{\theta_{n-1}=0}^{\pi} \int_{\alpha=0}^{1} \Big[\left(d_{lx}^{\theta\alpha} - d_{ly}^{\theta\alpha}\right)^2 + \left(d_{rx}^{\theta\alpha} - d_{ry}^{\theta\alpha}\right)^2 + $$
$$+ 2d_{xy} \prod_{i=1}^{n-1} \cos(\theta_i) \left(d_{lx}^{\theta\alpha} - d_{ly}^{\theta\alpha} - d_{rx}^{\theta\alpha} + d_{ry}^{\theta\alpha}\right)\Big] \, d\alpha d\theta_{n-1} \ldots d\theta_1. \tag{26}$$

The above equation shows that the proposed Euclidean fuzzy distance is the classical Euclidean distance between the centers of two ADFVs $\mathbf{X}, \mathbf{Y}$, modified by a factor that depends on the fuzziness that every ADFV holds. The Euclidean fuzzy distance can be considered as a generalized Euclidean distance, since Eq. (26) becomes equal to 0 when the ADFVs are crisp vectors.

1.5 Fuzzy vector median definition

Based on the previously defined distance of ADFVs, we extend the classical definition of the vector median as follows.

The Fuzzy Vector Median (FVM) of $\mathbf{X}_1, \mathbf{X}_2, \ldots, \mathbf{X}_n$ ADFVs is the ADFV $\mathbf{X}_{\text{FVM}}$ such that $\mathbf{X}_{\text{FVM}} \in \{\mathbf{X}_i, i = 1, 2, \ldots, n\}$ and for all $j = 1, 2, \ldots, n$, $j \neq k$:

$$\sum_{i=1}^{n} D_n[\mathbf{X}_{\text{FVM}}, \mathbf{X}_i] < \sum_{i=1}^{n} D_n[\mathbf{X}_j, \mathbf{X}_i]. \tag{27}$$

A straightforward algorithm to find the FVM of a set of fuzzy vectors is the following:

- For each fuzzy vector $\mathbf{X}_i$ compute the sum of the distances g_i to all other vectors:

$$g_i = \sum_{j=1}^{n} D_n[\mathbf{X}_j, \mathbf{X}_i]. \tag{28}$$

- Find k such that g_k is the minimum of g_i, $i = 1, 2, \ldots, n$.
- The Fuzzy Vector Median is $\mathbf{X}_k$.

When the Euclidean fuzzy distance is used the Euclidean FVM is defined. Similarly to the classical vector median, the Euclidean FVM $\mathbf{X}_{\text{FVM}}$ does not minimize the unconditional expression:

$$\sum_{i=1}^{n} D_{e_n}[\mathbf{X}_i, \mathbf{Y}] \tag{29}$$

but, by definition, it minimizes the same expression, when $\mathbf{Y}$ is chosen to be one of $\mathbf{X}_i$.

1.6 Fuzzy vector median application in image filtering

Vector median filters are used to remove impulses from noisy color images. In the following, we present such a filtering application of the FVM filters for image *Lenna* (256×256 pixels), corrupted by mixed impulsive and Gaussian noise. The FVM is applied using fuzzy Euclidean distance and a 3×3 window. Fuzziness is inserted to the problem by using the information of the neighbouring pixels. The chromatic RGB values of the nine pixels of each window are ranked and the average of the differences, between each pixel chromatic value and its two closest values, is used as a measure of the pixel fuzziness. The fuzziness of a pixel changes as the window moves, and when the size of the window is modified. By using this definition of fuzziness, the Fuzzy Vector Median filter removes the impulsive noise, preserves the edges and reduces the local variances of the filtered image in homogeneous regions.

Figure 2a shows the original image *Lenna*. This image corrupted by mixed impulsive (percentage $p = 0.2$) and Gaussian (standard deviation $s = 10$) noise is shown in Fig. 2b. Figure 2c shows the result of the FVM filter applied on the corrupted image. The filtered image has very good perceptual quality.

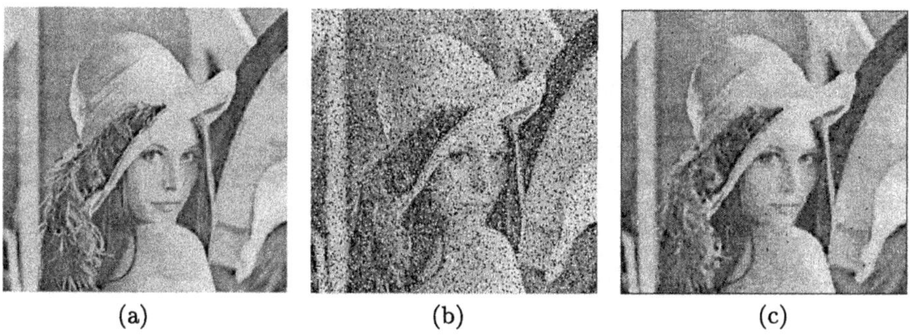

Fig. 2. (a) The original image *Lenna*. (b) The noisy image corrupted by impulsive (p=0.2) and Gaussian (s=10) noise. (c) The FVM filtered image.

2 Fuzzy cell Hough transform

2.1 Description of the classical Hough transform

The Hough Transform (HT) [11] is of fundamental importance for many applications in image analysis and computer vision, since it can be used to detect curves described by a number of parameters. The main disadvantage of HT is its large storage and computational time requirements that depend on the number of parameters used and on the split of the parameter space [12,13]. Many HT variations have been proposed in the literature, mainly regarding the reduction of its computational complexity [14]-[19]. Most of them can be modified to incorporate the use of fuzzy cells that will be described subsequently.

Let us suppose that the curve to be detected can be described by an equation with p parameters as follows:

$$f(a_1, a_2, \dots, a_p, x, y) = 0. \tag{30}$$

Using the classical HT, each parameter direction is limited to a range of values and each range is split into N_{a_i} intervals. Therefore, the parameter space is split into $N_{a_1} \times N_{a_2} \times \dots \times N_{a_p}$ crisp cells. An accumulator array M with $N_{a_1} \times N_{a_2} \times \dots \times N_{a_p}$ elements is needed for the calculation of the classical HT. The voting process is performed on a p-dimensional hypersurface for every contour point (x_j, y_j), described by the following equation:

$$M(a_1, a_2, \dots, a_p) = \sum_j v_{a_1, a_2, \dots, a_p}(x_j, y_j), \tag{31}$$

where:

$$v_{a_1, a_2, \dots, a_p}(x_j, y_j) = \begin{cases} 1 & \text{if } f(a_1, a_2, \dots, a_p, x_j, y_j) = 0 \\ 0 & \text{elsewhere.} \end{cases} \tag{32}$$

Then, the cell with the maximum value M corresponds to the detected curve. Local maxima in the array correspond to less significant curves which can also be detected. The computational complexity of the HT depends heavily on the number of parameters p and on the number of partitions in each parameter, something that also effects the accuracy of the detection.

When an image containing a given contour is corrupted by noise, some contour points are moved away from the actual contour. If this deviation is significant, the contour points may not contribute to the cell which corresponds to the actual contour during the voting process of the HT. The contour points that are distant from the actual contours may cause local maxima in the accumulator array and lead to incorrect curve detections.

The quantization steps of the parameters affect the tolerance of the Hough Transform when it is applied in noisy images. If small cells are used, relatively small deviation of contour points may throw them out of the cell which contributes to the actual curve. On the contrary, if large cells are used, the erroneous contour points may still contribute to the actual curve but the accuracy of the detection may not be satisfactory.

2.2 Definitions of fuzzy cells

Fuzzy Cell Hough Transform (FCHT) is based on a fuzzy split of the Hough Transform parameter space in fuzzy cells [20]. The use of fuzzy cells reduces false detections and improves the accuracy of the curve estimations, especially in noisy images. Accuracy can be traded with computational time, if needed.

Fuzzy Cell Hough Transform uses a fuzzy split of the Hough Transform parameter space which leads to a fuzzy voting process. The parameter space is split into fuzzy cells, which are considered as fuzzy vectors with p-dimensional membership functions $\mu(a_1, a_2, \dots, a_p)$. The values of the membership functions are limited in the range $[0, 1]$. The interval of confidence of each membership function can intersect with the neighboring cell intervals. Each contour point (x_j, y_j) in the spatial domain contributes to more than one fuzzy cell in the parameter space. Moreover, the contribution of each point is not constant, but depends on the value of the cell membership function at the specific transformed point $(a_1, a_2, \dots, a_p)$. Then, the contributions are added in an accumulator array. Finally, the local maxima in the array are found and the corresponding contours are detected.

Straight lines in an image are usually detected using the polar parameter space of Hough Transform. The transform equation is given by:

$$\rho = x \cos\theta + y \sin\theta, \tag{33}$$

where the parameter $\theta \in [0, \pi)$ is an angle, $\rho \in [-\frac{\sqrt{2}N}{2}, \frac{\sqrt{2}N}{2}]$ is the algebraic distance to the origin and $N \times N$ is the image size as shown in Fig. 3a. Then, the straight line given by Eq. (33) is transformed to a point (ρ, θ) in the parameter space, as shown in Fig. 3b.

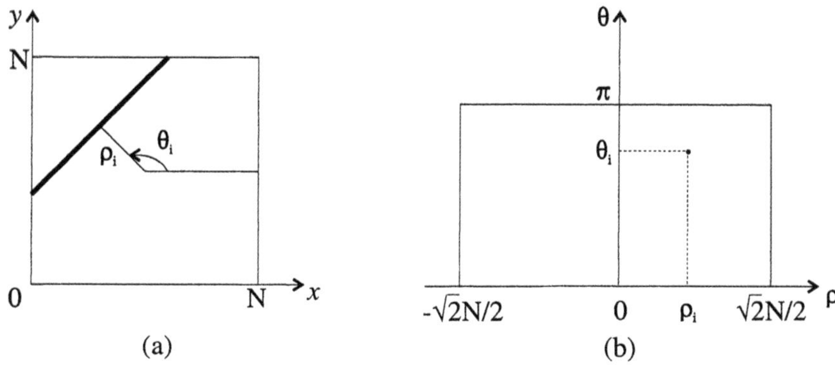

Fig. 3. (a) A straight line in an image $N \times N$. (b) The parameter space ρ, θ and the transformed straight line ρ_i, θ_i.

In classical HT the parameters θ, ρ are split in N_θ and N_ρ crisp sets (intervals) respectively, symbolized as Θ_i, P_i. The crisp cell C_{ij} where classical Hough Transform is applied, can now be defined as:

$$C_{ij} = \{(\rho, \theta), \rho \in P_i, \theta \in \Theta_j\}. \tag{34}$$

Thus, a couple (ρ, θ) belongs to a cell C_{ij}, if $\rho \in P_i$ and $\theta \in \Theta_j$, as shown in Fig. 4a.

Fuzzy cells are defined when one or more parameters are split in fuzzy sets. By using the same number of partitions for each parameter, the fuzzy sets in the θ coordinate are defined by the following equation:

$$\Theta_j^f = \left\{ \left(\theta, \mu_{\Theta_j^f}(\theta)\right), \theta \in \mathbb{R} \right\}, \tag{35}$$

where $\mu_{\Theta_j^f}(\theta)$ is a membership function. The interval of confidence of each fuzzy set Θ_j^f is assumed to be the crisp set $(l_{\Theta_j^f}, r_{\Theta_j^f})$. This fuzzy split of the parameter θ provides the ability to use overlapped sets with different membership values for every θ. Then, by using this fuzzy split in θ coordinate, a fuzzy-θ cell can be defined as a fuzzy vector:

$$\mathbf{C}_{ij}^\theta = \left\{ \left((\rho, \theta), \mu_{\mathbf{C}_{ij}^\theta}(\rho, \theta)\right), (\rho, \theta) \in \mathbb{R}^2 \right\}, \tag{36}$$

where:

$$\mu_{\mathbf{C}_{ij}^\theta}(\rho, \theta) = \begin{cases} \mu_{\Theta_j^f}(\theta) & \text{if } \rho \in P_i \\ 0 & \text{elsewhere.} \end{cases} \tag{37}$$

The same technique can be used to split ρ coordinate domain and define fuzzy-ρ cells $\mathbf{C}_{ij}^\rho$. If the fuzzy split in two coordinates is combined, then a fuzzy-$\rho\theta$ cell $\mathbf{C}_{ij}^{\rho\theta}$ is defined. Its membership function is given by:

$$\mu_{\mathbf{C}_{ij}^{\rho\theta}}(\rho,\theta) = \begin{cases} \min\left\{\mu_{P_i^f}(\rho), \mu_{\Theta_j^f}(\theta)\right\} & \text{if } \rho \in P_i \text{ and } \theta \in \Theta_j \\ 0 & \text{elsewhere.} \end{cases} \tag{38}$$

Examples of fuzzy cells are shown in Fig. 4 when triangular membership functions are used.

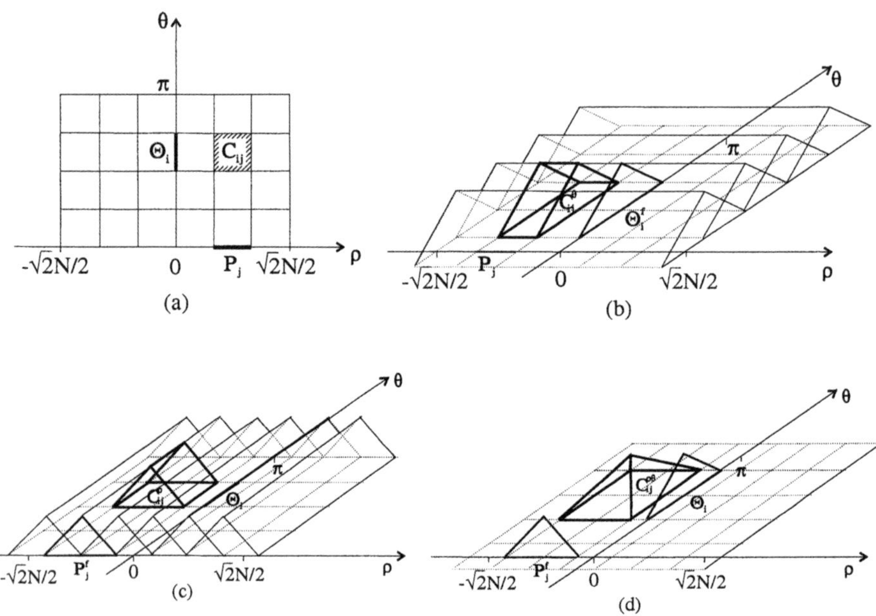

Fig. 4. (a) A classical split of the parameter space ρ, θ. (b) Fuzzy split of θ coordinate of the parameter space and the corresponding fuzzy-θ cells. (c) Fuzzy split of ρ coordinate of the parameter space and the corresponding fuzzy-ρ cells. (d) Fuzzy split of both coordinates ρ, θ and the corresponding fuzzy-$\rho\theta$ cells.

Circles in an $N \times N$ image can also be detected by using the conventional HT. The parameters used are the coordinates of the circle center a, b and the radius r. When HT is applied, the three dimensional parameter space is split in $N_a \times N_b \times N_r$ cells. The crisp cell C_{ijk} is defined by:

$$C_{ijk} = \{(a,b,r), a \in A_i, b \in B_j, r \in R_k\} \tag{39}$$

where A_i, B_j and R_k are crisp sets. In order to define fuzzy cells, each point, which belongs to the interval of confidence of the fuzzy cell, corresponds to a value in the range $[0,1]$ through its membership function. The fuzzy sets in coordinate a can be defined as $A_i^f = \{(a, \mu_{A_i^f}(a)), a \in \mathbb{R}\}$, where $\mu_{A_i^f}(a)$ is a membership function. Then, by using this fuzzy split, a fuzzy-a cell $\mathbf{C}^a_{ijk}$ can be defined as a 3-dimensional fuzzy vector:

$$\mathbf{C}^a_{ijk} = \left\{ \left((a,b,r), \mu_{\mathbf{C}^a_{ijk}}(a,b,r) \right), (a,b,r) \in \mathbb{R}^3 \right\}, \tag{40}$$

where

$$\mu_{\mathbf{C}^a_{ijk}}(a,b,r) = \begin{cases} \mu_{A^f_i}(a) & \text{if } b \in B_j \text{ and } r \in R_k \\ 0 & \text{elsewhere.} \end{cases} \tag{41}$$

The same technique can be used to split b and r coordinate in fuzzy sets and define the corresponding fuzzy-b $\mathbf{C}^b_{ijk}$ and fuzzy-r $\mathbf{C}^r_{ijk}$ cells as 3-dimensional fuzzy vectors. If the fuzzy split in two coordinates, for example a and b coordinates, are combined, then a fuzzy-ab cell can be defined as the fuzzy vector $\mathbf{C}^{ab}_{ijk}$ with membership function:

$$\mu_{\mathbf{C}^{ab}_{ijk}}(a,b,r) = \begin{cases} \min \left\{ \mu_{A^f_i}(a), \mu_{B^f_j}(b) \right\} & \text{if } r \in R_k \\ 0 & \text{elsewhere.} \end{cases} \tag{42}$$

Similarly, fuzzy-ar and fuzzy-br cells can be defined. Finally, if the fuzzy splits in three coordinates are combined, then a fuzzy-abr cell can be defined as the fuzzy vector $\mathbf{C}^{abr}_{ijk}$ with membership function:

$$\mu_{\mathbf{C}^{abr}_{ijk}}(a,b,r) = \min \left\{ \mu_{A^f_i}(a), \mu_{B^f_j}(b), \mu_{R^f_k}(r) \right\}. \tag{43}$$

The concept of a fuzzy cell can be easily generalized to p dimensions.

2.3 Description of the fuzzy voting processes

Let us assume that a pixel x, y is a contour point in an image, where straight lines will be detected and that the parameter space is split into $N_\theta \times N_\rho$ fuzzy-ρ cells $\mathbf{C}^\rho_{kl}$. By following the same process as in conventional Hough Transform, the centers θ_i of the crisp sets Θ_i are used to compute the distances ρ_i by using the following equation:

$$\rho_i = x \cos \theta_i + y \sin \theta_i. \tag{44}$$

Each couple (θ_i, ρ_i) belongs to more than one fuzzy cells $\mathbf{C}^\rho_{kl}$. The corresponding elements of the accumulator array $M(k,l)$ are increased by the values $V_{kl}(x,y) = \mu_{\mathbf{C}^\rho_{kl}}(\theta_i, \rho_i)$. The pixels of an image that vote in a certain fuzzy-ρ cell and the corresponding voting values are shown in Fig. 5a. Finally, the local maxima in the array M must be detected. The array that is created after the fuzzy voting process is smoother than in the crisp case. Thus, the fuzzy voting process reduces the local maxima which correspond to the effect of noise in an image. The straight lines can now be detected with better accuracy. However, this method slightly increases the computation time.

When fuzzy-$\rho\theta$ cells are used, the parameter θ is also split into fuzzy sets Θ_i^f as in Eq. (35). The parameter space is split into fuzzy-$\rho\theta$ cells $\mathbf{C}_{kl}^{\rho\theta}$ defined in Eq. (38). The fuzzy numbers Θ_i^f are used to compute the fuzzy distances P_i by using the following fuzzy equation:

$$P_i = x\mathrm{Cos}\Theta_i^f + y\mathrm{Sin}\Theta_i^f, \tag{45}$$

where Cos, Sin are the fuzzy extensions of the crisp functions cos, sin and the addition and multiplications are supposed to be fuzzy operators [4,5,21]. Then the inclusions of each fuzzy couple (Θ_i^f, P_i^f) in every fuzzy cell $\mathbf{C}_{kl}^{\theta\rho}$ have to be computed. Since Θ_i^f is equal to Θ_k^f, the inclusion of each fuzzy couple to a fuzzy cell is simplified to the inclusion of the computed fuzzy number P_i^f in the fuzzy set P_l^f, given by the following equation [21]:

$$\mu_{T_{il}}(u) = \sup_{\rho: u=\mu_{P_l^f}(\rho)} \mu_{P_i^f}(\rho), \qquad u \in [0,1]. \tag{46}$$

The compatibilities T_{il} of each fuzzy couple (Θ_i^f, P_i^f) in every fuzzy cell $\mathbf{C}_{kl}^{\theta\rho}$ are computed and added to form the corresponding accumulator array $M^f(k,l)$. Note that, since the compatibilities are fuzzy numbers with intervals of confidence in $[0,1]$, the array M^f is a two dimensional array of fuzzy numbers and the additions of compatibilities are additions of fuzzy numbers. When the voting process is completed, the fuzzy local maxima are detected. The pixels of an image that vote to a certain fuzzy-$\rho\theta$ cell and the corresponding voting values are shown in Fig. 5b.

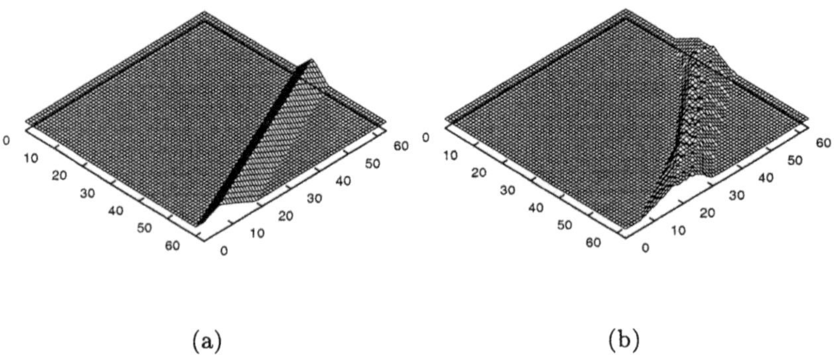

Fig. 5. The voting values of every pixel of a 64×64 image to: (a) a certain fuzzy-ρ cell and (b) a certain fuzzy-$\rho\theta$ cell.

Let us assume that a pixel (x, y) is a contour point in an image where circles will be detected and that the parameter space is split into fuzzy-r

cells $\mathbf{C}^r_{lmn}$. By following a similar procedure to the two parameter case, the centers a_i of the crisp sets A_i and the centers b_j of the crisp sets B_j are used to compute the distances r_{ij} by solving the equation:

$$(x - a_i)^2 + (y - b_j)^2 = r_{ij}^2. \tag{47}$$

Each point (a_i, b_j, r_{ij}) belongs to more than one fuzzy cell. The corresponding elements of the accumulator array $M(l, m, n)$ are increased by the values $V_{lmn}(x, y) = \mu_{\mathbf{C}^r_{lmn}}(a_i, b_j, r_{ij})$. The pixels of an image that vote to a certain fuzzy-ρ cell and the corresponding voting values are shown in Fig. 7a. Finally, the local maxima in the array M have to be detected. The voting process is similar when fuzzy-a or fuzzy-b cells are used. By using fuzzy instead of crisp cells, the accumulator array is smoother and the number of local maxima is reduced. An example of a two coordinate (a, r) accumulator array that is created after a classical voting process and after a fuzzy voting process is shown in Figs. 6[1]a, b respectively, when the methods are applied on an artificially generated image with one circle in it. Clearly the fuzzy approach produces less noise in the parameter space. When fuzzy-ab cells $\mathbf{C}^{ab}_{lmn}$ are

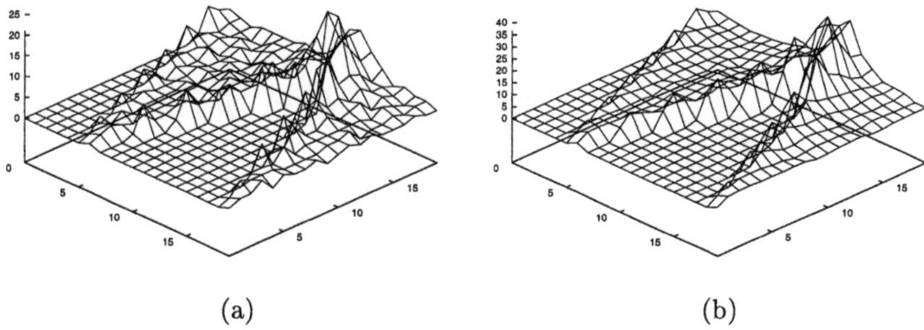

(a) (b)

Fig. 6. Two coordinate (a, r) domain of a three dimensional accumulator array after a classical voting process (a) and after a fuzzy voting process (b).

used defined as in Eq. (42), the voting process is more complicated but still similar to the one used on fuzzy-$\rho\theta$ cells for the detection of straight lines. The fuzzy numbers B_j^f and the crisp centers r_k of the crisp sets R_k are used to compute the fuzzy distances A_{jk}^f by solving the following fuzzy equation:

$$\left(x - A_{jk}^f\right)^2 + \left(y - B_j^f\right)^2 = r_k^2, \tag{48}$$

where all the operators are the extended fuzzy operators, through the extension principle. The first root of Eq. (48) is the fuzzy number A_{jk}^f, which is

[1] Figures 6, 8 and 9 were reprinted from Pattern Recognition, Vol. 30, V.Chatzis and I.Pitas, Fuzzy Cell Hough Transform for Curve Detection, pp. 2031-2042, Copyright 1997, with permission from Elsevier Science.

symbolized as:

$$A_{jk}^f = \bigcup_\alpha \alpha \cdot \left[a_{jk_l}^{(\alpha)}, a_{jk_r}^{(\alpha)}\right], \qquad \alpha \in [0,1] \tag{49}$$

and calculated by the equations:

$$a_{jk_l}^{(\alpha)} = x - \sqrt{r_k^2 - \min\left\{\left(y - b_{j_r}^{(\alpha)}\right)^2, \left(y - b_{j_r}^{(\alpha)}\right)\left(y - b_{j_l}^{(\alpha)}\right), \left(y - b_{j_l}^{(\alpha)}\right)^2\right\}}, \tag{50}$$

$$a_{jk_r}^{(\alpha)} = \begin{cases} x - \sqrt{r_k^2 - \max\left\{\left(y - b_{j_r}^{(\alpha)}\right)^2, \left(y - b_{j_l}^{(\alpha)}\right)^2\right\}} \\ \qquad \text{if } r_k^2 \geq \max\left\{\left(y - b_{j_r}^{(\alpha)}\right)^2, \left(y - b_{j_l}^{(\alpha)}\right)^2\right\} \\ x - \dfrac{r_k^2 - \max\left\{\left(y - b_{j_r}^{(\alpha)}\right)^2, \left(y - b_{j_l}^{(\alpha)}\right)^2\right\}}{\sqrt{r_k^2 - \min\left\{\left(y - b_{j_r}^{(\alpha)}\right)^2, \left(y - b_{j_r}^{(\alpha)}\right)\left(y - b_{j_l}^{(\alpha)}\right), \left(y - b_{j_l}^{(\alpha)}\right)^2\right\}}} \\ \qquad \text{if } r_k^2 < \max\left\{\left(y - b_{j_r}^{(\alpha)}\right)^2, \left(y - b_{j_l}^{(\alpha)}\right)^2\right\}. \end{cases} \tag{51}$$

The second root can be calculated by similar equations. Then, the inclusions of each fuzzy triple (A_{jk}^f, B_j^f, r_k) in every fuzzy cell $\mathbf{C}_{lmn}^{ab}$ have to be computed. They can be simplified to the inclusion of the computed fuzzy number A_{jk}^f in the fuzzy set A_l^f given by the following equation [21]:

$$\mu_{T_{jkl}}(u) = \sup_{a: u = \mu_{A_l^f}(a)} \mu_{A_j^f k}(a), \qquad u \in [0,1]. \tag{52}$$

The compatibilities T_{jkl}, which are fuzzy numbers, are added to form the corresponding accumulator array $M^f(l,m,n)$ of fuzzy numbers. The pixels of an image that vote to a certain fuzzy-*ab* cell and the corresponding voting values are shown in Fig. 7b. Fuzzy-*abr* cells can also be used to insert fuzziness to the detection of the centers and the radii of the circles. The voting process is similar and equations similar to Eqs. (50) and (51) can be used.

2.4 Simulation results

In the following, simulation results of the use of Fuzzy Cell Hough Transform (FCHT) for straight line and circle detection will be presented. The polar parameters θ and ρ were used for straight lines detection. The intervals $\theta \in [0, 2\pi)$ and $\rho \in [10, 70]$ where the parameters were restricted, were split in 30

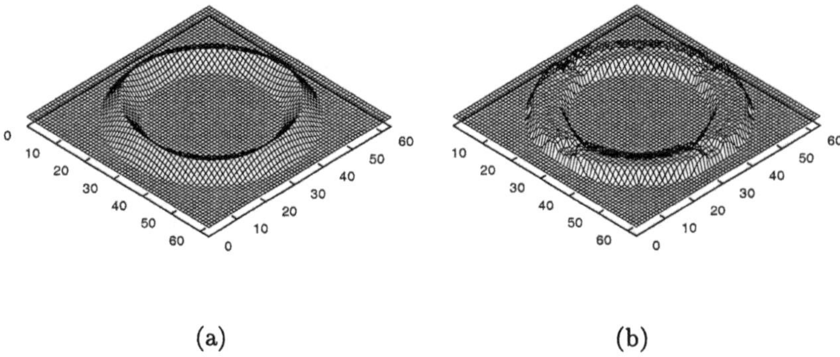

Fig. 7. The voting values of every pixel of a 64×64 image to: (a) a certain fuzzy-r cell. (b) a certain fuzzy-ab cell.

partitions. In the FCHT case, only the ρ parameter was fuzzily split and the fuzzy sets were chosen to have triangular slope.

Artificially generated images with five straight lines at random places were used. The images were corrupted by uniform noise having range $\pm d$ pixels added to the y-axis of a contour point (x, y). An example of such an image corrupted by uniform noise with range $d = \pm 10$ pixels is shown in Fig. 8a. Five straight lines were detected by HT and FCHT in each image. Figure 8b shows the straight lines that were detected by classical HT, when Fig. 8a was used as an input image. Figure 8c shows the corresponding straight lines that were detected by FCHT. In this example, FCHT detects correctly the four of the five straight lines, whereas the classical HT detects correctly only the two straight lines.

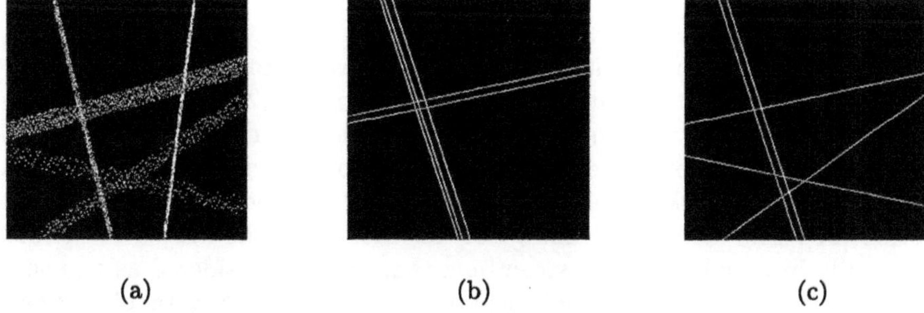

Fig. 8. An example of straight line detection. (a) An artificially generated 256×256 image with five straight lines, corrupted by uniform noise with range $d = \pm 10$ pixels. (b) The detected lines by using classical HT. (c) The detected lines by using FCHT.

Regarding circle detection, the parameters were restricted in the intervals $a \in [-128, 128]$, $b \in [-128, 128]$ and $r \in [10, 70]$, which were split into 30 partitions. In the FCHT case, only the r parameter was split in a fuzzy way. Fuzzy sets were chosen to have triangular slope.

Artificially generated images with five circles at random positions were used. The images were corrupted by uniform noise having range $\pm d$ pixels added to the ρ-coordinate of a contour point (x, y). An example of such an image corrupted by uniform noise with range $d = \pm 10$ pixels is shown in Fig. 9a. Five circles were detected by HT and FCHT in each image. Figure 9b shows the circles that were detected by classical HT, when Fig. 9a was used as an input image. Figure 9c shows the corresponding circles that were detected by FCHT. In this example, FCHT detects correctly the four of the five circles, whereas the classical HT detects correctly only the two circles.

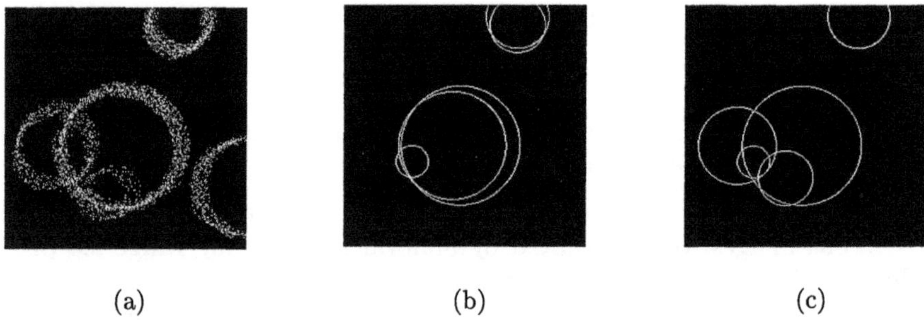

(a) (b) (c)

Fig. 9. An example of circle detection. (a) An artificially generated 256×256 image with five circles, corrupted by uniform noise with range $d = \pm 10$ pixels. (b) The detected circles by using classical HT. (c) The detected circles by using FCHT.

3 Generalized fuzzy mathematical morphology

3.1 Introduction

Mathematical morphology is a very rich and powerful tool used for the representation and analysis of binary and grayscale images [22]-[26]. The morphological image representation has been used for the description of the geometrical characteristics of image objects as well as for binary image compression. The morphological skeleton is the very popular method for shape representation [27]. The main disadvantage of this method is the lack of robustness especially in impulsive noise, something that can be considered as a general characteristic of all the morphological operations [27,28]. Several efforts have been done to reduce the sensitivity of morphological operations to impulses. Soft morphological operations have been proposed in [29], which can restrict the problem in some cases. The properties of soft mathematical morphology have been investigated in [30,31]. Fuzzy morphological operations have also been proposed and investigated [32]-[37]. Yet the fuzzy and soft morphologies are not efficient in binary object representation, since such a skeletonization still has strong dependency on the presence of outliers.

This section describes the Generalized Fuzzy Mathematical Morphology (GFMM). It is based on a Fuzzy Inclusion Indicator (FII), which is a fuzzy set used as a measure of the inclusion of a fuzzy set into another. The FII obeys a set of axioms, extensions of the known axioms that any inclusion indicator should obey. Each axiom corresponds to desirable characteristics that any mathematical morphology operation should have. The Binary and Grayscale mathematical morphologies can be considered as special cases of the proposed GFMM. The GFMM provides a very powerful and flexible tool in image analysis and filtering. An application to robust skeletonization of 2D and 3D objects will be presented in the following.

3.2 Relations between fuzzy sets

Several definitions can been found in the literature to extend the classical crisp set relation *subset* ($\subset$) and operations *union* ($\cup$) and *intersection* ($\cap$) to fuzzy sets. The most frequently used definitions follow. Let A, B be two fuzzy sets with membership functions μ_A, μ_B. Their union and intersection are the fuzzy sets $A \cup B$ and $A \cap B$ with membership functions $\mu_{A\cup B}$ and $\mu_{A\cap B}$ given by:

$$\mu_{A\cup B}(x) = \max\{\mu_A(x), \mu_B(x)\}, \tag{53}$$

$$\mu_{A\cap B}(x) = \min\{\mu_A(x), \mu_B(x)\}. \tag{54}$$

A fuzzy set A is considered to be subset of the fuzzy set B if and only if the membership degree $\mu_A(x)$ of A is less than or equal to the membership degree $\mu_B(x)$ of B for all x:

$$A \subset B \Leftrightarrow \mu_A(x) \leq \mu_B(x), \qquad \forall x, \tag{55}$$

or equivalently by using the α-cuts:

$$A \subset B \Rightarrow A^{(\alpha)} \subseteq B^{(\alpha)}, \qquad \forall \alpha \in [0,1]. \tag{56}$$

The complement of a fuzzy set A is usually defined as the fuzzy set A^c having membership function:

$$\mu_{A^c}(x) = 1 - \mu_A(x). \tag{57}$$

The translation of a fuzzy set A by a crisp vector v is the fuzzy set $T(A; v)$ with membership function:

$$\mu_{T(A;v)}(x) = \mu_A(x - v). \tag{58}$$

Similarly, the centrally symmetric fuzzy set $-A$ has membership function $\mu_{-A}(x) = \mu_A(-x)$.

3.3 The fuzzy inclusion indicator

An *inclusion indicator* or *inclusion grade operator* I measures the belief in the proposition "A is a subset of B", where A, B are fuzzy sets. It has been defined in the literature as a two argument function $I(A, B)$ that maps any pair of fuzzy sets A, B into the interval [0,1], and satisfies a number of axioms (properties) [32–34]:

- A1: $I(A, B) = 0$ if and only if $\{x : \mu_A(x) = 1\} \cap \{x : \mu_B(x) = 0\} \neq \emptyset$,
- A2: $B \subset C \Rightarrow I(A, B) \leq I(A, C)$,
- A3: $B \subset C \Rightarrow I(C, A) \leq I(B, A)$,
- A4: $I(A, B) = I(T(A; v), T(B; v))\ \forall v$, $I(A, B) = I(-A, -B)$,
- A5: $I(A, B) = I(B^c, A^c)$,
- A6: $I(\bigcup_i B_i, A) = \inf_i I(B_i, A)$,
- A7: $I(A, \bigcap_i B_i) = \inf_i I(A, B_i)$,
- A8: $I(A, \bigcup_i B_i) \geq \sup_i I(A, B_i)$.

The above mentioned axioms are not independent, since, for example, A3 can be derived by applying A5 to A2 and A7 can be derived by applying A5 to A6. Equivalences between these nine axioms and the desired properties of a fuzzy mathematical morphology are extensively investigated in [34]. It is just mentioned here that A4 is strongly related to the translation invariance property, A5 to the principle of duality, A1 and A9 lead to the compatibility of the constructed fuzzy mathematical morphology with the binary and grayscale one, A2 and A3 are related to the increasing (decreasing) property of erosion with respect to the reference (structuring element) set, and A6, A7 and A8 are related to the compatibility with union and intersection in classical morphology.

In the following, a generalized fuzzy mathematical morphology (GFMM) will be constructed based on the definition of an inclusion indicator as a fuzzy set rather than a number, defined in the interval $[0, 1]$. Large membership values of the inclusion indicator fuzzy set will strengthen the belief that "A is a subset of B". The use of a fuzzy inclusion indicator in the construction of a fuzzy mathematical morphology guarantees that fuzzy mathematical operations on fuzzy signals result to fuzzy signals.

A *fuzzy inclusion indicator* (FII) will be any two argument function $\mathbf{I}$, that maps a pair of fuzzy sets A, B into another fuzzy set with domain of definition the interval $[0, 1]$, and satisfies the following axioms:

- F1: $\mathbf{I}(A, B) = O$ if and only if $\forall \alpha \in [0, 1]$,
  $\{x : \mu_A(x) = \alpha\} \cap \{x : \mu_B(x) = 0\} \neq \emptyset$,
- F2: $B \subset C \Rightarrow \mathbf{I}(A, B) \subset \mathbf{I}(A, C)$,
- F3: If A is convex, then $B \subset C \Rightarrow \mathbf{I}(C, A) \subset \mathbf{I}(B, A)$,
- F4: $\mathbf{I}(A, B) = \mathbf{I}(T(A; v), T(B; v)) \qquad \forall v$,
- F5: In the general case: $\mathbf{I}(\bigcup_i B_i, A) \supset \bigcap_i \mathbf{I}(B_i, A)$,
  If A is convex then: $\mathbf{I}(\bigcup_i B_i, A) = \bigcap_i \mathbf{I}(B_i, A)$,

- F6: $\mathbf{I}(A, \bigcap_i B_i) = \bigcap_i \mathbf{I}(A, B_i)$,
- F7: $\mathbf{I}(A, \bigcup_i B_i) \supset \bigcup_i \mathbf{I}(A, B_i)$,

where O is the zero fuzzy set with membership function $\mu_O(x) = 0$ defined on its interval of confidence. These seven axioms F1 - F7 are extensions of the previously reported axioms A1 - A4 and A6 - A8 that are fundamental for the construction of any fuzzy mathematical morphology. Thus, the GFMM that will be constructed based on these seven axioms will have the desired properties of any mathematical morphology operation: translation invariance (F4), compatibility with the binary and grayscale mathematical morphology (F1), increasing (or decreasing) property of erosion with respect to the reference (structuring element) set (F2, F3), compatibility with union and intersection in classical morphology (F5, F6, F7). The principle of duality (A5) will be taken into account in the construction of the GFMM, by defining the dilation as the dual operation of the erosion.

An operation that is similar to the extension principle and defines the *compatibility index* τ will help us to define the fuzzy inclusion index of a fuzzy set A with respect to another fuzzy set B. The compatibility index has been defined as [21]:

$$\mu_\tau(u) = \sup_{x:\mu_A(x)=u} \mu_B(x), \qquad \forall u \in [0,1]. \tag{59}$$

However, since the compatibility index applies a $T - conorm$ (sup) operator, it cannot be used for the definition of an erosion operation. We can reach the same conclusion, by observing that the compatibility index does not satisfy the axioms F5 and F6.

The fuzzy inclusion indicator is defined by applying a $T - norm$ (inf) operator on the operation given in Eq. (59) as follows: Let A, B be two fuzzy sets. A fuzzy inclusion indicator $\mathbf{I}(A, B)$ is the fuzzy set with membership function:

$$\mu_{\mathbf{I}(A,B)}(u) = \inf_{x:\mu_A(x)=u} \mu_B(x), \qquad \forall u \in [0,1]. \tag{60}$$

If there is no x such that $\mu_A(x) = u$, then the inclusion index is *not* defined for the certain u. However, the fuzzy set A corresponds to the structuring element of the mathematical morphology and can be selected by the user. Thus, it can be selected to define a fuzzy inclusion indicator for all $u \in [0,1]$. The proposed fuzzy inclusion indicator given in Eq. (60) satisfies the axioms F1 - F7 [38].

The main difference between the proposed fuzzy inclusion indicator and other known indicators [32–34,36,39–41] is that the proposed indicator is generally a fuzzy set defined in $[0,1]$ instead of a crisp number in $[0,1]$. Although this characteristic provides fuzziness propagation through the inclusion operation, a crisp value could be desirable in some cases. Then, a defuzzification process should be followed, symbolized as $D(\mathbf{I})$. The resulting crisp value of inclusion I should vary from 0 to 1:

$$I(A, B) = D(\mathbf{I}(A, B)). \tag{61}$$

A variety of defuzzification processes have been proposed in the literature [5,21,42]. The most usual ones are the center of gravity method, the center of maxima method and the mean of maxima method as well as generalizations or weighted versions of them.

Some examples of fuzzy inclusion indicators between fuzzy sets and crisp sets are given in Fig. 10. We can observe in Fig. 10a that the domain of definition of the inclusion operation is set by the included set A (structuring element). Subsets of A, defined from its membership function through the equation $x : \mu_A(x) = \alpha$, provide the subdomains where the inf operator is applied on the values of the membership function $\mu_B(x)$ of the reference set B (signal or image). The correspondence of the values α to the values $\inf\{\mu_B(x)\}$ constructs the fuzzy set that is the fuzzy inclusion indicator $\mathbf{I}(A, B)$. It is also shown in Fig. 10b that the fuzzy inclusion of a crisp set into a fuzzy set is the minimum membership value of B in the crisp set, corresponding to $u = 1$. It is also easy to show that the fuzzy inclusion of a crisp set or a crisp value into a crisp set are extensions of classical set theory relations.

3.4 The generalized fuzzy mathematical morphology definition(GFMM)

Based on the fuzzy inclusion indicator that was previously defined in Eq. (60), the Generalized Fuzzy Mathematical Morphology framework will be constructed. In the following, the four basic operations of any morphology, erosion, dilation, opening and closing will be defined. Let A, B be two fuzzy sets. The erosion of the fuzzy set A by the *fuzzy structuring element* B, is defined as the fuzzy set $\mathcal{E}(A, B)$ having membership function given by :

$$\mu_{\mathcal{E}(A,B)}(x) = \mu_{\mathbf{I}(T(B;x),A)}(D(\mathbf{I}(T(B;x), A)), \tag{62}$$

where $\mathbf{I}$ denotes the fuzzy inclusion indicator and D is a defuzzification procedure. The dilation $\mathcal{D}(A, B)$ is then defined to obey to the duality principle :

$$\mathcal{D}(A, B) = \mathcal{E}(A^c, -B)^c, \tag{63}$$

where $-B$ denotes the centrally symmetric of the fuzzy structuring element B and A^c the complement of the fuzzy set A as in Eq. (57). Opening and closing are defined in terms of erosion and dilation:

$$\mathcal{O}(A, B) = \mathcal{D}(\mathcal{E}(A, B), B), \tag{64}$$

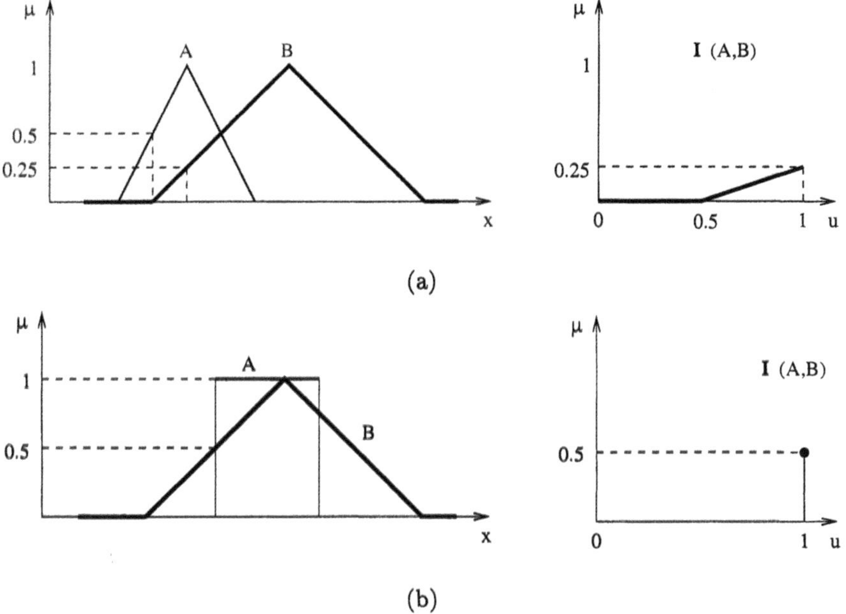

Fig. 10. Examples of fuzzy inclusion indicators, (a) of two fuzzy sets and (b) of a crisp and a fuzzy set. (©2000 IEEE)

$$\mathcal{C}(A,B) = \mathcal{E}(\mathcal{D}(A,-B),-B). \tag{65}$$

It can be observed in Eq. (62) that the membership value of the eroded set is the value of the fuzzy inclusion indicator membership function at the point resulting from the defuzzification procedure of the fuzzy inclusion indicator. Since the fuzzy inclusion indicator membership function takes its values from the membership function of the reference set, the eroded set takes also values coming from the reference set. The neighbor of the reference set that takes part in the erosion procedure is defined by the fuzzy structuring element. The above mentioned characteristics are similar with the classical morphological ones. The fuzziness of the structuring element forms sets with equal membership, which are responsible for the topological characteristics and special features of the corresponding morphological operations.

In the following, it will be shown that the classical Binary Mathematical Morphology (BMM) can be considered as a special case of the proposed Generalized Fuzzy Mathematical Morphology. Let A_c, B_c be two classical sets defined in a subset of R^n and any $x \in A_c$ correspond to a value in $\{0,1\}$ through a function $f_{A_c}(x)$. The above sets are considered as a reference and a structuring element set respectively of the classical BMM. In the GFMM framework, the bi-valued reference set A is equivalent to a fuzzy set with a bi-valued membership function $\mu_A(x) = f_{A_c}(x)$. Let also B be a fuzzy set defined in the same domain as the classical set B_c, having a membership value $\mu_B(x) = 1, \forall x$ in the domain of definition. For this special case of A, B, the fuzzy inclusion indicator membership function given in Eq. (60) is simplified as follows:

$$\mu_{\mathbf{I}(B,A)}(u) = \begin{cases} 0 & \text{if } \exists x \in B : \mu_A(x) = 0 \\ 1 & \text{otherwise,} \end{cases} \tag{66}$$

i.e., it is defined only in the set $u = 1$. Then, any defuzzification process in Eq. (62) leads to $D = 1$ and the corresponding fuzzy erosion operation is simplified to:

$$\mu_{\mathcal{E}(A,B)}(x) = \begin{cases} 0 & \text{if } \exists y \in T(B;x) : \mu_A(y) = 0 \\ 1 & \text{otherwise,} \end{cases} \tag{67}$$

which is equivalent to the definition of the classical binary erosion.

The Grayscale Mathematical Morphology (GMM) assumes that the reference set A_c takes values through a function $f_{A_c}(x)$, in a certain convex set. By normalizing the values of the convex set in $[0, 1]$, any membership function of a certain fuzzy set A can substitute the function f. The fuzzy structuring element is defined as in the previous case. For this special case of A, B, the fuzzy inclusion indicator membership function given in Eq. (60) is simplified as follows:

$$\mu_{\mathbf{I}(B,A)}(u) = \inf_{x \in B} \mu_A(x), \qquad u = 1, \tag{68}$$

i.e., it is defined only in the set $u = 1$. Then, any defuzzification process in Eq. (62) leads to $D = 1$ and the corresponding fuzzy erosion operation is simplified to:

$$\mu_{\mathcal{E}(A,B)}(x) = \inf_{x \in T(B;x)} \mu_A(x), \tag{69}$$

which is equivalent to the definition of the classical grayscale erosion. It can be concluded from Eqs. (67) and (69) that BMM and GMM can be considered as special cases of the proposed GFMM.

By using a crisp set as a fuzzy structuring element, the GFMM can include the BMM and the GMM as special cases and inherit the properties of the corresponding morphological operations. The GFMM characteristics can be handled by the membership function of the fuzzy structuring element. The role of the structuring elements membership function is important during the calculation of the fuzzy inclusion indicator. The membership function of the fuzzy structuring element B defines classical sets of points with equal degree of membership in B. The meaning of the membership equality is that it defines a locus of topological similarity. Then, the inf operation of the inclusion indicator takes place among the points that belong to each classical set of equal membership separately, and the infimum values are used to construct the fuzzy inclusion indicator set. Similarly to the definition of a fuzzy structuring element in other fuzzy mathematical morphologies, the fuzzy structuring element can be separated in two parts, the core where the membership function equals unit, and the fuzziness, which is the rest of the structuring element. The core is responsible for the properties which come from the classical morphological operation. The fuzziness of the membership function of the fuzzy structuring element can cause interesting behavior of the GFMM operations, and its suitable selection, can lead to desirable characteristics.

3.5 Application in robust morphological skeletonization

The morphological skeleton is a popular method for shape representation. It represents an object X as a number of components. Algebraic combinations of the components reconstruct the object. The representation of X using the morphological skeleton is [27]:

$$X = \bigcup_{k=0}^{N} S(k) \oplus kB, \tag{70}$$

where the sets $S(k)$ known as *skeletal subsets* are:

$$S(k) = (X \ominus kB) - (X \ominus kB)_B, \tag{71}$$

$\ominus$ and $\oplus$ are morphological erosion and dilation, X_B is the morphological opening of X by B, $X - Y$ denotes set difference, kB is the kth-order homothetic of B and N is the largest integer such that $X \ominus NB \neq \emptyset$.

A skeletonization procedure based on the GFMM morphology can be applied by substituting the classical erosion, dilation and opening operations that are used during their processes, with the corresponding operations of the GFMM defined in Eqs. (62), (63) and (64). Since X is a binary image, the simplified erosion equation given in Eq. (67) can be used. Simplified versions of dilation and opening are derived easily. The fuzzy structuring element was chosen to have a size of 9×9 pixels, a core set of 3×3 pixels where $\mu_B = 1$, and pyramidal fuzziness. Extensive simulations were applied by using two reference images. The first one contains an orthogonal spiral and it is shown in Fig. 11a. The second one contains a puzzle piece and it is shown in Fig. 12a. Both images were contaminated by impulsive (salt and pepper) noise with probability 1% and 5% respectively for positive and negative impulses.

Generally, the object reconstruction from skeletal subsets, that can be achieved by using the GFMM is better than the reconstruction obtained by the BMM. However, the most important characteristic of the use of GFMM is the preservation of the shape and the location of the skeletal subsets. Figures 11 and 12 illustrate this property. The original spiral image, the BMM skeleton, the reconstructed from the BMM skeleton object, the GFMM skeleton and the GFMM reconstructed object are shown in Figs. 11a to 11e. In

Figs. 11f to 11o, the skeletonization process is applied on the noisy spiral images shown, and the corresponding BMM and GFMM skeletons and reconstructed objects are presented. The corresponding results when skeletonization is applied on the original and two noisy puzzle images are presented in Fig. 12.

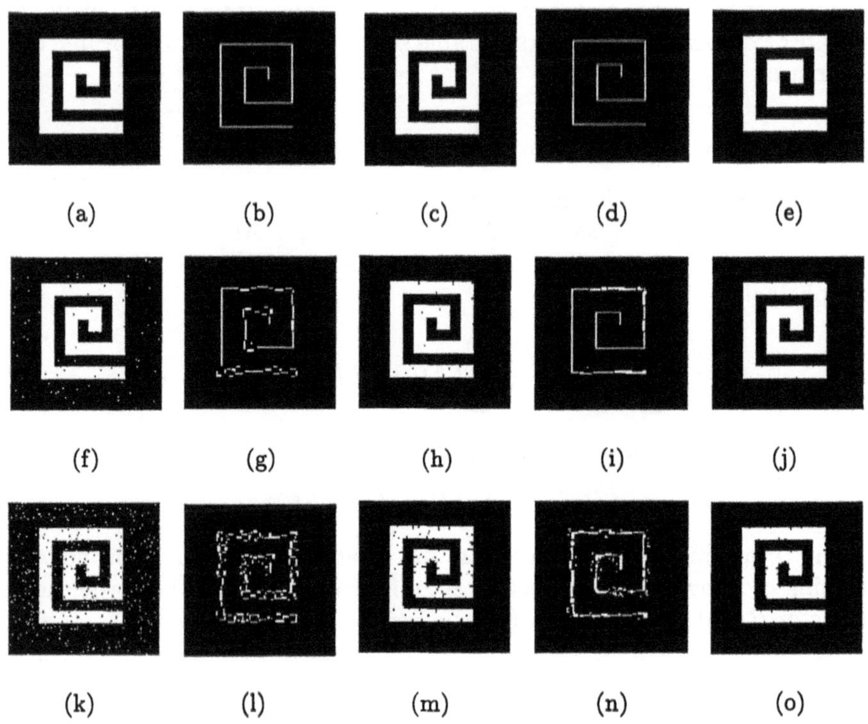

Fig. 11. BMM and GFMM skeletons: (a) The original *spiral* image, (b) the BMM skeleton, (c) the reconstructed object from the BMM skeleton, (d) the GFMM skeleton, (e) the reconstructed object from the GFMM skeleton, (f) the *spiral* image contaminated with impulsive noise (1%), (g) the BMM skeleton, (h) the reconstructed object from the BMM skeleton, (i) the GFMM skeleton, (j) the reconstructed object from the GFMM skeleton, (k) the *spiral* image contaminated with impulsive noise (5%), (l) the BMM skeleton, (m) the reconstructed object from the BMM skeleton, (n) the GFMM skeleton, (o) the reconstructed object from the GFMM skeleton. (©2000 IEEE)

The GFMM can also be applied on 3D binary data. The BMM and GFMM skeletonization were applied on the binary volume of a human body, consisted of 187 frames, each frame of 76×128 pixels, shown in Fig. 13a [43],[44]. The visualization of the morphological skeleton of the body by using either BMM or GFMM are shown in Figs. 13b and 13c. The skeletonization procedure

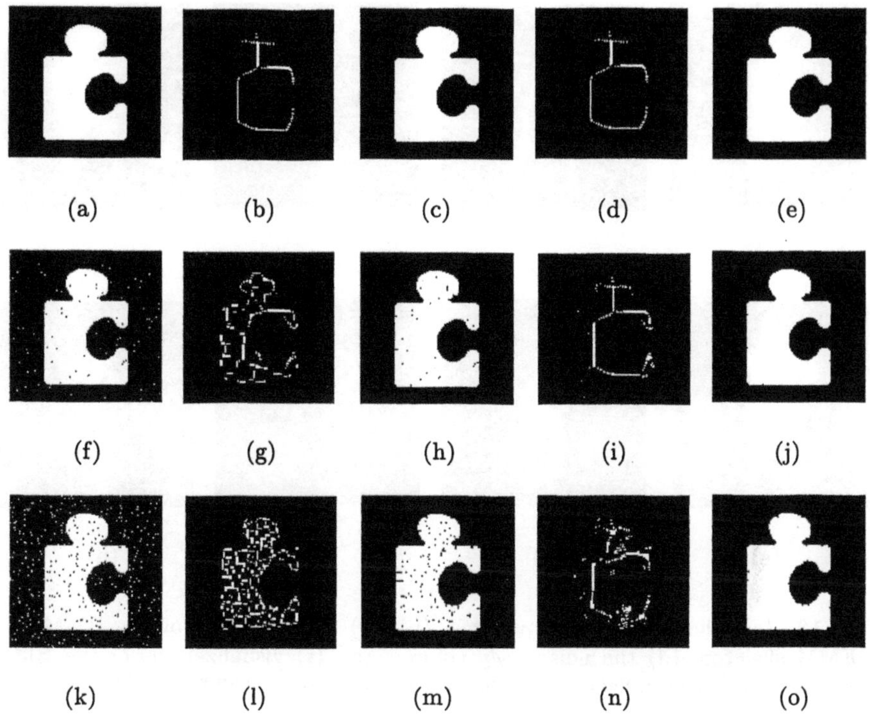

Fig. 12. BMM and GFMM skeletons: (a) The original *puzzle* image, (b) the BMM skeleton, (c) the reconstructed object from the BMM skeleton, (d) the GFMM skeleton, (e) the reconstructed object from the GFMM skeleton, (f) the *puzzle* image contaminated with impulsive noise (1%), (g) the BMM skeleton, (h) the reconstructed object from the BMM skeleton, (i) the GFMM skeleton, (j) the reconstructed object from the GFMM skeleton, (k) the *puzzle* image contaminated with impulsive noise (5%), (l) the BMM skeleton, (m) the reconstructed object from the BMM skeleton, (n) the GFMM skeleton, (o) the reconstructed object from the GFMM skeleton. (©2000 IEEE)

were also applied on a noisy body volume contaminated with impulsive noise (0.1%) shown in Fig. 13d. The visualization of the morphological skeleton of the noisy body by using BMM are shown in Figs. 13e and 13f. The same projections by using GFMM are shown in Figs. 13g and 13h. It can be observed that, as in the 2D case, the location and the shape of the morphological skeleton are better preserved by using the GFMM.

4 Conclusions

Fuzzy sets model uncertainty in the observation data. When data are available in fuzzy sets, fuzzy location and scale estimators are useful especially in

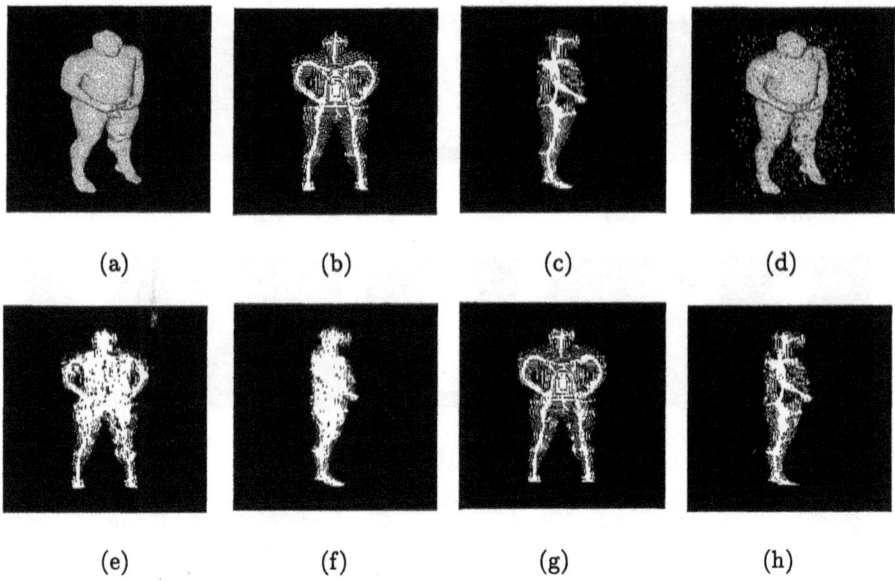

Fig. 13. (a) The original *body* volume, (b), (c) visualizations of the BMM and GFMM skeleton, (d) the noisy *body* volume, (e), (f) visualizations of the BMM skeleton, (g), (h) visualizations of the GFMM skeleton. (©2000 IEEE)

image processing applications. In this chapter fuzzy location and scale estimators based on the extension principle were presented. The fuzzy nonlinear means, the fuzzy location and scale estimators based on fuzzy order statistics and other fuzzy scale estimators, e.g., fuzzy sample standard deviation, were addressed. Equivalent relations that can be used to calculate the fuzzy estimators using classical arithmetic were also given. The Fuzzy Vector Median was also presented, defined as an extension of classical Vector Median based on a distance definition between fuzzy vectors. An application of Fuzzy Vector Median used for filtering images corrupted by mixed Gaussian and impulsive noise, was also given.

Fuzzy Cell Hough Transform was presented as an extension of classical Hough transform that uses a fuzzy split parameter space in order to detect curves in binary images. The Hough transform parameter space is split into fuzzy cells and the method is implemented by a fuzzy voting process. Therefore, the array that is created after the fuzzy voting process is smoother than in the classical case. Simulations on straight line and circle detections in artificially generated images showed that erroneous curve detections, that correspond to the effect of noise or any kind of contour point displacement, were drastically reduced. Moreover, the curves were detected with better accuracy in comparison to classical Hough Transform.

Finally, the Generalized Fuzzy Mathematical Morphology (GFMM) was presented. It is based on a Fuzzy Inclusion Indicator, which is a fuzzy set defined as a measure of the inclusion of one fuzzy set into another one. The classical Binary and Grayscale mathematical morphologies can be considered as special cases of the GFMM. The GFMM provides a very powerful and flexible tool for morphological operations. An application for robust 2D and 3D object representation using skeletonization was investigated. Simulations showed that the reconstruction of noisy objects from their skeletal subsets that can be achieved by using the GFMM is better than by using the classical Binary Mathematical Morphology in most cases. Besides, the use of the GFMM for skeletonization preserves the shape and the location of the skeletal subsets and, therefore, can be efficiently used for object representation especially in cases of impulsive noise.

References

1. Zadeh L., *Fuzzy Sets*, IEEE Inform. and Control, 8, 1965, 338–353.
2. Tzafestas S. G., Venetsanopoulos A. N., *Fuzzy Reasoning in Information, Decision and Control Systems*, Kluwer Academic Publ., Dordrecht, 1994.
3. Kruse R. , Meyer K. D., *Statistics with Vague Data*, D. Reidel Publishing Company, Dordrecht, 1987.
4. Tsoukalas L. H., Uhrig R. E., *Fuzzy and Neural Approaches in Engineering*, Wiley, New York, 1996.
5. Kaufmann A., Gupta M. M., *Introduction to Fuzzy Arithmetic: Theory and Applications*, Van Nostrand Reinhold, New York, 1985.
6. Chatzis V., Pitas I., *Nonlinear Location and Scale Estimators of Fuzzy Numbers*, in: IEEE Trans. on Signal Processing, 46, 1998, 231–234.
7. Chatzis V., Pitas I., *Mean and Median of Fuzzy numbers*, in: Proc. of IEEE Workshop on Nonlinear Signal and Image Processing, 1995, 297–300.
8. Pitas I., Venetsanopoulos A. N., *Nonlinear Digital Filters: Principles and Applications*, Kluwer Academic Publ., Boston, 1990.
9. Chatzis V., Pitas I., *Fuzzy Scalar and Vector Median Filters based on Fuzzy Distances*, in: IEEE Trans. on Image Processing, 8, 1999, 731–734.
10. Karayiannis N. B., Pai P. I., *Fuzzy Vector Quantization Algorithms and their Application in Image Compression*, in: IEEE Trans. on Image Processing, 4, 1995, 1193–1201.
11. Hough P. V. C., *Method and means for recognizing complex pattern*, U.S.Patent 3069654, 1962.
12. Kalviainen H., Hirvonen P., Xu L., Oja E., *Probabilistic and non-probabilistic hough transforms: Overview and comparisons*, in: Image and Vision Computing, 13, 1995, 239–252.
13. Leavers V. F., *Shape Detection in Computer Vision Using the Hough Transform*, Springer-Verlag, 1992.
14. Atiquzzaman M., *Multiresolution hough transform - an efficient method of detecting patterns in images*, in: IEEE Trans. on Pattern Analysis and Machine Intelligence, 14, 1992, 1090–1095.

15. Chang D., Hashimoto S., *An inverse voting algorithm for Hough transform*, in: Proc. ICIP'94, 1, 1994, 223–227.
16. Illingworth J., Kittler J., *The adaptive Hough transform*, in: IEEE Trans. on Pattern Analysis and Machine Intelligence, 9, 1987, 690–698.
17. Li H., Lavin M. A., Master R. J. L., *Fast hough transform: A hierarchical approach*, in: Computer Vision, Graphics, and Image Processing, 36, 1986, 139–161.
18. Tsuji S., Matsumoto F., *Detection of ellipses by a modified Hough transform*, IEEE Trans. Comput., 27, 1978, 777–781.
19. Xu L., Oja E., *Randomized hough transform (rht): Basic mechanisms, algorithms, and computational complexities*, in: CVGIP:Image Understanding, 57, 1993, 131–154.
20. Chatzis V., Pitas I., *Fuzzy Cell Hough Transform for Curve Detection*, in: Pattern Recognition, 30, 1997, 2031-2042.
21. Dubois D., Prade H., *Fuzzy Sets and Systems, Theory and Applications*, Academic Press, Boston, 1980.
22. Haralick R. M., *Image analysis using mathematical morphology*, in: IEEE Trans. on Pattern Analysis and Machine Intelligence, 9, 1987, 532–550.
23. Serra J., *Image Analysis and Mathematical Morphology*, Academic Press, London, 1982.
24. Serra J. (editor), *Image Analysis and Mathematical Morphology, Part II: Theoretical Advances*, Academic Press, London, 1988.
25. Serra J., Soille P., *Mathematical Morphology and its Applications to Image Processing*, Kluwer Academic Publ., Boston, 1994.
26. Maragos P., Schafer R. W., Akmal M., *Mathematical Morphology and its Applications to Image and Signal Processing*, Kluwer Academic Publ., Boston, 1996.
27. Maragos P. A., Schafer R. W., *Morphological skeleton representation and coding of binary images*, in: IEEE Trans. on Acoustics, Speech, and Signal Processing, 34, 1986, 228–244.
28. Wang D., Haese-Coat V., Bruno A., Ronsin J., *Some statistical properties of mathematical morphology*, in: IEEE Trans. on Signal Processing, 43, 1995, 1955–1965.
29. Koskinen L., Astola J., Neuvo Y., *Soft morphological filters*, in: Proc. SPIE Symp. on Image Algebra and Morphological Image Processing, 1568, 1991, 262–270.
30. Koskinen L., Astola J., *Statistical properties of soft morphological filters*, in: Proc. SPIE Symp. on Nonlinear Image Processing, 1658, 1992, 25–36.
31. Shih F. Y., Pu C. C., *Analysis of the properties of soft morphological filtering using threshold decomposition*, in: IEEE Trans. on Signal Processing, 43, 1995, 539–544.
32. Popov A. T., *Convexity indicators based on fuzzy morphology*, in: Pattern Recognition Letters, 18, 1997, 259–267.
33. Sinha D., Dougherty E. R., *A general axiomatic theory of intrinsically fuzzy mathematical morphologies*, in: IEEE Trans. on Fuzzy Systems, 3, 1995, 389–403.
34. Bloch I., Maitre H., *Fuzzy mathematical morphologies: A comparative study*, in: Pattern Recognition, 28, 1995, 1341–1387.
35. Daneshgar A., *Residuated semigroups and morphological aspects of translation invariant systems*, in: Fuzzy Sets and Systems, 90, 1997, 69–81.

36. Grossert S., Koppen M., Nickolay B., *A new approach to fuzzy morphology based on fuzzy integral and its application in image processing*, in: Proc. of ICPR, 1996.
37. Gader P. D., *Fuzzy spatial relations based on fuzzy morphology*, in: Proc of the Sixth IEEE International Conference on Fuzzy Systems, 1997.
38. Chatzis V., Pitas I., *A Generalized Fuzzy Mathematical Morphology and its Application in Robust 2D and 3D Object Representation*, in: IEEE Trans. on Image Processing, 9, 2000, 1798-1810.
39. Popov A. T., *Fuzzy morphology and fuzzy convexity measures*, in: Proc. of ICPR, 1996.
40. Heijmans H. J. A. M., *Morphological Image Operators*, Academic, Boston, 1994.
41. Werman M., Peleg S., *Min-max operators in texture analysis*, in: IEEE Trans. on Pattern Analysis Machine Intelligence, 7, 1985, 730–733.
42. Klir G. J., Yuan B., *Fuzzy Sets and Fuzzy Logic, Theory and Applications*, Prentice Hall, New Jersey, 1995.
43. Ackerman M. J., *The visible human project*, in: IEEE Proceedings, 86, 1998, 504–511.
44. U. S. National Library of Medicine, *The visible human project*, http://www.nlm.nih.gov/research/visible/visible_human.html, 1996.

Chapter 13

Adaptive Fuzzy Filters and Their Application to Online Maneuvering Target Tracking

Mohammad B. Menhaj

AmirKabir University
Electrical Engineering Department
Hafez Ave. 529, Tehran, Iran
email: menhaj@cic.aku.ac.ir

Currently: Oklahoma State University, Department of Computer Science
219 MSCS, Stillwater, Ok. 74078, USA
menhaj@okstate.edu

Summary. This chapter is devoted to present two adaptive fuzzy filters with online structure and parameter learning ability with an important feature that they can dynamically partition the input and output spaces using a modified FCM (Fuzzy C-Means) clustering algorithm according to the input-output data distribution. These filters are also able to tune membership functions and find fuzzy logic rules in an on-line manner. This chapter also introduces a new evolutionary algorithm called OGA as a learning system for the adaptive fuzzy filters. Finally the adaptive fuzzy filters are applied to maneuvering target tracking problem and their performance is compared with that of the classical Kalman Based techniques (Interacting Multiple Model).

1 Introduction

Knowing that the term filter represents a device processing information and filtering is nothing but extracting information about a desired quantity from a set of measured data gathered from operations in an environment of unknown statistics, it was always desirable to use all the information sources including numerical (quantitative) data and qualitative data in terms of linguistic values coming from human expert in the filter design. Adaptive fuzzy filters as a special kind of nonlinear adaptive filters consist of two main parts: a set of tunable if-then rules and an adaptive mechanism that uses the empirical data for adjusting the parameters of the memberships functions corresponding to the linguistic variables defined on the input-output reference sets.

This chapter presents adaptive fuzzy filters with online structure and parameter learning ability. These filters have an important feature that they can dynamically partition the input and output spaces using a modified FCM (Fuzzy C-Means) clustering algorithm according to the input-output data distribution. The introduced filters are also able to partition dynamically input-output spaces, tune membership functions and find fuzzy logic rules in an on-line manner. This chapter also introduces a new evolutionary algorithm called OGA as a learning system for the adaptive fuzzy filters. Section 4 fully develops these filters. The maneuvering target tracking problem has been considered as a case study. This problem has attracted many researchers attention in the last three decades due to its importance in operation of military surveillance systems. Section 2.1 presents the target model used in this study.

The most commonly used technique to the target tracking problem is the Kalman filtering. Sections 2.2-2.3 review Kalman and extended Kalman filtering. The standard Kalman filter fails to perform satisfactory for many target tracking, however it can optimally estimate the target motion from noisy radar data. In the case of non-maneuvering, the standard Kalman filter has good performance, but when the target starts maneuvering the performance of standard Kalman filter degrades and a large error bias occurs. In this case, it is necessary to modify the standard Kalman filter algorithm. There exist many maneuvering target tracking algorithms[1]-[20] which are discussed in section 3. Among these techniques the Interacting Multiple Model (IMM) method provides a better performance with efficient computations. In this method, a bank of Kalman filters run in parallel, and the final output is the weighted sum of all filters. The performance of this method depends on: the number of models selected, the appropriate transition probability matrix of the models, and the knowledge of statistics of measurement noise. It is not possible to know, in advance, the values of these factors. Consequently, the performance of this method will degrade.

By considering the maneuver as an inherent part of the target dynamics leading to a non-stationary target model, we can use an adaptive filter to track maneuvering target. The structure of adaptive fuzzy systems (LAFF and SAFF) along with an on-line structure-parameter learning algorithm which combines modified FCM clustering algorithm, BP and OGA learning algorithm are presented in section IV. In this section the proposed filters are also applied to maneuvering target tracking problem and their performance are compared with that of IMM method. Finally, section 5 concludes the chapter.

2 Preliminaries

In this section we want to present the target tracking problem, the Kalman and extended Kalman filtering that are used in the remainder of this chapter.

2.1 Target modeling

The ability of an adaptive filter to operate satisfactorily in an unknown environment and track time variations of input statistics make them a powerful device for signal processing and control applications. Among the various applications of adaptive filters, the identification shown in figure 1 is of great interest. The desired response d(k), is the summation of actual output , s(k), and measurement noise, n(k). This paper considers the following target model given in[19] as a benchmark. The dynamic equation of the maneuvering target is :

$$\underline{\mathbf{x}}(k+1) = \mathbf{A}(\omega)\underline{\mathbf{x}}(k) + \mathbf{G}\underline{\mathbf{w}}(k), \tag{1}$$

$$\mathbf{A}(\omega) = \begin{bmatrix} 1 & \frac{\sin \omega T}{\omega} & 0 & -\frac{1-\cos \omega T}{\omega} \\ 0 & \cos \omega T & 0 & -\sin \omega T \\ 0 & \frac{1-\cos \omega T}{\omega} & 1 & \frac{\sin \omega T}{\omega} \\ 0 & \sin \omega T & 0 & \cos \omega T \end{bmatrix}, \mathbf{G} = \begin{bmatrix} \frac{T^2}{2} & 0 \\ T & 0 \\ 0 & \frac{T^2}{2} \\ 0 & T \end{bmatrix}, \underline{\mathbf{x}}(k) = \begin{bmatrix} x(k) \\ \dot{x}(k) \\ y(k) \\ \dot{y}(k) \end{bmatrix}. \tag{2}$$

where $\underline{\mathbf{w}}(k)$ is the process noise and the measurement equation is:

$$\underline{\mathbf{z}}(k) = \underline{\mathbf{h}}(\underline{\mathbf{x}}(k)) + \underline{\mathbf{v}}(k). \tag{3}$$

In the polar coordinates, we have:

$$\underline{\mathbf{h}}(\underline{\mathbf{x}}(k)) = \begin{pmatrix} \sqrt{x^2(k) + y^2(k)} \\ \arctan(\frac{x(k)}{y(k)}) \end{pmatrix}. \tag{4}$$

This system, which represents a model of MA-2D radar with unknown turn rate ω and radar sampling period T, can be viewed as the plant block in figure 1, with input $\underline{\mathbf{w}}(k)$, output s(k) and measurement noise $\underline{\mathbf{v}}(k)$. Because ω is unknown, we can consider it as the uncertainty that causes the non-stationarity of the plant.

2.2 Kalman filtering

The Kalman filter represents a unified state space-based recursive minimum mean-squared estimation (MSSE) algorithm ideally suited for implementation on digital computers with a plethora of industrial applications in real-world engineering problems. Consider the systems shown in Fig. 1 with the following stochastic dynamic equation

$$\underline{\mathbf{y}}_t = \mathbf{C}_t\underline{\mathbf{x}}_t + \underline{\mathbf{d}}_t + \mathbf{F}_t\underline{\mathbf{v}}_t, \tag{5}$$

$$\underline{\mathbf{x}}(t+1) = \mathbf{A}_t\underline{\mathbf{x}}_t + \mathbf{B}_t\underline{\mathbf{u}}_t + E_t\underline{\mathbf{v}}_t. \tag{6}$$

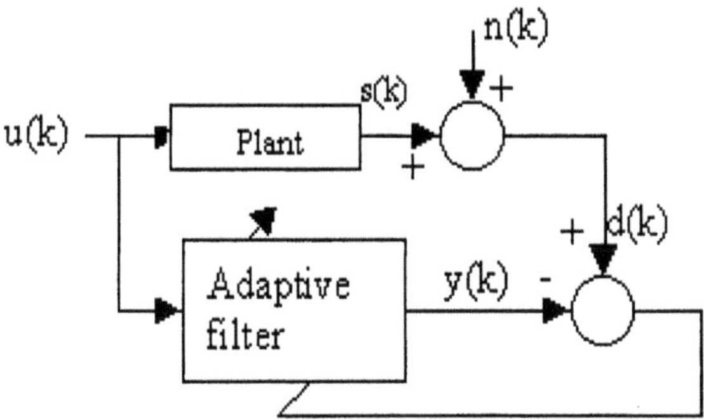

Fig. 1. Block diagram of identification mechanism.

where $\underline{\mathbf{x}}_t \in \Re^n$ represents non-observable state vector, $\underline{\mathbf{y}}_t \in \Re^m$ constitutes the observable data, $\underline{\mathbf{u}}_t \in \Re^r$ is deterministic control input vector, and the process noise $\underline{\mathbf{w}}_t \in \Re^l$ and the measurement noise $\underline{\mathbf{v}}_t \epsilon \Re^q$ are assumed to be stationary white with mean values and covariance matrices given below.

$$\begin{aligned} \mathbf{E}(\underline{\mathbf{w}}_t) = 0, \quad \mathbf{E}(\underline{\mathbf{w}}_t \underline{\mathbf{w}}_t^T) = \mathbf{Q}_t, \quad \mathbf{Q_t} = \mathbf{Q_t}^T \geq 0, \\ \mathbf{E}(\underline{\mathbf{v}}_t) = 0, \quad \mathbf{E}(\underline{\mathbf{v}}_t \underline{\mathbf{v}}_t^T) = \mathbf{R}_t, \quad \mathbf{R_t} = \mathbf{R_t}^T \geq 0. \end{aligned} \tag{7}$$

It is further assumed that the initial sate is unknown with a priori available knowledge of $\mathbf{E}(\underline{\mathbf{x}}_0) = \underline{x}_0$, $\mathbf{E}(\underline{\mathbf{x}}_0 \underline{\mathbf{x}}_0^T) = \mathbf{P}_{x_0}$ with $\mathbf{P_{x_0}} = \mathbf{P}_{x_0}^T \geq 0$ and the processes $\underline{x}_0$, $\underline{\mathbf{w}}_t$ and $\underline{\mathbf{v}}_t$ are mutually uncorrelated for all t and all s so that $\mathbf{E}(\underline{\mathbf{v}}_t \underline{\mathbf{w}}_t^T) = 0$, $\mathbf{E}(\underline{\mathbf{x}}_0 \underline{\mathbf{v}}_t^T) = 0$, and $\mathbf{E}(\underline{\mathbf{x}}_0 \underline{\mathbf{w}}_t^T)$.

The estimate $\widehat{\underline{\mathbf{x}}}_t \equiv \mathbf{E}(\underline{\mathbf{x}}_t \mid \mathbf{Y}_t)$ of $\underline{\mathbf{x}}_t$ based on the information up to time t, i.e. $\mathbf{Y}_t = \{\underline{\mathbf{y}}_1, \cdots, \underline{\mathbf{y}}_t\}$ is known as filtering. The filtering algorithm is known as the Kalman filter described by the following equations[21]. This algorithm can be derived using many approaches such as normality assumption for the process noises, mixed estimation approach and minimum mean squared estimation.

$$\begin{aligned} \widehat{\underline{\mathbf{x}}}_t^- &= \mathbf{A}_t \widehat{\underline{\mathbf{x}}}_{t-1} + \mathbf{B}_t \underline{\mathbf{u}}_t, \\ \mathbf{P}_t^{x-} &= \mathbf{A}_t \mathbf{P}_{t-1}^x + E_t \mathbf{Q}_t E_t^T, \\ \widehat{\underline{\mathbf{y}}}_t^- &= \mathbf{C}_t \widehat{\underline{\mathbf{x}}}_t^- + \underline{\mathbf{d}}_t, \\ \mathbf{G}_t^- &= \mathbf{C}_t \mathbf{P}_t^{x-} \mathbf{C}_t^T + \mathbf{F}_t \mathbf{R}_t \mathbf{F}_t^T, \\ \mathbf{K}_t &= \mathbf{P}_t^{x-} \mathbf{C}_t^T (\mathbf{G}_t^-)^{-1}, \\ \mathbf{P}_t^x &= \mathbf{P}_t^{x-} - \mathbf{K}_t \mathbf{G}_t^- \mathbf{K}_t^T, \\ \widehat{\underline{\mathbf{x}}}_t &= \widehat{\underline{\mathbf{x}}}_t^- + \mathbf{K}_t (\underline{\mathbf{y}}_t - \widehat{\underline{\mathbf{y}}}_t^-), \end{aligned} \tag{8}$$

where $\widehat{\underline{\mathbf{x}}}_t^- \equiv \mathbf{E}(\widehat{\underline{\mathbf{x}}}_t \mid \mathbf{Y}_{t-1})$, $\mathbf{P}_t^x \equiv Cov(\underline{\mathbf{x}}_t \mid \mathbf{Y}_t)$- Conditional covariance matrix of $\underline{\mathbf{x}}_t$ given $\mathbf{Y}_t$, and $\mathbf{P}_t^{x-} \equiv Cov(\underline{\mathbf{x}}_t \mid \mathbf{Y}_{t-1})$.

2.3 The extended Kalman filter

The EKF [21] is a special kind of nonlinear filtering where the state space model is specified as:

$$\begin{aligned} \underline{\mathbf{y}}_t &= \underline{\mathbf{g}}_t(\underline{\mathbf{x}}_t, \underline{\mathbf{v}}_t), \quad \underline{\mathbf{g}}_t \text{ -- known and invertible} \\ \underline{\mathbf{x}}_{t+1} &= \underline{\mathbf{h}}_t(\underline{\mathbf{x}}_t, \underline{\mathbf{w}}_{t+1}), \quad \underline{\mathbf{h}}_t \text{ -- known and invertible} \end{aligned} \tag{9}$$

and it is based on the normality and independence assumption for the process noises. In this case in order to apply the standard Kalman filtering algorithm, the two nonlinear functions $\underline{\mathbf{g}}_t(\underline{\mathbf{x}}_t, \underline{\mathbf{v}}_t)$ and $\underline{\mathbf{h}}_t(\underline{\mathbf{x}}_{t-1}, \underline{\mathbf{w}}_t)$ are approximated around $(\underline{\mathbf{x}}_t, \underline{\mathbf{v}}_t) = (\widehat{\underline{\mathbf{x}}}_t^-, 0)$ and $(\underline{\mathbf{x}}_{t-1}, \underline{\mathbf{w}}_t) = (\widehat{\underline{\mathbf{x}}}_{t-1}, 0)$ by first order Taylor series expansion to obtain the following state-space model:

$$\begin{aligned} \underline{\mathbf{x}}_t &= \mathbf{A}_t^-(\underline{\mathbf{x}}_{t-1} - \widehat{\underline{\mathbf{x}}}_{t-1}) + E_t^- \underline{\mathbf{w}}_t + \underline{\mathbf{b}}_t, \\ \underline{\mathbf{y}}_t &= \mathbf{C}_t^-(\underline{\mathbf{x}}_t - \widehat{\underline{\mathbf{x}}}_t^-) + \mathbf{F}_t^- \underline{\mathbf{v}}_t + \underline{\mathbf{d}}_t^-, \end{aligned} \tag{10}$$

where all the functions $\mathbf{C}_t^-$, $\underline{\mathbf{d}}_t^-$, $\mathbf{F}_t^-$, $\mathbf{A}_t^-$, E_t^-, and $\underline{\mathbf{b}}_t$ defined below and $\widehat{\underline{\mathbf{x}}}_t^-$, $\widehat{\underline{\mathbf{x}}}_{t-1}$) are dependent on the information available at time t-1.

$$\begin{aligned} \mathbf{C}_t^- &\equiv \frac{\partial \underline{\mathbf{g}}_t(\underline{\mathbf{x}}_t, \underline{\mathbf{v}}_t)}{\partial \underline{\mathbf{x}}_t}\Big|_{\substack{\underline{\mathbf{x}}_t = \widehat{\underline{\mathbf{x}}}_t^- \\ \underline{\mathbf{w}}_t = 0}}, \quad \mathbf{F}_t^- \equiv \frac{\partial \underline{\mathbf{g}}_t(\underline{\mathbf{x}}_t, \underline{\mathbf{v}}_t)}{\partial \underline{\mathbf{v}}_t}\Big|_{\substack{\underline{\mathbf{x}}_t = \widehat{\underline{\mathbf{x}}}_t^- \\ \underline{\mathbf{v}}_t = 0}}, \\ \mathbf{d}_t^- &\equiv \underline{\mathbf{g}}_t(\widehat{\underline{\mathbf{x}}}_t^-, 0), \quad \mathbf{A}_t^- \equiv \frac{\partial \underline{\mathbf{h}}_t(\underline{\mathbf{x}}_{t-1}, \underline{\mathbf{w}}_t)}{\partial \underline{\mathbf{x}}_{t-1}}\Big|_{\substack{\underline{\mathbf{x}}_{t-1} = \widehat{\underline{\mathbf{x}}}_{t-1} \\ \underline{\mathbf{w}}_t = 0}}, \\ E_t^- &\equiv \frac{\partial \underline{\mathbf{h}}_t(\underline{\mathbf{x}}_{t-1}, \underline{\mathbf{w}}_t)}{\partial \underline{\mathbf{w}}_t}\Big|_{\substack{\underline{\mathbf{x}}_{t-1} = \widehat{\underline{\mathbf{x}}}_{t-1} \\ \underline{\mathbf{w}}_t = 0}}, \quad \mathbf{b}_t^- \equiv \underline{\mathbf{h}}_t(\widehat{\underline{\mathbf{x}}}_{t-1}, 0). \end{aligned} \tag{11}$$

The following equations describe the EKF:

$$\begin{aligned} \hat{\underline{\mathbf{x}}}_t^- &= \mathbf{E}(\underline{\mathbf{h}}_t(\underline{\mathbf{x}}_{t-1}, \underline{\mathbf{w}}_t) \mid \mathbf{Y}_{\mathbf{t-1}}) = \underline{\mathbf{b}}_t^-, \\ \mathbf{P}_t^{x-} &= \mathbf{E}((\underline{\mathbf{x}}_t - \hat{\underline{\mathbf{x}}}_t^-)(\underline{\mathbf{x}}_t - \hat{\underline{\mathbf{x}}}_t^-)^T \mid \mathbf{Y}_{\mathbf{t-1}}) = \mathbf{A}_t^- \mathbf{P}_{t-1}^x \mathbf{A}_t^{-T} + E_t^- \mathbf{Q}_t E_t^{-T}, \\ \hat{\underline{\mathbf{y}}}_t^- &== \mathbf{E}(\underline{\mathbf{g}}_t(\underline{\mathbf{x}}_t, \underline{\mathbf{v}}_t) \mid \mathbf{Y}_{\mathbf{t-1}}) = \underline{\mathbf{d}}_t^-, \\ \mathbf{G}_t^- &= \mathbf{E}((\underline{\mathbf{y}}_t - \hat{\underline{\mathbf{y}}}_t^-)(\underline{\mathbf{y}}_t - \hat{\underline{\mathbf{y}}}_t^-)^T \mid \mathbf{Y}_{\mathbf{t-1}}) = \mathbf{C}_t^- \mathbf{P}_t^{x-} \mathbf{C}_t^{-T} + \mathbf{F}_t^- \mathbf{R}_t \mathbf{F}_t^{-T}, \\ \mathbf{L}_t^- &= \mathbf{E}((\underline{\mathbf{y}}_t - \hat{\underline{\mathbf{y}}}_t^-)(\underline{\mathbf{x}}_t - \hat{\underline{\mathbf{x}}}_t^-)^T \mid \mathbf{Y}_{\mathbf{t-1}}) = \mathbf{C}_t^- \mathbf{P}_t^{x-}, \\ \mathbf{K}_t &= \mathbf{L}_t^{-T}(\mathbf{G}_t^-)^{-1}, \\ \mathbf{P}_t^x &= \mathbf{P}_t^{x-} - \mathbf{K}_t \mathbf{G}_t^- \mathbf{K}_t^T, \\ \hat{\underline{\mathbf{x}}}_t &= \hat{\underline{\mathbf{x}}}_t^- + \mathbf{K}_t(\underline{\mathbf{y}}_t - \hat{\underline{\mathbf{y}}}_t^-). \end{aligned} \tag{12}$$

3 Kalman-Based filter designs

3.1 A short review

In general, the transient equation of the target motion equations is the same as given in equation (1) with states of target positions and speeds in the x and y directions, and the input vector u represents the acceleration inputs. The Kalman filter, because of its recursive structure and real-time capability of sequential processing of data has been frequently employed to estimate the target states. Such a tracking filter was not efficiently suited to handle the maneuvering target. [1] introduced a target model by considering the maneuvering motion as a first-order Markov process. This filter performs well for low maneuvering targets. This filter has been improved by [2–4]. These filters all require the availability of the maneuvering model of the target either as an autoregressive or semi-Markov process. Furthermore, because the Kalman estimate of the state vector requires the target acceleration inputs which are not available to the Kalman filter given in (2), in [5–7] (extended) Kalman filter based tracking filters have been developed using estimates of the acceleration inputs; this technique does not require the knowledge of target type and its maneuvering characteristics.

In short, the non-maneuvering based filters devised on the non-maneuvering models show poor performances as the target maneuvers. The reason for that is the Kalman filtering gain is too small to contribute the maneuver measurement data. The maneuvering target based filters though improve the tracking filtering performance during the period of maneuvering they do not outperform the non-maneuvering based filters when the target has a constant velocity. The use of these filters becomes more complicated if there is not enough information about the maneuvering. To resolve this, [8] proposed Variable Dimension filters. In these filters both of maneuvering based and non-maneuvering filters are used. The poor performance of the filter devised in [1] for high maneuvering targets has been enhanced in [9]. Interested readers may refer to [10] for a comparative study of these algorithms. [11] has further developed the filter devised in [5] by introducing a revised input estimation method that estimates the onset time and magnitude of the maneuver. In 1988, [12] enhanced the filter devised in [9] by proposing an acceleration model by augmenting a mean target acceleration term into that model. These filter require a big number of filters that represents a time-consuming task.

By using a better method of hypothesis management, Bar-Shalom [13,14] developed the Integrating Multiple Model (IMM) shown to have a good performance with less amount of computational complexity. Because of its wide applications to maneuvering target tracking problems [15] and using this technique as a benchmark study for our technique, the IMM will be reviewed later in this chapter. In [16] the enhanced variable dimension filter using measurement concatenation for maneuvering target tracking has been presented.

In [17], there has been designed an adaptive tracking filter based on a modified input estimation using pseudoresiduals and it has been shown that the proposed method takes less computation time in some cases compared with some IMM algorithms.

3.2 Integrating multiple model method

Figure 2 summarizes the IMM algorithm indexIMM algorithm for the system defined below.

$$\begin{aligned} \underline{\mathbf{x}}_{t+1}^{m_t} &= \mathbf{A}_t^{m^t} \underline{\mathbf{x}}_t^{m_t} + \mathbf{B}_t^{m^t} \underline{\mathbf{w}}_t^{m_t}, \\ \underline{\mathbf{y}}_t^{m_t} &= \mathbf{C}_t^{m^t} \underline{\mathbf{x}}_t^{m_t} + \underline{\mathbf{v}}_t^{m_t}. \end{aligned} \tag{13}$$

with Markovian transition modes probabilities $p(m_{t+1} = i \mid m_t = j) = c_{ji}, i, j \epsilon I \equiv$ the set of all modes , $\sum_{i\epsilon\imath} c_{ij} = 1$, and m_t is the mode index indicating that the current mode in effect during the sampling period beginning at time t, the process noise and measurement noise are mutually uncorrelated nonzero mean Gaussian with positive definite covariance matrices $\mathbf{Q}_t^{m_t}$ and $\mathbf{R}_t^{m_t}$, respectively; they are uncorrelated with the initial state vector which is a zero mean Gaussian process.

As you may easily see from Fig. 2, the IMM is a recursive and modular algorithm with the main steps of interaction, filtering and combination.

4 The adaptive fuzzy filters

Fuzzy Computation (FC), Neural Computation (NC), and Evolutionary Computation (EC) represent the main components of the Computational Intelligence (CI). One may easily comprehend the FC based methods because of its perception based if-then structure. The calculus of Fuzzy if-then rules mimicking human mind and reasoning plays a central role in adaptive system design, i.e., filtering and control. A general form, linguistic type, of fuzzy if-then-rules is expressed as "if Input is A then Output is B" in which Input and Output are I/O variables and A and B represent the linguistic terms like Big and more and less tall; they are defined as fuzzy sets in the respective universal sets. How to build up these rules systematically (establishment of the rules from experimental data) was a main concern of researchers and has a central role in calculus of fuzzy systems.

NC and EC constituting two main components of model-free based learning systems and gradient based techniques representing classical learning systems can be employed as numerical optimization methods in a mixture with fuzzy if-then rules to furnish us the necessary equipment for establishing a common framework know as supervisory or adaptive fuzzy systems; this is nothing but the mixture of fuzzy if-then rule systems with a learning algorithm to tune the parameters of the rules, shaping the membership functions

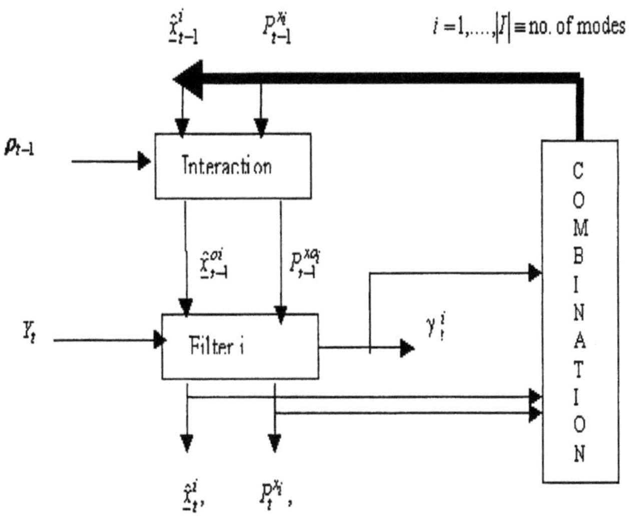

$$\rho_i^- \equiv \text{predicted mode probabilities} = p\left(I_t^i \middle| Y_t\right) = \sum_j c_{ji}\rho_{t-1}^j, \quad \rho_t \equiv \text{mode probabilities at time t},$$

$$C = \left[c_{ji}\right] \equiv \text{Markov Transition Matrix}, \; I_t^j \equiv \text{the event that mode j is in effect at time t},$$

$$\rho_{j|i} \equiv \text{mixing probability} \equiv p\left\{I_{t-1}^j \middle| I_t^i, Y_{t-1}\right\} = \left(\rho_i^-\right)^{-1} c_{ji}\rho_{t-1}^j,$$

$$\underline{\hat{x}}_{t-1}^{oi} \equiv E\left(\underline{x}_{t-1} \middle| I_{t-1}^i, Y_{t-1}\right) = \sum_j \underline{\hat{x}}_{t-1}^j \rho_{j|i},$$

$$P_{t-1}^{x_{oi}} = \sum_j P_{t-1}^{x_j}\rho_{j|i} + \sum_j\sum_n \left(\underline{\hat{x}}_{t-1}^n - \underline{\hat{x}}_{t-1}^j\right)\left(\underline{\hat{x}}_{t-1}^n - \underline{\hat{x}}_{t-1}^j\right)^T \rho_{n|j}\rho_{j|i},$$

$$\underline{\hat{x}}_t^{i-} = A_{t-1}^i \underline{\hat{x}}_{t-1}^{oi} + B_{t-1}^i \underline{\overline{w}}_{t-1}^i; \qquad \underline{\overline{w}}_{t-1}^i \equiv \text{mean of the mode i process noise at time t - 1},$$

$$P_t^{x_i-} = A_{t-1}^i P_{t-1}^{x_{oi}}\left(A_{t-1}^i\right)^T + B_{t-1}^i Q_{t-1}^i\left(B_{t-1}^i\right)^T,$$

$$e^i \equiv \text{residual} = y - C^i\underline{\hat{x}}_t^{i-}, \text{ with covariance } D^i = C^i P_t^{x_i-}\left(C^i\right)^T + R^i,$$

$$G^i = \text{Kalman Filter gain} = P_t^{x_i-}\left(C^i\right)^T\left(D^i\right)^{-1},$$

$$\underline{\hat{x}}_t^i = \underline{\hat{x}}_t^{i-} + G^i e^i,$$

$$P_t^{x_i} = P_t^{x_i-} - G^i D^i\left(G^i\right)^{-1},$$

$$\gamma^i = \text{Likelihood function} \sim \aleph\left(e^i; 0, D^i\right) - \text{Gaussian with mean 0 and covariance } D^i, \; \rho^i = \frac{\rho_i^-\gamma^i}{\sum_j \rho_i^-\gamma^j},$$

$$\underline{\hat{x}}_t \equiv E\left(\underline{x}_t \middle| Y_t\right) = \sum_j \underline{\hat{x}}_t^j \rho_j,$$

$$P_t^x = E\left(\left(\underline{x}_t - \underline{\hat{x}}_t\right)\left(\underline{x}_t - \underline{\hat{x}}_t\right)^T\right) = \sum_j P_t^{x_i}\rho_j + \sum_j\sum_n \left(\underline{\hat{x}}_t^n - \underline{\hat{x}}_t^j\right)\left(\underline{\hat{x}}_t^n - \underline{\hat{x}}_t^j\right)^T \rho_n\rho_j.$$

Fig. 2. The IMM Baseline Structure and its Algorithm.

of linguistic terms, fixing the number of rules based on numerical (empirical) pairs of I/O observations. So an adaptive fuzzy system is defined as a system that employs a mixture of FC and learning algorithms. The interested readers may refer to [22,23] to see more about adaptive fuzzy systems.

In this section the Linguistic-type Adaptive Fuzzy Filter (LAFF) and Sugeno-type Adaptive Fuzzy Filter (SAFF) are presented[70].

4.1 The LAFF

The main objective of a fuzzy system is a linguistic description prescribing qualitative (approximate) actions for a given state of the system under study. These descriptions consist of associations of fuzzy variables and procedures for inference. In contrast to conventional filtering in which the physical process is modelled, in fuzzy based systems the objective is to involve expert human knowledge (to incorporate expert human operator) in the design process.

The basic structure of a fuzzy-in-loop-system is outlined in Fig. 3. The fact that the measuring device gives us quantitative (non-fuzzy) measurements and actuators require quantitative inputs justifies the two additional considerations when one employs qualitative data (linguistic descriptions). They are fuzzifier that fuzzifies the input of the fuzzy system and defuzzifier that defuzzifies its output; it makes the output non-fuzzy.

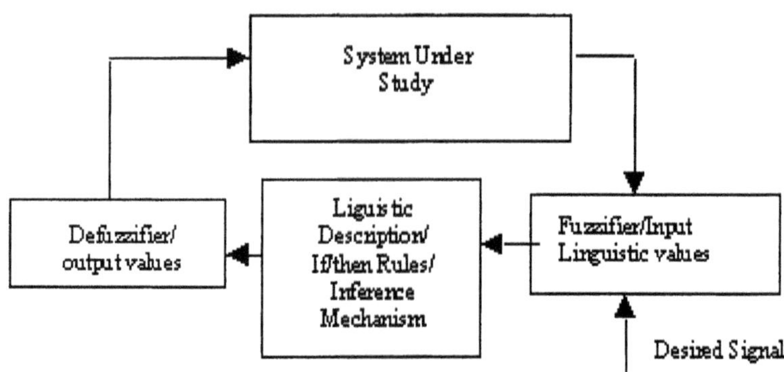

Fig. 3. Block diagram of the Fuzzy-in-loop-System.

As seen in the above figure a set of fuzzy if-then rules with linguistic values in both if-parts and then-parts referred to as system's knowledge base is to be evaluated at time instants in parallel (i.e. several rules can be performed together). These rules relate the input and output variables of the fuzzy system. The fuzzy values for the I/O variables share the same universe of discourses with some common parametric shapes of the types: S-shape,

Z-shape, and bell-shape or trapezoidal shape. These allow for flexible categorization of a variable's universe of discourse; i.e, slow, fast and medium as the linguistic values of fuzzy variable velocity. Indeed a unique membership function for each linguistic value can be sought that describes the degree to the fuzzy value of each individual ordinary element of the universe of discourse. This can be done enumeratively using learning capability of neural networks as a powerful learning mechanism followed by a fuzzification process in which to each training data in the set of classified clusters fuzzy membership values are assigned.

Here the fuzzy system represents an adaptive filter with the ability of operating in a satisfactory fashion in an unknown environment and tracking time variations of input statistics; these make them very powerful in the field of signal processing and control theory. So the desired signal may represent the summation of actual system output and measurement noise. The system under study is the maneuvering target tracking problem described in section 2.

The LAFF system consists of rules with the following generic form:

$\mathbf{R}^j$: if $\mathbf{x}_1$ is $\mathbf{A}_1^j$ and ... and $\mathbf{x}_n$ is $\mathbf{A}_n^j$
then $\mathbf{y}$ is $\mathbf{B}^j$ j=1,..., M.

where $\mathbf{A}_i^j$'s and $\mathbf{B}^j$'s are fuzzy sets in input and output spaces, respectively.

Using the algebraic product for T-norm operator, the operator max for S-norm, and the centroid defuzzification method, the final crisp value of fuzzy inference system is:

$$\mathbf{y} = \frac{\sum_{j=1}^{M} w_j \bar{m}_j}{\sum_{j=1}^{M} w_j}, \quad w_j = \prod_{i=1}^{n} \mathbf{A}_i^j(x). \tag{14}$$

where $\bar{m}_j$ is the mean of $\mathbf{B}^j$.

Fig. 4 shows the structure of the LAFF. In the structure-learning phase, conflicting rules (i.e. rules with the same if-parts and different then-parts) may be created. This situation is represented by dashed line-links between layer-2 and layer-3 nodes. We will next describe the functions of the nodes in each of the three layers of the LAFF.

Layer 1: Each processing element in this level acts as a one-dimensional membership function. The following Gaussian membership function is used:

$$\begin{aligned} \mathbf{n}_{ij}^{(1)} &= -\left(\frac{(\mathbf{x}_i - \mathbf{m}_{ij}^1)}{\sigma_{\mathbf{ij}}^1}\right)^2, \quad i = 1, 2, \ldots, n, \\ \mathbf{a}_{ij}^{(1)} &= -\exp\left(\mathbf{n}_{ij}^{(1)}\right), \qquad j = 1, 2, \ldots, M. \end{aligned} \tag{15}$$

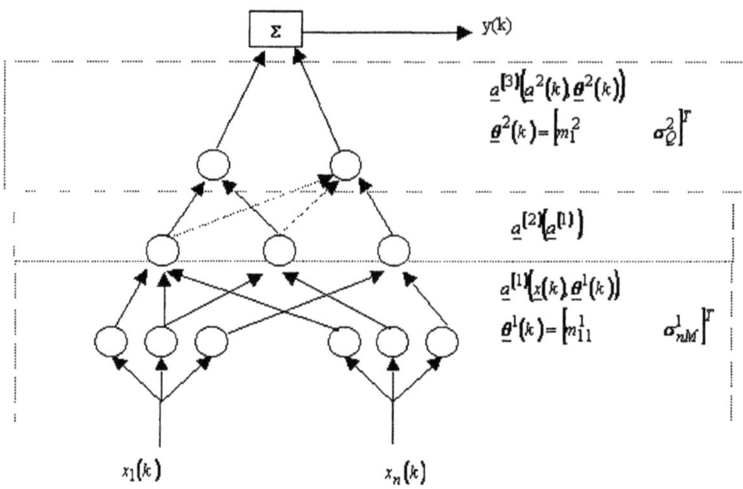

Fig. 4. Structure of the LAFF.

where n is the number of inputs, M is the number of current rules, $\mathbf{m}_{ij}^1$ and σ_{ij}^1 are respectively the mean and variance of gaussian membership function of the j-th fuzzy term set of the i-th input variable $\mathbf{x}_i$; they represent the adjustable parameters of the nodes in layer 1.

Layer 2: A node in this layer represents one fuzzy logic rule. By applying the algebraic product, we have:

$$\mathbf{n}_j^{(2)} = \exp\left(-\left[\mathbf{D_j}\left(\underline{\mathbf{x}} - \underline{\mathbf{m}}_j^1\right)\right]^T \left[\mathbf{D_j}\left(\underline{\mathbf{x}} - \underline{\mathbf{m}}_j^1\right)\right]\right), \quad j = 1, 2, \ldots, M$$

where $\mathbf{D_j} = diag\left(\frac{1}{\sigma_{1j}^1}, \ldots, \frac{1}{\sigma_{nj}^1}\right)$ and $\underline{\mathbf{m}}_j^1 = \left(m_{11}^1, \ldots, m_{nj}^1\right)^T$. The outputs of this layer represent the firing strength of the corresponding fuzzy rules. The strength values are then normalized as:

$$\mathbf{a}_j^{(2)} = \frac{\mathbf{n}_j^{(2)}}{\sum_{j=1}^{M} \mathbf{n}_j^{(2)}}, \quad j = 1, 2, \ldots, M \tag{16}$$

Layer 3: Each node in this layer represents the membership function in output space and has two operation modes: structure learning mode and normal mode. The structure-learning mode is used in the structure-learning phase. In this mode, the membership value of d(k), the desired response, in each current node is calculated as follows:

$$\mathbf{n}_q^{s(3)} = -\frac{\left(d(k) - \mathbf{m}_q^{(2)}\right)^2}{\sigma_q^{(2)}},$$
$$\mathbf{a}_q^{s(3)} = \exp\left(\mathbf{n}_q^{s(3)}\right), \quad q = 1, \dots, Q. \tag{17}$$

where Q is the number of fuzzy sets in the output space. In the normal mode, the nodes in this layer perform the defuzzification operation. Because of the presence of conflicting rules in the LAFF, the inference engine is based on a combination of composition-based inference and individual-rule-based inference. The composition-based inference is applied to these rules resulting in one rule, and then the individual rule is applied to the remaining rules. So we may write:

$$\mathbf{a}_q^{n(3)} = m_q \sum_j p_j^{(3)}, \quad q = 1, 2, \dots, Q \tag{18}$$

where c_q is the center of fuzzy set $\mathbf{B}^q$ and $p_j^{(3)}$is the j-th input to the node q in layer 3, and it is calculated as follows:
If rule j is a conflicting rule, then:

$$p_j^{(3)} = \begin{cases} 0 & \text{if } \mathbf{a}_j^{(2)} \mathbf{a}_q^{s(3)} < \mathbf{a}_j^{(2)} \mathbf{a}_{q'}^{s(3)}, \ q' = 1, 2, \dots Q, \ q' \neq q \\ \mathbf{a}_j^{(2)} & \text{otherwise,} \end{cases} \tag{19}$$

else:

$$p_j^{(3)} = \mathbf{a}_j^{(2)}. \tag{20}$$

The parameters of these nodes are m_q 's. Finally the output of the system is obtained as

$$\mathbf{y} = \sum_{q=1}^{Q} \mathbf{a}_q^{n(3)}. \tag{21}$$

4.2 The LAFF structure learning algorithm

The structure learning in the LAFF includes input-space clustering, output-space clustering, and mapping process.

A. Input-space clustering: For input-space clustering, a Modified FCM (MFCM) clustering algorithm is presented. The FCM algorithm developed by Bezdek [24] generalizes the work of Dunn [25] and it is guaranteed to converge [26]. This algorithm is not appropriate for on-line operation and dynamic clustering. Below, a new algorithm, which is suitable for both on-line operation and dynamic clustering, is presented.

Noting that the input clusters represent the precondition of fuzzy rules, therefore we can consider μ_{ik} , the membership value of the data x(k) in the i-th cluster, as firing strength of the i-th rule associated with x(k). Thus, we have:

$$\mu_{ik} = \prod_{j=1}^{n} \mathbf{A}_j^i \left(\mathbf{x}_j(k)\right), \quad i = 1, 2, \ldots, c(k). \tag{22}$$

where c(k) is the total number of clusters at the time instant k. By assuming that the membership functions are Gaussian, we obtain:

$$\mu_{ik} = \prod_{j=1}^{n} \exp - \left(\frac{\mathbf{x}_j(k) - m_{ij}^1(k)}{\sigma_{ij}^1} \right)^2 = \exp\left(\underline{\mathbf{s}}_{ik}^T \underline{\mathbf{s}}_{ik}\right), \tag{23}$$

with $\underline{\mathbf{s}}_{ik} = \mathbf{D}_i \left(\underline{\mathbf{x}}(k) - \underline{\mathbf{m}}_i^1\right)$ and $\mathbf{D}_i = diag\left(\frac{1}{\sigma_{i1}^1}, \ldots, \frac{1}{\sigma_{in}^1}\right)$.

Now we define the following objective function:

$$\mathbf{J} = 0.5 \sum_{i=1}^{c(k)} (\mu_{ik})^p \left\|\underline{\mathbf{x}}(k) - \underline{\mathbf{m}}_i\right\|^2 . \tag{24}$$

where c(k) as defined before is the number of clusters up to time k and μ_{ik} is given in (22). The goal is to adjust $\underline{\theta}_i^1 = \left[\underline{\mathbf{m}}_i^1, \underline{\sigma}_i^1\right]$ such that the objective function J is minimized. Using the steepest descent method, we may have:

$$\begin{aligned}
\underline{\mathbf{m}}_i^1(k+1) &= \underline{\mathbf{m}}_i^1(k) - \alpha \cdot p \cdot (\mu_{ik})^p \cdot \\
&\cdot \left(\mathbf{D}_i^2 \cdot \left(\underline{\mathbf{x}}(k) - \underline{\mathbf{m}}_i^1(k)\right) \cdot \left(\underline{\mathbf{x}}(k) - \underline{\mathbf{m}}_i^1(k)\right)^T - \mathbf{I}\right) \times \left(\underline{\mathbf{x}}(k) - \underline{\mathbf{m}}_i^1(k)\right), \\
\underline{\sigma}_i^1(k+1) &= \underline{\sigma}_i^1(k) - \alpha \cdot p \cdot (\mu_{ik})^p \cdot \\
&\cdot \left(\mathbf{D}_i^2 \cdot diag\left(\underline{\mathbf{x}}(k) - \underline{\mathbf{m}}_i^1(k)\right) \cdot \mathbf{D}_i \cdot \left(\underline{\mathbf{x}}(k) - \underline{\mathbf{m}}_i^1(k)\right) \cdot \left(\underline{\mathbf{x}}(k) - \underline{\mathbf{m}}_i^1(k)\right)^T - \mathbf{I}\right) \times \\
&\times \left(\underline{\mathbf{x}}(k) - \underline{\mathbf{m}}_i^1(k)\right).
\end{aligned} \tag{25}$$

Figure 5 lists general steps of the structure-learning algorithm. In the structure-learning algorithm, $\bar{\mu}$ is a threshold which controls the number of rules; i.e. for a high value of $\bar{\mu}$ more rules are generated.

B. Output-space clustering: The goal here is to generate fuzzy term sets in the output space. The clustering algorithm is summarized in Fig. 6. In this algorithm, $\bar{\mathbf{V}}$ is a threshold which controls the number of output clusters, and y(k-1) is the output of system at time k-1.

IF there are no input clusters,

then

generate the first cluster with

$$\underline{m}_1^1 = \underline{x}(k) \quad \text{and} \quad D_1 = diag\left[\frac{1}{\sigma_{init}^1}, \ldots, \frac{1}{\sigma_{init}^1}\right],$$

assign the means and variances of the Gaussian membership functions as:

$$m_{i1}^1 = x_i, \quad \sigma_{i1}^1 = \sigma_{init}, \quad i = 1, \ldots, n$$

ELSE

calculate μ_{ik} from equation (23),

calculate $D_{ik} = \left\| \underline{x}(k) - \underline{m}_i^1(k) \right\|^2$,

find $\mu_{\max} = \max\limits_{1 \le i \le c(k)} (\mu_{ik})$,

$$D_{\min} = \min_{1 \le i \le c(k)} (D_{ik})$$

if $(\mu_{\max} \ge \bar{\mu})$, then

adjust $\underline{m}_i, \underline{\sigma}_i$ using (25) for the existing rules.

Else

$c(k+1) = c(k) + 1$,

generate a new fuzzy rule with

$$\underline{m}_{c(k+1)} = \underline{x}(k),$$

$$D_{c(k+1)} = diag\left[\frac{1}{\sqrt{D_{\min}}}, \ldots, \frac{1}{\sqrt{D_{\min}}}\right]$$

$$m_{ic(k+1)}^1 = x_i(k), \quad \sigma_{ic(k+1)}^1 = \sqrt{D_{\min}}$$

End

END

Fig. 5. The structure learning steps.

IF there is no output membership function

generate the first membership function with

$m_1^{[2]} = d(k)$, $\sigma_1^{(2)} = \sigma_{init}$

where σ_{init} is a predetermined constant.

ELSE

calculate $V_{qk} = \exp - \left(\frac{d(k) - m_q^{[2]}(k)}{\sigma_q^{[2]}(k)} \right)^2 , q = 1, \ldots, Q(k)$

where Q(k) is the current number of output clusters.

Calculate $D_{qk} = \left(d(k) - m_q^{[2]}(k)\right)^2 , q = 1, \ldots, Q(k)$.

Find $V_{\max} = \max_{1 \le q \le Q(k)} \{V_{qk}\}$, $D_{\min} = \min_{1 \le q \le Q(k)} \{D_{qk}\}$.

if $\{V_{\max} < \bar{V}\}$, $Q(k) = Q(k) + 1$, then

generate a new output cluster with

$m_Q^{(2)} = y(k-1)$, $\sigma_Q^{[2]} = \sqrt{D_{\min}}$

end

END

Fig. 6. The output clustering algorithm.

C. Input-output Mapping: In this phase, the fuzzy rules are then constructed by the following mapping process shown in Fig. 7. After the mapping process, the system starts normal operation to calculate the output y(k). Then the parameter-learning algorithm begins to tune the membership functions. In the parameter-learning phase, we adopt the back-propagation algorithm and an evolutionary algorithm will be introduced in section 4.5.

4.3 The adaptive filter based on fuzzy rules with functional consequents

Sugeno's type fuzzy systems [27], fuzzy rules with functional consequents can be considered as a special class of fuzzy systems that will be briefly described. Figure 8 shows the basic configuration of Sugeno-type fuzzy systems.

In contrast to the linguistic-type fuzzy systems in which the then-parts in fuzzy rule base are also in the form of linguistic variables, here the outputs of the rules are a function of the inputs and consequently their inputs and outputs are real-valued variables (no need of fuzzifier and defuzzifier blocks). As seen in the above figure a set of fuzzy-if-then rules referred to as system's

```
IF a new cluster is created in the input space,
  then
      Link this cluster to the selected or a new output cluster.
ELSE
     If a new cluster is created in the output space,
       then
           link the selected input cluster to this cluster; obviosly,
           conflicting rules may be created.
     Else
          If there exists no link between the selected input and
          output clusters,
           then
               connect the input cluster to that of the output.
          end
     End
END
```

Fig. 7. The main steps of mapping process.

knowledge base are evaluated at time instants in parallel (i.e. several rules can be performed together). These rules relate the input and output variables of the fuzzy system.

The general form of the fuzzy if-then rule is

$$if \quad \mathbf{x}_1 \quad is \quad \mathbf{P}^i_1, \ldots, \quad \mathbf{x}_n \quad is \quad \mathbf{P}^i_n, \quad then \quad \mathbf{y}^i =$$

$$\left[\theta^i_0(k), \ldots \; \theta^i_n(k)\right] \cdot \begin{bmatrix} 1 \\ \underline{\mathbf{x}}(k) \end{bmatrix} = \left(\underline{\theta}^i\right)^T \cdot \underline{\mathbf{q}}.$$

In the above $\mathbf{P}^i_j$'s are linguistic terms (fuzzy sets) with membership functions which are automatically tuned, and the θ^i_j's are real values adjustable parameters, $\underline{\mathbf{x}}$ is the input and y^i is the output of the fuzzy system due to the i-th rule with the if-part of fuzzy variables and then-part of non-fuzzy (a linear combination of the inputs.

Denoting the strength of rule i for the input $\underline{\mathbf{x}}$ by $s^i = \prod_{j=1}^n P^i_j(x_j)$, we may have the following equation relating the output y of the fuzzy system to the input vector $\underline{\mathbf{x}}(k)\epsilon\Re^n$.

$$y\left(\underline{\mathbf{x}}, \underline{\theta}\right) = \sum_{i=1}^{M} s^i_{(n)} \cdot y^i; \quad s^i_{(n)} \equiv \frac{s^i}{\sum_{i=1}^M s^i}.$$

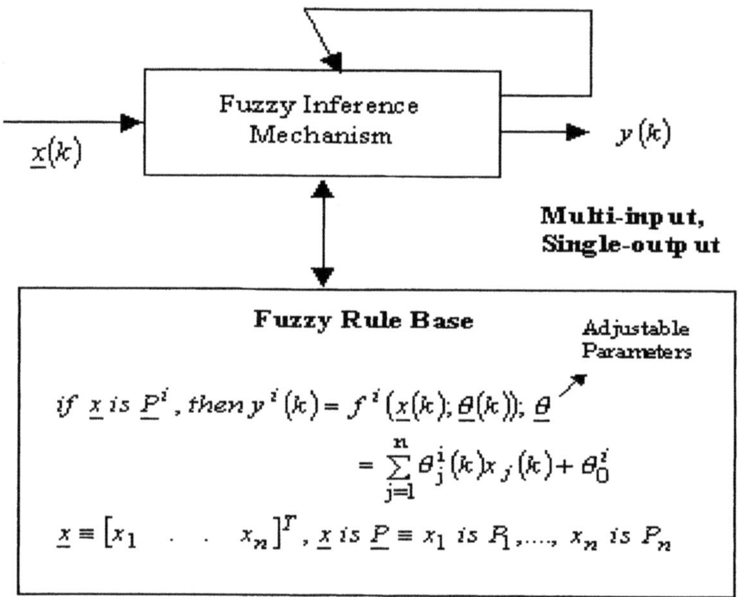

Fig. 8. Block Diagram of a Sugeno-type Fuzzy System.

It represents a parametric-based compact system in which the parameters can be tuned (estimated); note that the order of the system M must be first determined. It is worth mentioning that this freedom of tuning and shaping of the fuzzy system is obtained in the expense of losing the general and natural structure in which the then-part does not preserve. However, these systems (Sugeno- type fuzzy models) become well-defined for system modeling by interpolating linear models that in turn makes them very well-suited to mathematical analysis (i.e. convergent and stability) and further to the applications of classical adaptive (learning) methods such as recursive least squares and least mean squares as we will see in the next subsection.

The adaptive filter in Fig. 2 represents a fuzzy system whose structure is given in Fig. 9. The fuzzy system consists of a set of rules of the form:

$\mathbf{R}^j$: if $\mathbf{x}_1$ is $\mathbf{A}_1^j$ and ... and $\mathbf{x}_n$ is $\mathbf{A}_n^j$
then $y^j = \sum_{i=0}^{n} \theta_i^{2j} \cdot x_i$, j=1,..., M.

where $\mathbf{A}_i^j$'s are fuzzy sets in the input space. As explained in the previous subsection the final crisp value of the fuzzy inference system becomes:

$$\mathbf{y} = \frac{\sum_{j=1}^{M} w_j \sum_{j=1}^{M} w_j}{,} \quad w_j = \prod_{i=1}^{n} \mathbf{A}_i^j(x). \tag{26}$$

We call this filter Sugeno-type Adaptive Fuzzy Filter (SAFF).

Like the LAFF, this filter also is represented by a network with three

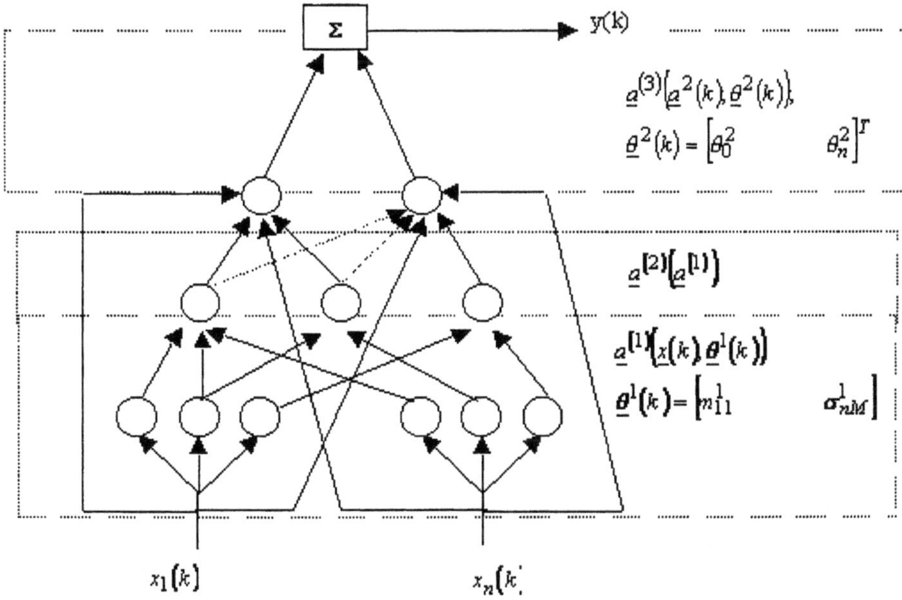

Fig. 9. Structure of the SAFF.

layers. Layers 1 and 2 act the same as those given in the LAFF.

Layer 3: This layer is called the consequent layer. The function of each node in this layer is formulated as:

$$\mathbf{n}_j^{(3)} = \sum_{i=0}^{n} \theta_i^{2j} \cdot \mathbf{x}_i, \quad \mathbf{x}_0 = 1,$$
$$\mathbf{a}_j^{(3)} = \mathbf{a}_j^{(2)} \cdot \mathbf{n}_j^{(3)}, \quad j = 1, 2, \dots, M.$$

where M is the current number of rules. The parameters of these nodes are θ_i^{2j} 's, i=1,2,..., n.

Output: This node integrates all the actions given by the nodes of the third layer and acts as a defuzzifier. That is, $y(k) = \sum_q^M \mathbf{a}_q^{(3)}$.

4.4 The SAFF structure-learning algorithm

The structure learning in the SAFF includes input-space clustering and consequent identification process.

A. Input-space clustering: It is the same as that of for the LAFF discussed in section 4.2.
B. Consequent identification: When a new cluster (new fuzzy rule) is created in the input space, we should decide about the consequent of the rule. The following algorithm does this.

If a new cluster is created in the input space,
then
set $b_0^{c(k)} = y(k-1)$,
$b_1^{c(k)},\ldots b_n^{c(k)}$: arbitrary initial values and y(k-1) is the output of the system at time (k-1).
End

Fig. 10. Consequent Identification.

After the structure learning process, the network starts normal operation to calculate the output y(k). Then, the parameter-learning algorithm, which will be discussed in the next section, begins to tune the membership functions.

4.5 Learning algorithms

The main objective of a fuzzy system is a set of verbal descriptions consisting of associations of fuzzy variables and procedures for inference prescribing qualitative (approximate) actions for a given state of the system under study. In contrast to conventional filtering in which the physical process is modelled, in fuzzy based systems the objective is to involve expert human knowledge (to incorporate expert human operator) in the design process.

The dream of using all information (qualitative and quantitative) in a systematic yet scientific manner came true through self-tuning (feedback) systems of a hybrid combination of fuzzy and learning systems (i.e., neural networks or evolutionary systems) in which verbal information (non-numerical information) provided by human experts is given in the form of (adjustable) parameterized fuzzy terms while numerical information is obtained from the sensors. However, in general and in most practical real- world engineering problems such as filtering, the system under study is modelled by a set of fuzzy if-then rules in terms of linguistic words using fuzzy logic principles (i.e. the extension principle). To make it self-fabricatory or adaptive, its structure including its rules are automatically tuned by employing some learning mechanism through learning training phase using the empirical input/output data measured by the sensors. Any learning system employs an optimization technique.

Most traditional optimization methods can be divided into two broad categories: direct search and gradient search. Direct methods require only the objective function values while the gradient methods need gradient information (exact or approximate) with the common characteristic that they all work on a point-by-point basis, i.e., they begin with an initial solution (usually applied by the user) and a new solution is calculated according to the steps of the algorithm. In general, direct search methods, however they require no gradient calculations, are computationally extensive and in most cases they work on some simple unimodal objective functions. On the other hand, gradient-based techniques require the knowledge of gradients of functions and constraints and in most cases they suffer from the problems of gradient calculations and local optimality. The learning systems used are the Back Propagation (BP) learning (gradient search) and evolutionary (direct search) algorithms. First a very brief review of the BP is given and a new evolutionary algorithm will be fully introduced.

4.5.1 A Short Review on Backpropagation: The BP as an approximated steepest descent numerical optimization method iteratively adjusts the tunable parameters of a layered structure neural networks. It updates the parameters in such a way that the networks (learning system) approaches a solution acceptable in some metric sense showing that successfully learns the task of training. Though there are many versions of BP [28], the standard error BP is described by the following generic rule:

$$\theta(k+1) = \theta(k) + \Delta\theta(k), \quad \Delta\theta(k) \equiv -\alpha \cdot g(k), \quad 0 < \alpha < 1.$$

At each iteration (or epoch for batch backpropagation), the gradient g(k) is calculated and the parameter $\theta(k)$ is updated based on the step size known as learning rate α. This is an approximate steepest descent guaranteeing that the objective function (the total error) is reduced at each iteration or epoch so long as the step size α is sufficiently small and the gradient g(k) is not sufficiently close to zero.

Now consider a feed-forward NN with L layers, like Fig. 4 and Fig. 9, the forward pass produces the outputs for a given input vector $\underline{\mathbf{x}}$ through the calculation of each element $\underline{\mathbf{a}_i^l(k;\mathbf{x})}$ for l=1,2,..., L≡number of layers, in our case L= 3. In the backward pass the instant error $\mathbf{J}\left(k,\underline{\theta};\underline{\mathbf{x}}\right) = 0.5\left(d\left(k;\underline{\mathbf{x}}\right) - \left(y\left(k;\underline{\mathbf{x}}\right)\right)\right)^2$ and the corresponding gradients $g\left(k;\underline{\mathbf{x}}\right)$ with elements $\frac{\partial J}{\partial \theta_{ij}^{lm}}$ as:

$$\frac{\partial J}{\partial \theta_{ij}^{lm}} = -s_i^l\left(k;\underline{\mathbf{x}}\right) \cdot a_i^{l-1}\left(k;\underline{\mathbf{x}}\right).$$

with the sensitivity recursive equations is given by

$$s_i^L(k;\underline{\mathbf{x}}) = (d(k;\underline{\mathbf{x}}) - (y(k;\underline{\mathbf{x}}))) \cdot \dot{f}\left(a_i^L(k;\underline{\mathbf{x}})\right),$$

$$s_j^q(k;\underline{\mathbf{x}}) = \dot{f}\left(a_j^q(k;\underline{\mathbf{x}})\right) \cdot \sum_{N(i;j)} s_i^l(k;\underline{\mathbf{x}})\,\theta_{ij}^{lq}, \quad q < L.$$

where $N(i;j) \equiv$ {all nodes in layer i which are connected to the neuron j in layer q}, $\theta_{ij}^{lq} \equiv$ The parameter connectes the i-th node in layer l to the node j in layer q, and $\dot{f}(x) = \frac{df}{dx}$; the first derivative of f w.r.t. x.

4.5.2 The evolutionary algorithm: Due to its ability to model genetic inheritance and Darwinian Strife for survival, the Evolutionary Computation (EC), as a modern heuristic random search (guided), plays an important role in computational intelligence-based techniques and is getting popular in real world engineering problems such as filtering. ECs consisting of the principal components such as Genetic Algorithms (GA) and Evolutionary Strategies (ES) are mainly based on principle of evolution and mechanism of fittest survival modelling the human mutation. The idea of EC has been emerged more than 30 years on the USA and Europe, independently. At the university of Michigan in 1975, the GA by Holland and the ES at the Berlin of technology in 1973 by Rechenberg have been developed [29–31]. They introduced a novel optimization algorithm known as Evolutional algorithm that is structurally different from the two major classes of classical calculus-based and enumerative techniques [32,33]. The EC is good for ill-defined or multi-modal cost functions in which local optima are frequently observed, since it tends to seek the global optimum solution without getting trapped at local minima; this is often claimed, but the truth is an EC can get trapped anywhere on the function.

The GA in EC appeared to become more popular. In 1985, a ten years perspective of GA along with several research directions has been summarized by DeJong [34]. Grefenstette studied and listed the effects of control parameters on the performance of GA [35]. The GA was then extensively explored by Goldberg [36]. As a type of structured random search, the GA is compared with Simulated Annealing, another type of random search method, by Davis and Steenstrap [37], in 1987. It has been shown that the GA functions as a totally different optimization algorithm than the simulated annealing [38], however there is mathathematical equivalence between GA, evolutionary strategies [39] and evolutionary programming [40]. Because of its ease of implementation procedure, the GA has been successfully used as an optimization tool for a variety of optimization problems such as machine learning [41], delay estimation of sampled signals [42,43], robotic trajectory planning [44], fuzzy controllers [45–47], neural networks [48]-[53], and linear transportable process [54].

Knowing that solving a problem is nothing but to seek the best solution in the space of potential solutions known as population, the following figure illustrates the general structure of any EC. As easily observed from Fig. 11,

```
Initialize population (at time =0).
Evaluate the population
Do
    time=time+1,
    choose population (time) from population (time-1),
    act on population (time) by some evolutionary operator(s),
    evaluate population (time),
until (stopping criteria).
```

Fig. 11. The general structure of EC.

the evolution is simulated as an optimizing process. The process of selection attempts to move the response as close to the best response as possible. The selection mechanism is merely probabilistic. Hoping that the best individual represents a reasonably suboptimal solution, the algorithm converges after some number of iterations. Different data structure, different sort of controlling the search process, different operators result in different ECs.

To produce a new generation, one may 1) to replace the whole generation by a population of children, 2) select the best individual from the two population of parents and children, 3) produce possibly few children replacing some worst individual, 4) keep the best individual from one generation to the next by using an elitist model. The most important components of any EC are the data structure employed for a particular problem and the set of operators.

In solving a given optimization task, the GA starts with a collection of solutions (i.e. parameter estimates) known as chromosomes. Each individual (chromosome) is evaluated for its fitness. In each iteration of the GA, the best fit-chromosomes (parents) are allowed to mate and bear offspring (produce new individuals using the operators reproduction, crossover and mutation). These individuals (children) or new parameter estimates provide the basis for the next generation. The population size plays a crucial role in GA and offers many considerations: optimal population size and quality of using individual's age in the genetic process. Adding to them, the considerations like efficient selection method of individuals in population, and efficient crossover operator, it becomes clear why the conventional GA is not guaranteed to yield an online optimum solution, and often performs rather poorly and there are few applications of GAs for online problems. There are many researchers who have employed different ways to resolve the aforementioned problems by proposing some corrections on the procedure of simple GA; i.e., the interested reader may refer to [55] for optimization of control Parameters, to [56–58] for population size, to [59] for fitness function, to [60] for aging of individuals, to [61–63] for mutation operator, to [64,65] for crossover operator, to [66,67] for migration operator, and to [68,69] for multi-population of GA.

In contrast to GA, the earliest ES is based on a population consisting of one individual. An individual is represented as a pair of float-valued vectors. In compared with the GA, the ES is indeed a real coded process. The ES in fact represents an individual-based optimization method that uses only the mutation operator to produce offspring from the current parent. Here the mutation is realized by replacing the current individual, q(t) with q(t)+N(0,s), where N is a vector of independent random Gaussian numbers with a mean of zero and standard deviations s. Note that the (m,l)-ES and (m+l)-Es have been developed [38] that produce new population from the present population. Next section offers modifications to the conventional GA that are designated to increase its suitability for online problems.

4.5.3 The OGA as an EC: To begin, first consider a conventional EC whereby a population of individuals is produced by a stochastic selection from the response space. Individuals are evaluated in light of a fitness function, and the best individual from the present population is identified. We call this elite individual the "queen". Individuals are chosen as parents by roulette wheel selection and then undergo recombination. Whereas individuals in a conventional EC are recombined two-by-two, in the OGA the queen recombines with all individuals. After the individuals recombine with the queen, mutation is also applied to each. This formula is then iterated over generation. By enforcing recombination with a single queen, we expect that the algorithm may be more quickly trapped in local optima. To prevent this, the probability of mutation is increased. This recombination process generates a set of offspring that are essentially the queen, with a part changed through recombination. The following figure shows the main steps of the OGA [71].
Convergence Theorem: For a regular function optimization problem with an objective function g, $prob\,(\lim_{t\to\infty}) = 1$ provided that the optimum g* is finite.
Proof: will be reported in near future.

Before presenting the application of the OGA, which will be discussed in section 4.5.5, to the fuzzy networks given in sections 4.3-4.4, we consider the following optimization problem in order to show the high performance of the OGA.

4.5.4. Function Optimization Example: Consider the following function. This function has a unique global maximum surrounded by so many local maxima. As one observes from Figure 13, this function possesses a unique global maximum of 1 at the origin surrounded by an army of local optima. Because of these oscillations many different optimization techniques

```
Select a coding scheme to represent adjustable parameters.
Initialize population (at time =0) by taking randomly a chromosome as a
queen.
Evaluate the chromosome.
Do
    time=time+1,
    perform mutation on chromosome (time),
    evaluate the chromosome (time),
until (stopping criteria).
```

Fig. 12. The structure OGA.

had hard time to find the maximum of this function. To show this, we apply two conventional hill-climbing methods (S.D. and CG), the CGA and OGA to find the optimum of this function. The results are summarized in Fig. 14

$$f\left(x_1, x_2\right) = \frac{\cos^2\left(\left(x_1^2 + x_2^2\right)^{0.5}\right)}{1 + 0.001\left(x_1^2 + x_2^2\right)}.$$

Because the classical methods, steepest descent and conjugate gradient,

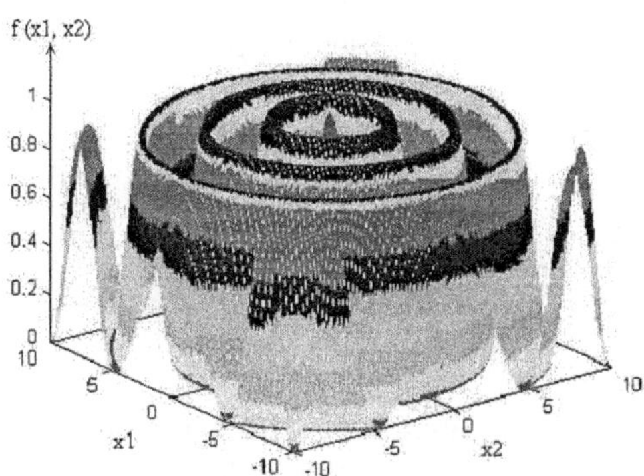

Fig. 13. 3-D plot of the function in the function optimization example.

act on individuals, for the purpose of comparison, we set the initial conditions to four different values as (x_1, x_2) = (32,32), (x_1, x_2)=(10.7,10.7), (x_1, x_2)=(32,10.7) and (x_1, x_2) = (-3.5,-3.5). For each initial condition, we performed the methods 40 runs (times) on a Pentium III PC and their average values of running times are listed in Table 1.

	SD	CG	OGA
Initial Point	(31.9844, 31.9844)	(31.9844, 31.9844)	(31.9844, 31.9844)
Final Point	(31.0896, 31.0896)	(31.0976, 31.0976)	(0, 0)
Mean of Time (Sec.)	Not Converged	Not Converged	6.9702
Initial Point	(106562, 106562)	(106562, 106562)	(106562, 106562)
Final Point	(11.0983, 11.0983)	(11.1060, 11.1060)	(0, 0)
Mean of Time (Sec.)	Not Converged	Not Converged	5.1053
Initial Point	(31.9844, 10.6562)	(31.9844, 10.6562)	(31.9844, 10.6562)
Final Point	(32.7708, 10.9182)	(32.7828, 10.9222)	(0, 0)
Mean of Time (Sec.)	Not Converged	Not Converged	6.3110
Initial Point	(-3.5312, -3.5312)	(-3.5312, -3.5312)	(-3.5312, -3.5312)
Final Point	(-4.4386, -4.4386)	(-4.4424, -4.4424)	(0, 0)
Mean of Time (Sec.)	Not Converged	Not Converged	5.5655

Fig. 14. Simulation Results for the function approximation example.

As we could expect the hill climbing methods starting from a randomly generated point in the two dimensional parameter space and following the steepest descent gradient directions get trapped in one of the local optima. In contrast, the evolutionary algorithms (both OGA and CGA) show robust performance though the OGA presents a more robust performance than the CGA. This means that both CGA and OGA reach the maximum value above the 901 at (0,0) within a few ten evaluations and the CGA reaches the maximum value above the 99the performance of the CGA for a run in which a population of 100 strings has been generated in 10000 epochs with run time of 5175 seconds. In compared with the OGA, the CGA could reach the maximum value of 0.9902 instead of 1; remember that the value of 1 is the global maximum.

5.5.5. Fuzzy Network Design by the OGA: This section is devoted to show how one may design a fuzzy network like the networks given in figures 4 and 9. Each fuzzy network consists of two parts: Topology and Rule set. The networks Topology covers the number of membership functions (term sets) corresponding to each system's input/output, their locations on the universal discourse and the parameters representing the shape of each membership function, while the number of rules and their forms make the network's Rule set. First, the design of fuzzy network topology is introduced and then its rule set design will be discussed.

A-Fuzzy Network Topology
The goal is to use the OGA to determine the best number of membership functions, and their shapes for each system's input and output. To do so, we may use a bit string to model the number of membership functions (term sets) and their location on the given I/O reference sets, then apply the OGA to obtain the best arrangement. For example, consider the Gaussian membership function. We may use the bit string 1001101, to represent the number of term sets and their distributions on the universal discourse depicted in Fig. 15.

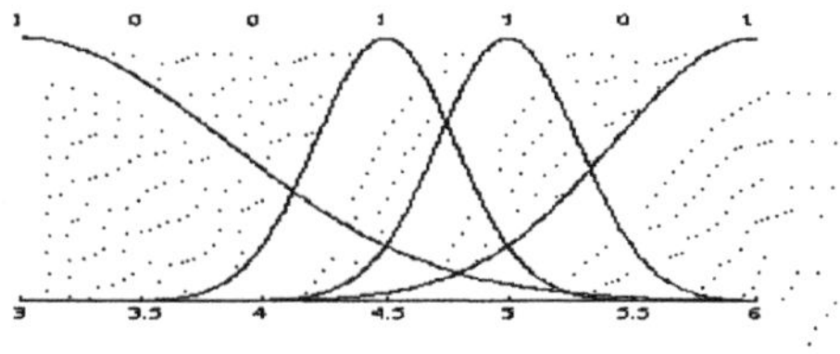

Fig. 15. Representation of term sets by Bit strings.

The procedure may be summarized as follows. First, the universal discourse of each fuzzy variable is divided into some, say N, parts each containing the center of one membership function. For example, in Fig. 15, the universal discourse of fuzzy variable x is divided into 7 parts. On the other hand, each bit of the string shows one end/first part of the universal discourse. It is assumed that a membership function (term set) exists if its end/first part bit is 1, and there will be no term set if the corresponding bit 0. In Figure 14 there are four "1" bits; hence this fuzzy variable will have 4 membership functions (term sets) with the center points 3, 4.5, 5, 5.5, and 6. The standard deviation of Gaussian membership function can be determined by K-means rule. You may use other rules for determination of the standard deviations. Of course, the designer is free to choose the kind of membership function such as triangular membership functions, see Fig. 16.

B-The Rule Set
The objective here is to use the OGA to find the best set of rules along with their forms known as the network Rule set for a given optimization problem. For example, if the system has m inputs and each one has m_i membership functions (term sets), then the maximum non-conflict accessible fuzzy rules are $\prod_{i=1}^{m} m_i$. Now if the system has one output with n membership functions on its universal discourse, then each rule from the accessible rule space has

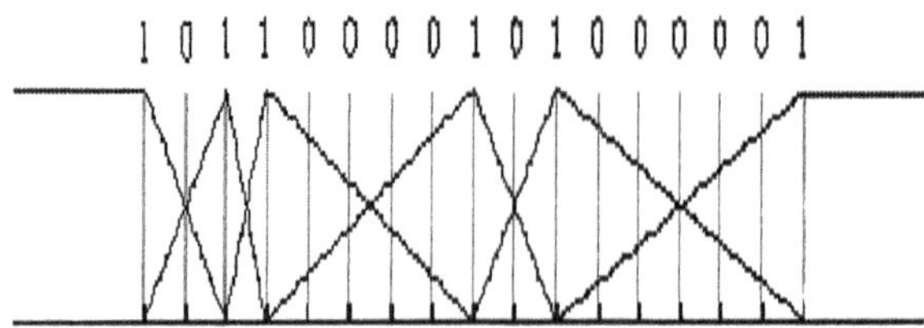

Fig. 16. Triangular membership functions.

n choices. There are two ways to choose the rule-base:
1) The OGA acts on the total rule space by a bit string with length $\prod_{i=1}^{m} m_i \cdot n$; if the bit is one, the rule selected, and if it is zero, the rule is disregarded. Here we may end up with some some conflict rules, which must be identified, and take only one of them randomly.
2) The OGA acts on the accessible rule space by a bit string with length $\prod_{i=1}^{m} m_i$. In this case, the OGA selects the if-part of the rule base if the bit is one and reject it, if the bit is zero. When an if-part of the rule base is chosen, the then-part rule is randomly chosen from the set of n output membership functions.

4.6 Simulation results

Using the learning algorithms discussed in section 4.5, the performances of the LAFF and the SAFF have been evaluated by Monte Carlo simulations over the test trajectory given in Fig. 17, and the results were compared with those of the IMM algorithms, FGIMM, SGIMM and AGIMM [20]. The position RMSE for LAFFF and IMM- algorithms are shown in figures 18-20 and those of the SAFF are given in figures 21-23. These results show the good performance of proposed adaptive filters in comparision with the IMM algorithms. Noting that the perfomance of the LAFF outperforms that of the SAFF as expected. In the simulations we applied the BP learning algorithm. The application of OGA for this case study is under investigation and will be reported later.

5 Conclusion

After briefly reviewing the classical adaptive filtering for maneuvering target tracking problem, this chapter introduced the LAFF and the SAFF adaptive fuzzy filters. It has also discussed the training of the fuzzy filter using gradient technique and a proposed evolutionary method. Finally the classical and fuzzy filters have been applied to the target-tracking problem.

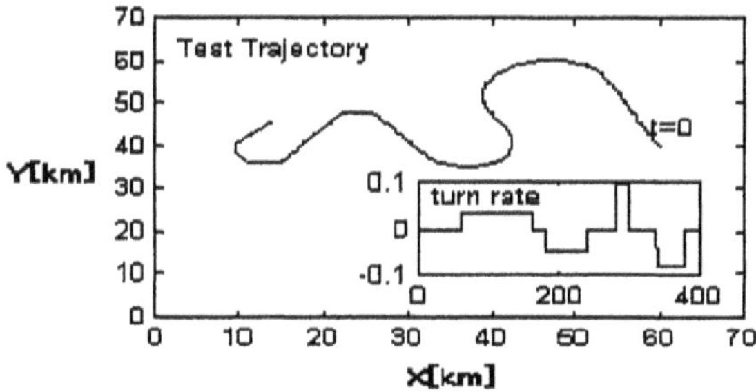

Fig. 17. Test Trajectory.

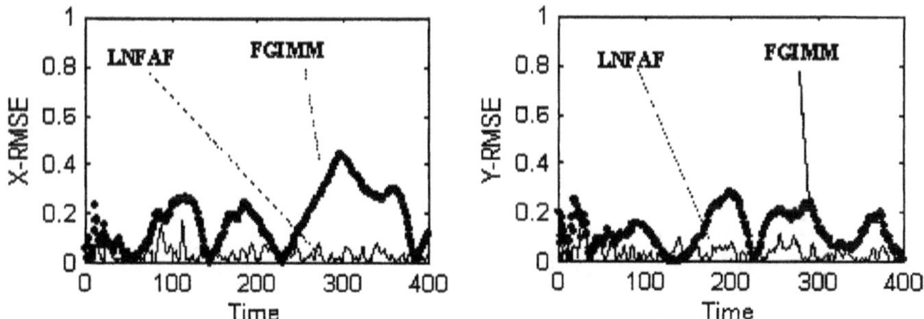

Fig. 18. RMSE of the LNFAF and the FGIMM.

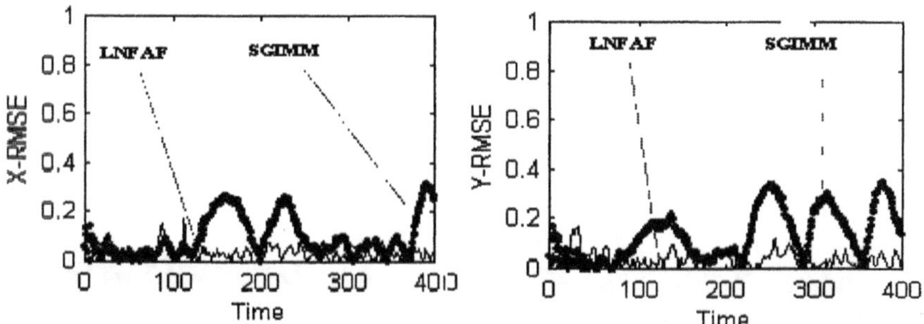

Fig. 19. RMSE of the LNFAF and the SGIMM.

References

1. Singer R. A. , *Estimating Optimal Tracking Filter Performance for Manned Maneuvering Targets*, IEEE Transactions on Aerospace, Electronic Systems, vol. AES-6, pp. 478-483, July 1970
2. McAulay R. J. , *A decision-directed Adaptive Tracker*, IEEE Transactions on Aerospace, Electronic Systems, vol. AES-9, pp. 229-236, March 1973
3. Gholson N. H. , Moose E. L. , *Maneuvering Target Tracking Using Adaptive State Estimation*, IEEE Transactions on Aerospace, Electronic Systems, vol. AES-13, pp. 310-317, May 1977

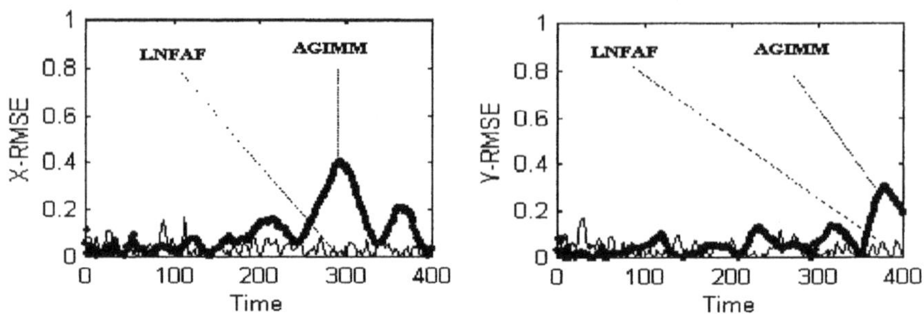

Fig. 20. RMSE of the LNFAF and the AGIMM.

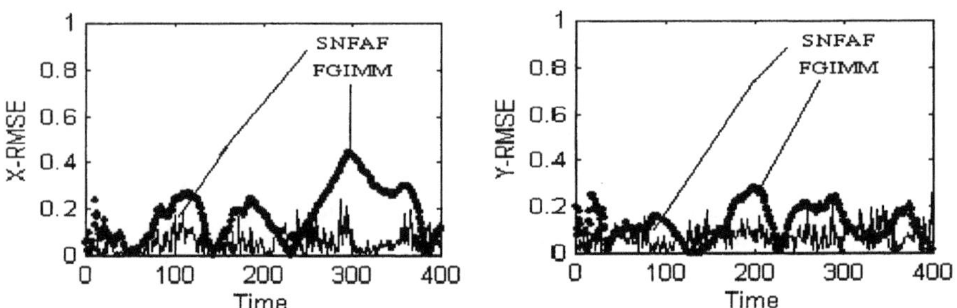

Fig. 21. RMSE of the SNFAF and the FGIMM.

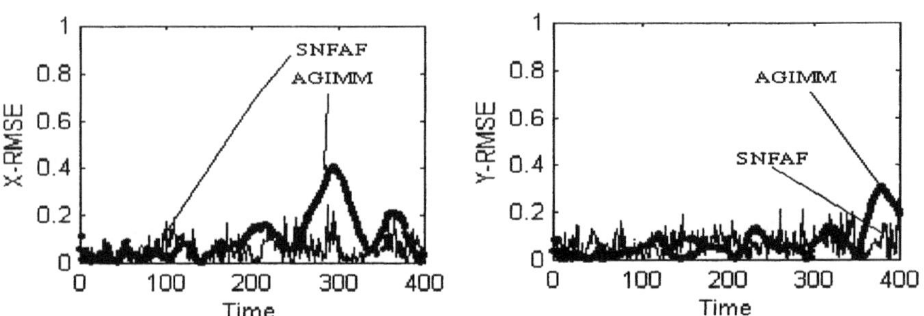

Fig. 22. RMSE of the SNFAF and the AGIMM.

4. Ricker G. G. ,Williams J. R. , *Adaptive Tracking Filter for Maneuvering Targets*, IEEE Transactions on Aerospace, Electronic Systems, vol. AES-14, pp. 185-193-317, January 1978
5. Chan Y. T. ,Hu A. G. C. and Plant J. B. , *A Kalman Filter Based Trackeing Scheme with Input Estimation*, IEEE Transactions on Aerospace, Electronic Systems, vol. AES-15, pp. 237-244, March 1979
6. Moose R. R. , *Modeling and estimation for tracking maneuvering targets*, IEEE Transaction on Aerospace and Electronic Systems, vol. AES-15, pp.448-456, 1979

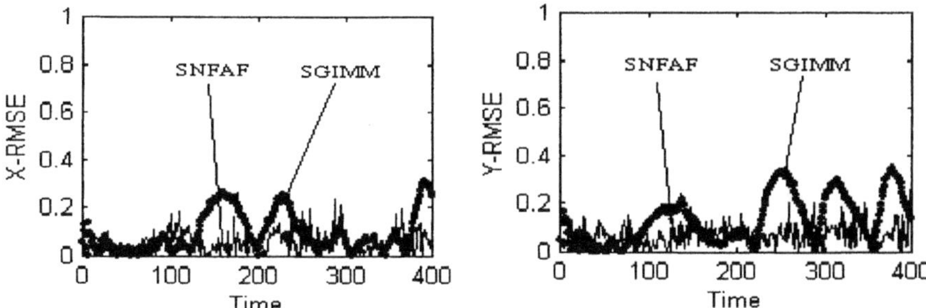

Fig. 23. RMSE of the SNFAF and the SGIMM.

7. Chan Y. T. , Plant J. B. , and Bottomley J. R. T. , *A Kalman Tracker with a Simple Input Estimator*, IEEE Transactions on Aerospace, Electronic Systems, vol. AES-18, pp. 235-241, March 1982
8. Speyer J. L. ,Song T. L. , *A Comparison between pseudomeasurements and extended Kalman Observers*, Proceedings of IEEE Conference on Decision and Control, pp. 324-329, December 1981
9. Bar-Shalom Y. , Birmiwal K. , *Variable Dimension Filter for Maneuvering Target Tracking*, IEEE Transactions on Aerospace, Electronic Systems, vol. AES-18, pp. 621-629, September 1982
10. Berg R. E. , *Estimation and Prediction for Maneuvering Target Trajectories*, IEEE Transactions on Automatic Control, AC-28, pp. 294-304, March 1983
11. Lin C. F. ,Shafroth M. W. , *A Comparative evaluation of some Maneuvering Target Tracking Algorithms*, Proceedings of the AIAA Guidance and Control Conference, pp. 39-56, August 1983
12. Bolger P. L. , *Tracking a Maneuvering Target Using Input Estimation*, IEEE Transactions on Aerospace, Electronic Systems, vol. AES-23, pp. 298-310, May 1987
13. Song T. L. , J. L. Ahn, and Park C. , *Suboptimal Filter Design with Pseudomeasurements for Target Tracking*, IEEE Transactions on Aerospace, Electronic Systems, vol. AES-24, pp. 28-39, January 1988
14. Blom H. A. P. , Bar-Shalom Y. , *The Interacting Multiple Model Algorithm for Systems with Markovian Switching Coefficients*, IEEE Transactions on Automatic Control, vol. 33, no. 8, pp. 780-783, August 1988
15. Bar-Shalom K. C. , Blom H. A. , *Tracking a Maneuvering Target Using Input Estimation Versus the Interacting Multiple Model Algorithm*, IEEE Transactions on Aerospace, Electronic Systems, vol. AES-25, pp. 296-300, March 1989
16. Rong Li X. ,Bar-Shalom Y. , *Performance Prediction of the Interacting Multiple Model Algorithm*, IEEE Transactions on Aerospace, Electronic Systems, AES-29, no.3, pp. 755-771, July 1993
17. Cloutier J. R. C. , Yang C. , *Enhanced Variable Dimension Filter for Maneuvering Target Tracking*, IEEE Transactions on Aerospace, Electronic Systems, AES-29, no. 3, pp. 786-797, July 1993
18. Whang I. H. ,Lee J. G. , Sungand T. K. , *A Modified Input Estimation Technique Using Pseudoresiduals*, IEEE Transactions on Aerospace, Electronic Systems, AES-30, no. 2, pp. 591-598, April 1994

19. Mazor E. ,Averbuch A. ,Bar-Sholaom Y. ,and Dayan J. , *Interacting Multiple Model Methods in Target Tracking*, Aerospace and Electronic Systems, Vol. 34, No.1, pp. 343-349, January 1998
20. Jilkov V. P. ,Angelova D. S. ,Semerdjiev TZ. A. , *Design and Comparison of Mode-set Adaptive IMM Algorithms for Maneuvering Target Tracking*, IEEE Trans. on Aerospace and Electronic Systems, Vol. 35, No.1, pp. 343-349, January 1999
21. Mendel J. M. , *Kalman-Bucy Filtering*, Lessons in Estimation Theory for Signal processing, Communications, and Control, Prentice Hall PTR, Englewood Cliffs, New Jersey 07632, 1995
22. Wang Li-Xin, *Training of Fuzzy Logic Systems Using Back-Propagation*, Adaptive Fuzzy Systems and Control: Design and Stability Analysis, Englewood Cliffs, New Jersey, 1994
23. Lin C. T. , *On-line Supervised Structure/Parameter Learning*, Neural Fuzzy Control Systems with Structure and Parameter Learning, World Scientific, London, 1994
24. Bezdek J.C. , Pattern recognition with fuzzy objective function algorithms, year of publication, Plenum Press, N.Y., 1981
25. Dunn J. C. , *A fuzzy relative of the Isodata process and its use in detecting compact, well separated clusters*, Journal of Cybernetics, Vol.3, pp.32-57, 1974
26. Bezdek J. C. and Hathaway R. J. , *Convergence theory for fuzzy c-means: Counterexamples and repairs*, IEEE Transaction Systems, Man and Cybernetics, vol. SMC-17, no. pp.873-877, 1987
27. Takagi T. ,Sugeno M. , *Fuzzy identification of systems and its application to modeling and control*, IEEE Transaction On Systems, Man and Cybernetics, SMC-15, pp. 116-132, 1985
28. Hagan M. T. ,Menhaj M. B. , *Training feedforward networks with the Marquardt algorithm*, IEEE Transactions on Neural Networks, vol. 5, No. 6, pp. 989-993, 1994
29. Fogel D. B. , Evolutionary computation: Toward a new philosophy of machine intelligence, IEEE Press , Piscataway, NJ, 1995
30. Holland J. H. , *Genetic algorithms and the optimal allocations of trails*, SIAM Journal of Computing, 2(2),pp. 88-105, 1973
31. Holland J. H. , Adaptation In Natural And Artificial Systems, Ann Arbor, MI: The University of Michigan Press, 1975.
32. Scales E. , Introduction to Nonlinear Optimization, Springer-Verlag, N. Y. 1985
33. Bellman R. E. , Dynamic Programming, Princeton Univ. Press, Princeton, N. J., USA, 1957
34. DeJong K. A. , *Genetic Algorithms: A 10 Years Perspective*, Proceedings of an International Conference on Genetic Algorithms and Their Applications, pp. 169-177, 1985
35. Grefenstetle J. J. , *Optimization of Control Parameters for Genetic Algorithms*, IEEE Trans. on Systems, Man and Cybernetic, SMC-16, No. l, pp. 122-128, Jun-Feb., 1986
36. Goldberg D. E. , Genetic Algorithms in Search, Optimization, and Machine Learning, Addison-Wesley, Publishing Company 1988
37. Davis L. Steenstrup M. , *Genetic Algorithms and Simulated Annealing*, In GAs and SA, L. Davis, Ed., Pitman, London, UK, pp. 1-11,1987
38. Kirkpatrick S. , Gellat Jo C. D. , and Vecchi M. P. , *Optimization by Simulated Annealing*, Science, Vol. 220, No. 4598, pp. 671-680, 1983

39. Back T. ,Hoffmeister F. , and Schwefel H. P. , *A Survey of Evolution Strategies,* in R. K. Belew, and L. B. Booker, Eds. Proc. 4th International Conference on Genetic Algorithms, pp. 2-9, San Mateo, CA: Morgan Kaufmann, 1991
40. Fogel L. J. , Owens A. J. , Walsh and M. J. , Artificial Intelligence Through Simulated Evolution, New York: John Wiley, 1966
41. Fogel D. B. , System Identification Through Simulated Evolution: A Machine Learning Approach to Modeling, Needham Heights, MA: Ginn Press, 1991
42. Chan Y. T. ,Riley J.M. and Plant J. B. , *Modeling of Time Delay and it's Application to Estimation of Nonstationary Delays*, IEEE Trans. Acoust. Speech, Signal Processing, ASSP-29, pp. 577-581, June 1981
43. Elter D. M. ,Masukawa M. M. , *A Comparison of Algorithms for Adaptive Estimation of The Time-Delay Between Sampled Signals*, proceedings of IEEE int. conf. on ASSP., pp. 1253-1256, 1981
44. Davidor Y. , Genetic Algorithms and Robotics, A Heuristic Strategy for Optimization, World Scientific, Singapore, 1991
45. Karr C. L. , *Genetic Algorithms for Fuzzy Controllers*, A.I. Expert, Vol. 6, No. 2, pp. 26-33, 1991
46. Park D. ,Kandel A. , and Langholz G. , *Genetic-Based New Fuzzy Reasoning Models with Application to Fuzzy Control*, IEEE Trans. on Sys., Man, and Cyb., Vol. 24, No. 1, January 1993
47. Nomura H. ,Hayashi I. and Wakami N. (editors), *A Self Method of Fuzzy Reasoning by Genetic Algorithms*, in Fuzzy Control System, Kandel and Langholz , CRC Press, Inc., London, pp337-354, 1994
48. Linkens D. A. ,Nyongesa H. O. , *Genetic Algorithms for Fuzzy Control Part 2: Online System Development and Application*, IEE Proc.-Control Theory Appl., Vol. 142, No. 3, May 1995
49. Montana D. J. ,Davis L. , *Training Feed-forward Neural Networks Using Genetic Algorithms*, Proc. of Int. Joint Conf. on Artificial Intelligence (Detroid), pp. 762-767, 1989
50. Brill F. Z. ,Brown D. E. , and Martin W. N. , *Fast Genetic Selection of Features for Neural Networks Classifiers*, IEEE Trans. on Neural Networks, Vol. 3, No. 2, March 1992
51. Maniezzo V. , *Genetic Evolution of the Topology and Weight Distribution of Neural Networks*, IEEE Transactions on Neural Networks, Vol. 5, No. 1, pp. 39-53, Jun 1994
52. Angeline P. J. , Saunders Gregory M. , and Pollack Jordan B. , *An Evolutionary Algorithm that Construct Recurrent Neural Networks*, IEEE Transactions on Neural Networks, Vol. 5, No.1, Jan. 1994
53. Tang K. S. ,Man K. F. ,and Chan C. Y. , *Genetic Structure for NN Topology and Weight Optimization*, 1st IEE/IEEE int. conf. on GAs in Engineering System, innovations and Applications, pp. 280-255, 12-14, Sept. 1995
54. Vignaun G. A. ,Michalewicz Z. , *A Genetic Algorithm for the Linear Transportation Problem*, IEEE Trans. on Systems Man and Cybernetics, Vol. 21, pp. 445-452, March-April, 1991
55. Grefenstette J. J. , *Optimization of Control Parameters for Genetic Algorithms*, IEEE Trans. on Sys., Man, and Cyb., Vol. SMC-16, No. 1, January/February 1986
56. Odetayo M. O. , *Optimal Population Size for Genetic Algorithms: An Investigation*, IEEE Colloquium on Genetic Algorithms for Control Systems Engineering, pp. 2/1-2/4, 1993

57. Alander J. T. , *On Optimal Population Size of Genetic Algorithms*, Proc. CompEuro '92 Computer Systems and Software Engineering, pp. 65-70, 1992
58. Arabas J. ,Michalewicz Z. and Mulawka J. , *GAVaPS - a Genetic Algorithms with Varying Population Size*, Evolutionary Computation, IEEE World Congress on Computational Intelligence, Proceedings of the IEEE Conf., Vol. 1, pp. 73-78, 1994
59. Lima J. A. ,Gracias N. ,Pereira H. and Rosa A. , *Fitness Function Design for Genetic Algorithms in Cost Evaluation Based Problems*, Proc. of IEEE, International Conf. on Evolutionary Computation, pp. 207-212, 1996
60. Ghosh A. ,Tsutsui S. and Tanaka H. , *Individual Aging in Genetic Algorithms*, Proc. Conf. on Intelligence Information System, Australian and New Zealand, pp. 276-279, 1996
61. Hesser J. ,Manner R. , *Towards an Optimal Mutation Probability for Genetic Algorithms*, Proc. First Int. Workshop on Parallel Problem Solving from Nature, Dortmund, 1990, Paper A-XII
62. Oi X. ,Palmieri F. , *Theoretical Analysis of Evolutionary Algorithms With an Infinite Population Size in Continuous Space Part I: Basic Properties of Selection and Mutation*, IEEE Trans. on Neural Networks, Vol. 5, No. 1, January 1994
63. Oi X. ,Palmieri F. , *Theoretical Analysis of Evolutionary Algorithms With an Infinite Population Size in Continuous Space Part II: Analysis of the Diversification Role of Crossover*, IEEE Trans. on Neural Networks, Vol. 5, No. 1, January 1994
64. Shang Y. ,Li G. J. , *New Crossover in Genetic Algorithms*, Proc. of IEEE, Third International Conf. on Tools for Artificial Intelligence, TAI '91, pp. 150-153, 1991
65. Coli M. , G. Gennuso, P. Palazzari, *A New Crossover Operator for Genetic Algorithms*, Proc. of IEEE, International Conf. on Evolutionary Computation, pp. 201-206, 1996
66. Potts J. C. ,Giddens T. D. , and Yadav S. B. , *The Development and Evaluation of an Improved Genetic Algorithms Based on Migration and Artificial Selection*, IEEE Trans. on Sys., Man, and Cyb., Vol. 24, No. 1, January 1994
67. Moed M. C. ,Stewart C. V. ,and Kelly R. B. , *Reducing the Search Time of a Steady State Genetic Algorithms Using the Immigration Operator*, Proc. of IEEE, Third International Conf. on Tools for Artificial Intelligence, TAI '91, pp. 500-501, 1991
68. Tsutsui S. , Fujimoto Y. , *Phenotypic Forking Genetic Algorithms (p-fGA)*, Proc. of IEEE, International Conf. on Evolutionary Computation, Vol. 2, pp. 566-572, 1995
69. Tsutsui S. ,Fujimoto Y. , and Hayashi I. , *Extended Forking Genetic Algorithms for Order Representation (o-fGA)*, Proc. of the First IEEE, Conf. on IEEE World Congress on Computational Intelligence, Vol. 2, pp. 566-572, 1994
70. Menhaj M. B. , Amani S. , *Online Maneuvering Target Tracking Using Neuro -Fuzzy Adaptive Filters*, Workshop "Fuzzy Filters for Image Processing" ,the 10th IEEE International Conference on Fuzzy Systems, Fuzz-IEEE 2001, Melborne, Australia.
71. Seifipour N. ,Menhaj M. B. ,*A GA-Based Algorithm with a very fast rate of convergence*, 7th Fuzzy days in Dortmund, Lecture Notes in Computer Science, Bernd Reusch (Ed.), Springer, Dortmund-Germany, October 01-03, 2001.

Chapter 14

Lossy Image Compression and Reconstruction Based on Fuzzy Relational Equation

Hajime Nobuhara, Yasufumi Takama, Witold Pedrycz, and Kaoru Hirota

Tokyo Institute of Technology
Department of Computational Intelligence and Systems Science
4259 Nagatsuta, Midori-ku, Yokohama-city 226-8502, Japan
{nobuhara, takama, hirota}@hrt.dis.titech.ac.jp

University of Alberta
Department of Electrical and Computer Engineering
Edmonton, T6R, 2G7, Canada
pedrycz@ee.ualberta.ca

Summary. A fast image reconstruction method for **I**mage **C**ompression method based on **F**uzzy relational equation (ICF) is proposed. Furthermore, in order to improve the quality of the image reconstructed by ICF, a new image reconstruction method is also proposed. In image compression and reconstruction experiments using 20 images (extracted from Standard Image DataBAse, SIDBA), we confirm that the decrease of the image reconstruction time to 1/132.02 and 1/382.29 are obtained when the compression rate is 0.0156 and 0.0625, respectively. Furthermore, it is confirmed that the quality of the reconstructed image obtained by the proposed method is better than that of the conventional one from the viewpoint of PSNR (Peak Signal to Noise Ratio).

1 Introduction

Due to the tremendous popularization of Internet, high performance personal computers, and large storage devices, people can treat digital image information as means of communication. These promote a development of lossy image compression method, leading to the proposal of many lossy image compression methods [1] [2] [10] [15] [16] [17] [19] particularly, Discrete Cosine Transform (DCT) used for JPEG [29].

Nowadays, lossy image compression is needed not only for the coding performance (e.g., computation time, compression rate) but also for other applications, e.g., a feature extraction based on image compression can be utilized to manage a huge image database system [12] [26] [31] [32], and a digital watermarking based on image compression [3] [4] [8] [13] [25] [28] is also used for protection of digital contents on Internet. An example of such image compression methods has been proposed by Hirota and Pedrycz, called **I**mage **C**ompression method based on **F**uzzy relational equation (ICF) [9].

In this chapter, a fast image reconstruction method for ICF is proposed. Furthermore, in order to improve the quality of the reconstructed image, a new image reconstruction method for ICF is also proposed. In addition, various features of ICF lacking in other image coding methods are introduced.

The ICF is formulated in Sect. 2 and a fast image reconstruction method for ICF is presented in Sect. 3. In Sect. 4, a new image reconstruction method for ICF is proposed in order to improve the quality of reconstructed images. Section 5 shows various properties of ICF for image compression-based applications.

2 Image compression method based on fuzzy relational equation (ICF)

2.1 Image compression

A method of lossy **I**mage **C**ompression and reconstruction based on **F**uzzy relational equation (ICF) has been proposed by Hirota and Pedrycz in 1999 [9]. In ICF, a still grayscale image of size $M \times N$ is expressed as a fuzzy relation $R \in F(\mathbf{X} \times \mathbf{Y})$, $\mathbf{X} = \{x_1, x_2, \dots, x_M\}$, $\mathbf{Y} = \{y_1, y_2, \dots, y_N\}$ by normalizing the intensity range of each pixel into $[0, 1]$. In the case of color image (of RGB color space), an original image can be expressed by three fuzzy relations, i.e., $\mathbf{P} = \{P^{(R)}, P^{(G)}, P^{(B)}\} \subset F(\mathbf{X} \times \mathbf{Y})$, where, $P^{(R)}$,$P^{(G)}$, and $P^{(B)}$ denote the red, green, and blue planes of the color image, respectively. Other color spaces can be also considered. Therefore, the ICF can also treat color images. For simplicity, in this chapter, we consider only the case of grayscale images.

The grayscale image $R \in F(\mathbf{X} \times \mathbf{Y})$ is compressed into $G \in F(\mathbf{I} \times \mathbf{J})$ by a max t-norm composite fuzzy relational equation [22][23],

$$G(i,j) = \max_{y \in \mathbf{Y}} \left\{ B_j(y) \circledt \max_{x \in \mathbf{X}} \{A_i(x) \circledt R(x,y)\} \right\} \qquad (\in [0,1]), \tag{1}$$

where $\circledt$ denotes a continuous t-norm, and $A_i \in \mathbf{A} = \{A_1, A_2, \dots, A_I\} \subset F(\mathbf{X})$, $B_j \in \mathbf{B} = \{B_1, B_2, \dots, B_J\} \subset F(\mathbf{Y})$, are called coders. The image compression process is shown in Fig. 1.

In this chapter, only the max t-norm composite fuzzy relational equation is considered. However, other types of fuzzy relational equation [18][24], e.g., min s-norm composite, adjoint of max t-norm, and adjoint of min s-norm can be also applied to ICF [21]. The coders $\mathbf{A}$ and $\mathbf{B}$ are defined by

$$\mathbf{A} = \{A_1, A_2, \dots, A_I\}, \tag{2}$$

$$A_i(x_m) = \exp\left(-Sh\left(\frac{iM}{I} - m\right)^2\right), \tag{3}$$

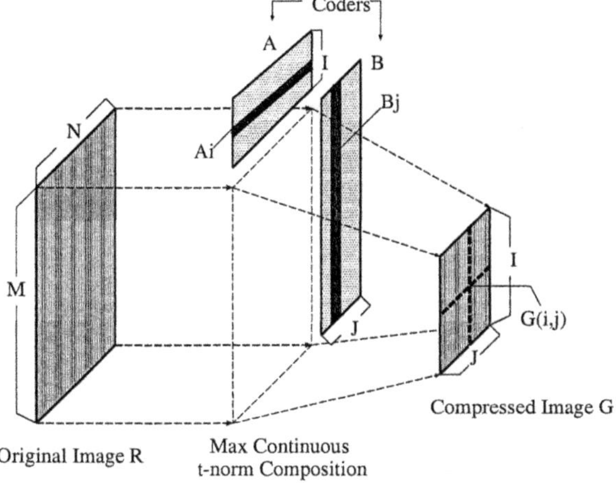

Fig. 1. Image compression process

$$(m = 1, 2, \ldots, M),$$

and

$$\mathbf{B} = \{B_1, B_2, \ldots, B_J\}, \tag{4}$$

$$B_j(y_n) = \exp\left(-Sh\left(\frac{jN}{J} - n\right)^2\right), \tag{5}$$

$$(n = 1, 2, \ldots, N),$$

where Sh denotes the sharpness of fuzzy sets A_i and B_j contained in coders **A** and **B**. A preferable range of Sh depends on the compression rate, and it will be mentioned in following image compression and reconstruction experiments. Notice that coders **A** and **B** have employed Sh. The shape of the fuzzy sets of coders are Gaussian-like shown in Fig. 2, and it is preferable for the ICF. (When the shape is triangle, a block noise will appear in the reconstructed images [21].) The compression rate of ICF can be adjusted by the number of fuzzy sets contained in coders **A** and **B**. The compression rate ρ is defined as

$$\rho = \frac{IJ}{MN}, \tag{6}$$

where IJ and MN denote the number of coefficients of compressed image and original one, respectively. In this chapter, we take Yager's t-norm [30] as the continuous t-norm appeared in fuzzy relational equation (1).

$$\text{(Yager's t-norm)} :$$

$$a \textcircled{t}^{(p)} b = 1 - \min\left\{1, \{(1-a)^p + (1-b)^p\}^{1/p}\right\}$$

$$(p \geq 1). \tag{7}$$

Yager's t-norm can cover all continuous t-norms by changing the parameter p. Zadeh's t-norm and others cannot cover all the continuous t-norms. Frank's t-norm can also cover all continuous t-norms, but it has high computational complexity compared with Yager's one [6].

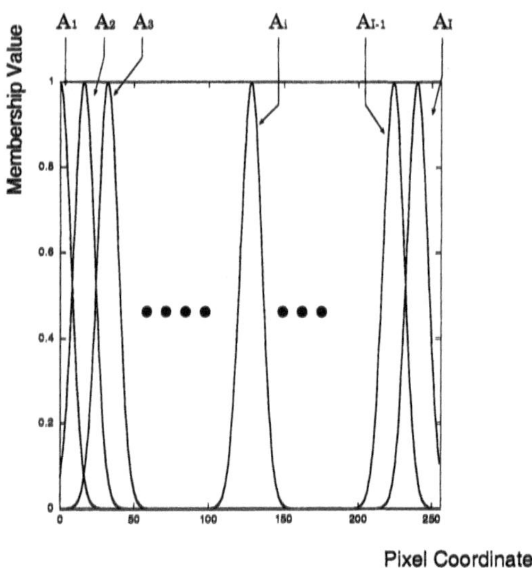

Fig. 2. Shape of fuzzy sets on coders

2.2 Image reconstruction

The reconstruction of an image $R' \in F(\mathbf{X} \times \mathbf{Y})$ from the compressed image G is considered to be an inverse problem, under the condition that $\mathbf{A}$, $\mathbf{B}$, and G are given. That is, the image reconstruction corresponds to the estimation of the solution R' that satisfies the fuzzy relational equation (1). In the case of max t-norm composite fuzzy relational equation, the structure of the solution set $\mathbf{R} \subset F(\mathbf{X} \times \mathbf{Y})$ is shown in Fig. 3 and the greatest and minimal solutions of $\mathbf{R}$ can be obtained analytically [5].

Examples of reconstructed images (the greatest and one of the minimal solutions) are shown in Fig. 5. The reconstructed images are obtained under the conditions that compression rate $= 0.0625$ ($I \times J = 64 \times 64$) and $Sh = 0.04$. As can be seen from Fig. 5, the minimal solutions present bright pixels scattered on a black plane (background), yielding a meaningless result from the viewpoint of reconstructed image, whereas the greatest solution (Fig. 5, left) has quality comparable to the original image. Therefore, only the greatest solution is considered as a reconstructed image in this chapter. The greatest solution has been presented by Hirota and Pedrycz [9], and will be

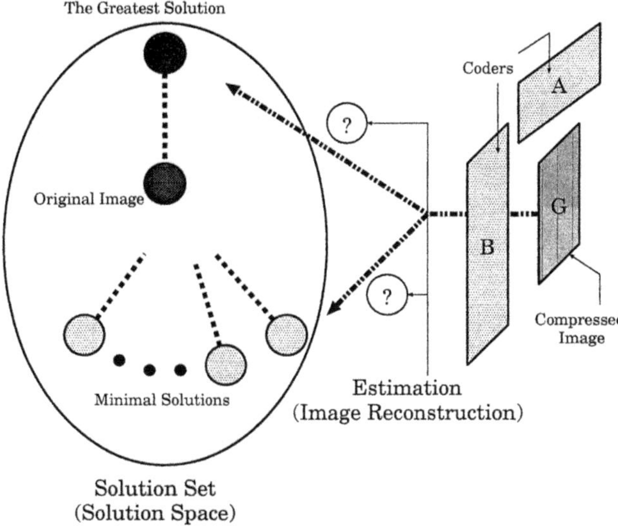

Fig. 3. Image reconstruction

referred as H-P method. In the case of H-P method, if Sh is a large value, the reconstructed image has high sharpness, but the white slit is more prominent in the image (see Fig. 6, left side image). Conversely, if Sh is a small value, then the white slit is decreased, but the reconstructed image is blurred (see Fig. 6, right side image). Therefore, we employ a compromise value between such opposite sides.

Fig. 4. Original image (Lenna)

In [9], the greatest solution $\hat{R}$ of fuzzy relational equation (1) is given by

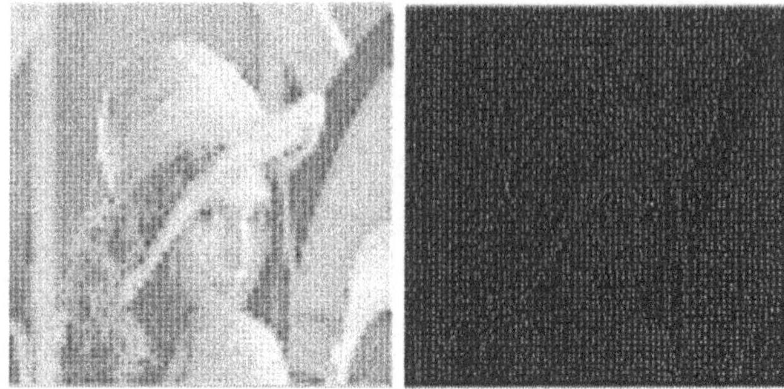

Fig. 5. The greatest (left) and one of minimal solutions (right) (Compression Rate = 0.0625)

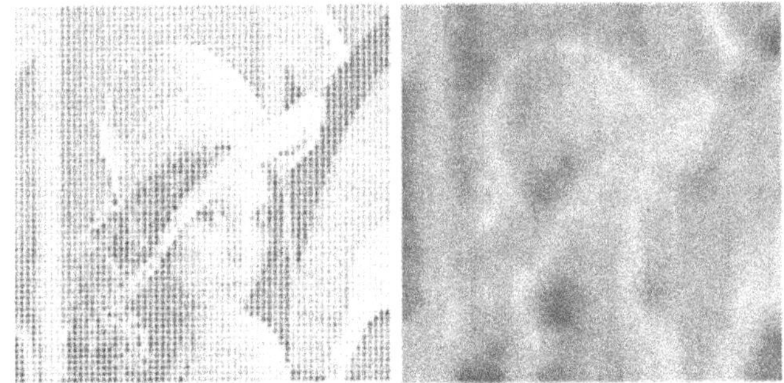

Fig. 6. Left side image : the greatest, Sh = 0.2 , right side image : the greatest, Sh = 0.001 (Compression Rate = 0.0625)

$$\hat{R}(x,y) = \min_{(i,j)\in \mathbf{I}\times\mathbf{J}} \{(A_i(x) \circledt B_j(y)) \ \varphi_t \ G(i,j)\}, \tag{8}$$

where,

$$a \ \varphi_t \ b = \sup\{c \in [0,1] | a \circledt c \leq b\}. \tag{9}$$

2.3 Image compression and reconstruction experiments

In order to investigate the computation time of ICF, an image compression and reconstruction experiment using 20 images (extracted from Standard Image Data BAse, SIDBA) is performed. The original images of size $M \times N = 256 \times 256$ are compressed into images $I \times J = 32 \times 32$ and $I \times J = 64 \times 64$ pixels. In the experiment, the parameter p of Yager's t-norm is set to 1.0 and Sh of coders $\mathbf{A}$ and $\mathbf{B}$ is set to 0.01 and 0.04, respectively, under the condition that the compression rate is 0.0156 ($I \times J = 32 \times 32$) and 0.0625 ($I \times J = 64 \times 64$), respectively.

The averages of image compression and reconstruction times of ICF are shown in Table 1. One example of the original image and the reconstructed one (H-P method, the greatest solution) are shown in Figs. 4 and 5, respectively. Table 1 shows that the image reconstruction time (H-P method) is large. Therefore, in the next section, we propose a fast image reconstruction method for ICF.

	Compression Time (Average)	Reconstruction Time (Average)
$\rho = 0.0156$	0.93(s)	168.98(s)
$\rho = 0.0625$	1.43(s)	546.67(s)

Table 1. Image compression and reconstruction time (H-P method), computed using a Sun Ultra 10 (440MHz) workstation

3 Fast image reconstruction method of ICF

3.1 Derivation of fast image reconstruction method

In order to reduce the image reconstruction time of ICF, a fast image reconstruction method is proposed [20]. The fast image reconstruction method is based on the following theorem:

[**Theorem 1**] ([20])

H-P method (Eq. (8)) is equivalent to

$$\hat{R}(x,y) = \min_{j \in J} \left\{ B_j(y) \ \varphi_t \ \left\{ \min_{i \in I} \{ A_i(x) \ \varphi_t \ G(i,j) \} \right\} \right\}. \tag{10}$$

Theorem 1 tells us the greatest solution obtained by the H-P method (Eq. (8)) is the same one obtained by the proposed method (Eq. (10)). The theoretical computation time of the H-P method and the proposed method correspond to

$$\text{H-P method} : IJMN(L+T+P), \tag{11}$$
$$\text{Proposed method} : JM(I+N)(L+P), \tag{12}$$

where, MN and IJ denote the size of original image R and compressed image G, respectively. The computation time of min, t-norm, and relative pseudocomplement (Eq. (9)) are denoted L, T, and P, respectively. The theoretical comparison shows that the computation time of the proposed method is shorter than that of the H-P method.

3.2 Experimental comparison of computation time

A comparison of the proposed method with the H-P method from the viewpoint of reconstruction time is presented by an image compression and reconstruction experiment using 20 images (extracted from SIDBA). The original images of size $M \times N = 256 \times 256$ pixels are compressed into images of size $I \times J = 32 \times 32$ ($\rho = 0.0156$) and $I \times J = 64 \times 64$ ($\rho = 0.0625$). The image reconstruction is performed by the H-P method and the proposed method, respectively. The result of comparison is shown in Table 2. The decrease of the computation time to 1/132.02 and 1/382.29 are obtained, respectively.

	H-P method (Average)	Proposed method (Average)
$\rho = 0.0156$	168.98(s)	1.28(s)
$\rho = 0.0625$	546.67(s)	1.62(s)

Table 2. Image reconstruction time comparison of H-P method with proposed method, Computed using Sun Ultra 10 (440MHz)

4 Improvement of quality of reconstructed image

4.1 Solution space structure and reformulation of image reconstruction problem

As shown in Sect. 2, the quality of the reconstructed image based on the greatest solution is low due to a white slits on the image (Fig. 5, left). In order to improve the quality of the reconstructed image, we should estimate a solution of fuzzy relational equation which is close to the original image as the ideal reconstructed one. A part of the original image and the corresponding part of the reconstructed image, given as the greatest solution, are shown in Fig. 7.

The solution structure of the fuzzy relational equation (Fig. 3) and Fig. 7 show that the original image is always located under the greatest solution in terms of order. Therefore, in order to improve the quality of the reconstructed image, a solution located under the greatest one should be estimated. However, the analytical solutions of $\mathbf{R}$ are the greatest and minimal solutions only, and other solutions can not be obtained analytically. That is, except for the greatest and minimal solutions, it is very difficult to find an analytical solution $\bar{R}$ that satisfies the fuzzy relational equation (1), given A, B and G. In order to find an other reconstructed image that is close to the original image, we reformulate the image reconstruction problem as to estimate

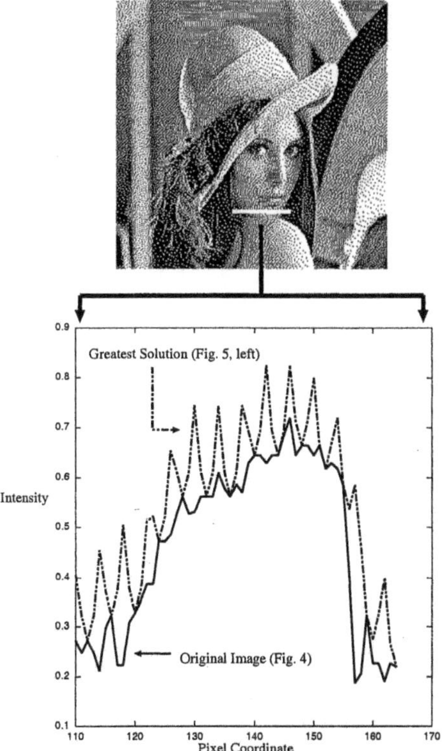

Fig. 7. Intensity of each image

a solution $\bar{R}$ which satisfies the **fuzzy relational inequality** [7],

$$G(i,j) \geq \max_{y \in \mathbf{Y}} \left\{ \left\{ \max_{x \in \mathbf{X}} \left(R(x,y) \circledt A_i(x) \right) \right\} \circledt B_j(y) \right\}, \tag{13}$$

given $\mathbf{A}$, $\mathbf{B}$, and G.

This reformulation corresponds to the extension of the solution space $\mathbf{R}$ of equation (1) to the solution space of equation (13) shown in Fig. 8, so that the reconstructed image with a good quality can be easily obtained. A solving method of the fuzzy relational inequality (13) is presented in [21].

[Solving Method of Fuzzy Relational Inequality] ([21])

Let $\circledt_1$ be a t-norm of inequality (13). We select a t-norm $\circledt_2$ such that

$$\forall a, b \in [0,1] \quad a \circledt_1 b \leq a \circledt_2 b. \tag{14}$$

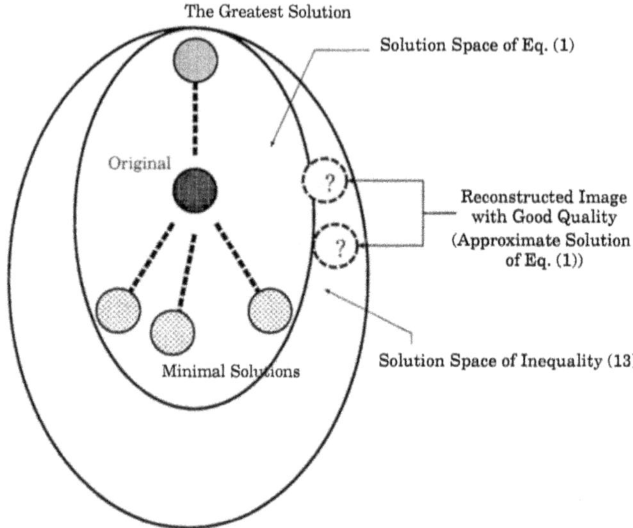

Fig. 8. Image reconstruction

Then, a fuzzy relation $\tilde{R}$ which is obtained by

$$\tilde{R}(x,y) = \min_{j \in \mathbf{J}} \left\{ B_j(y) \; \varphi_{t2} \left\{ \min_{i \in \mathbf{I}} \left(A_i(x) \; \varphi_{t2} \; G(i,j) \right) \right\} \right\}. \tag{15}$$

is a solution of the fuzzy relational inequality (13). Here φ_{t2} denotes the ⓣ-relative pseudocomplement of t-norm $ⓣ_2$.

The method is based on the fast solving method of fuzzy relational equation (Eq. (10)), particularly, the method is equivalent to Eq. (10) when $ⓣ_1$ is equal to $ⓣ_2$. Therefore, the computation time of the proposed method is comparable with the image reconstruction based on the greatest solution. An overview of the proposed method is shown in Fig. 9.

4.2 Image compression and reconstruction experiment

In order to evaluate the effectiveness of the proposed method, an image compression and reconstruction experiment using 20 test images of SIDBA is performed. In this experiment, we use Yager's t-norm (7) which can adjust the order of t-norm by changing the parameter p. The greatest solution [9] [20] performs the image compression and reconstruction under the condition that the parameter $p = 1.0$. On the other hand, the proposed method compresses an image under the condition that the parameter $p = 1.0$, while the image reconstruction is performed under the condition of $p = 2.1$.

The Peak Signal to Noise Ratio (PSNR) comparison of the proposed method with the greatest solution is shown in Fig. 10. The PSNR is defined

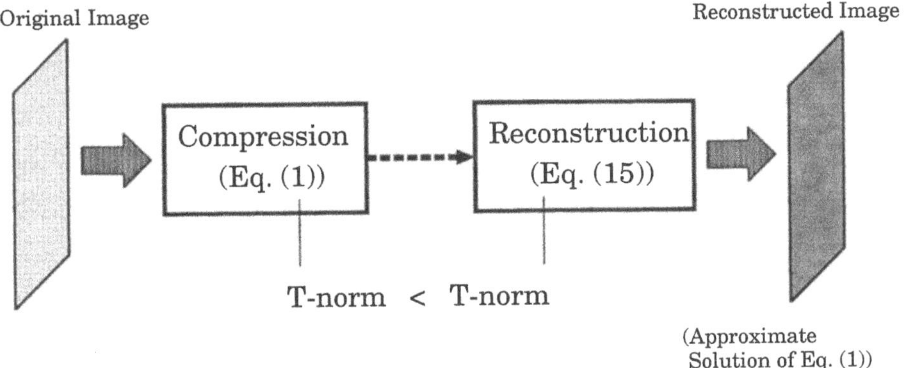

Fig. 9. Overview of the proposed method

as

$$PSNR = 20 \log_{10} \left(\frac{255}{RMSE} \right), \tag{16}$$

where, RMSE stands for root mean square error defined by

$$RMSE = \sqrt{\frac{\sum_{(x,y)\in \mathbf{X}\times\mathbf{Y}} (R(x,y) - \hat{R}(x,y))^2}{|\mathbf{X}\times\mathbf{Y}|}}. \tag{17}$$

The results of Fig. 10 correspond to the average value of PSNR for 20 images. Furthermore, under the condition that the compression rates are 0.0625 and 0.25, the improvements for each image are presented in Table 3.

	The Greatest (PSNR, Average)	Proposed Method (PSNR, Average)
$\rho = 0.0625$	15.69	21.97
$\rho = 0.25$	20.08	25.38

Table 3. PSNR comparison

Figure 10 and Table 3 show that the quality of the reconstructed image obtained by the proposed method is better than that of the greatest solution from the view point of PSNR. With respect to the aspect of the reconstructed image, the white slits on the reconstructed image of the original method can be reduced by applying the proposed method (Figs. 11 - 12).

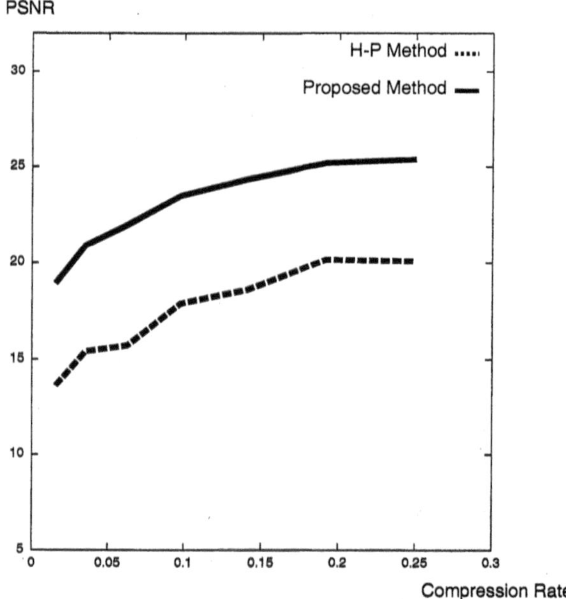

Fig. 10. PSNR comparison

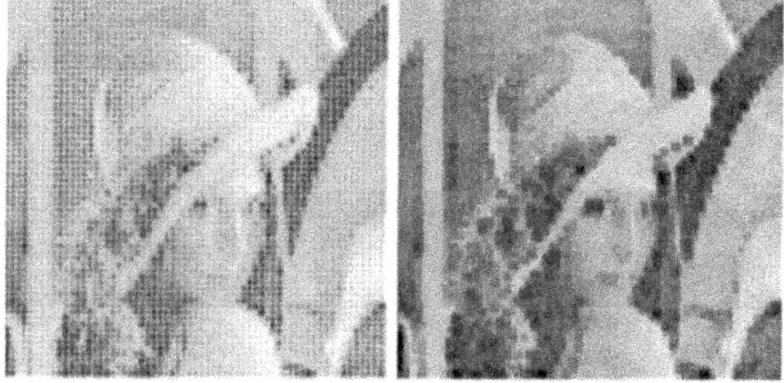

Fig. 11. Reconstructed images (Lenna), left : the great solution, right : proposed method

5 Additional features of ICF

In this section, two features of ICF are introduced, namely, the ability to deal with multi resolution images and image compression restricted to an image region. These features can be utilized to feature extraction for image database management systems [12] [26] [31] [32].

[Multi resolution images]

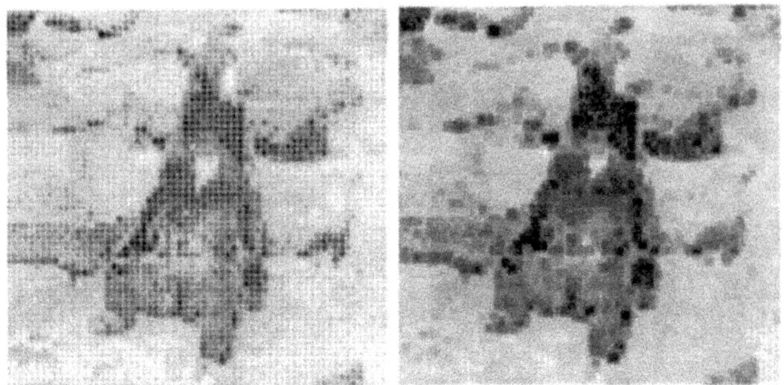

Fig. 12. Reconstructed images (Bear), left : the great solution, right : proposed method

Multi resolution images correspond to multiple images with different resolutions compressed from the same original image. In the case of ICF, multi resolution images can be obtained by adjusting the number of fuzzy sets of coders **A** and **B**. Examples of multi resolution images obtained by ICF are shown in Fig. 13.

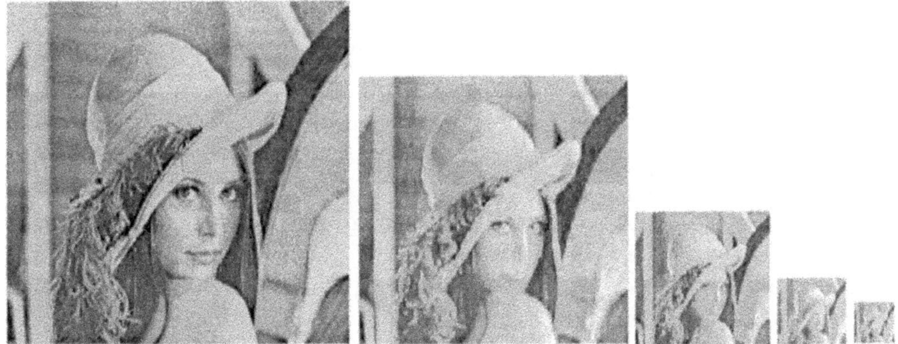

Fig. 13. Multi resolution images, $I \times J = 256 \times 256$, 200×200, 100×100, 50×50, and 30×30

The wavelet transform also has this capability, but the image resolution is lower than that of ICF's. On the other hand, DCT does not have this capability. In image databases, higher order local autocorrelation[14] is frequently employed as a feature for multi resolution images[11]. In this case, the higher order local autocorrelation is extracted from the multi resolution images. Therefore, ICF outperforms wavelet and DCT in multi resolution images function because it extracts more higher order local autocorrelation when compared with its competitors.

[Image compression restricted to an image region]

An example of image compression restricted to a region is shown in Fig. 14.

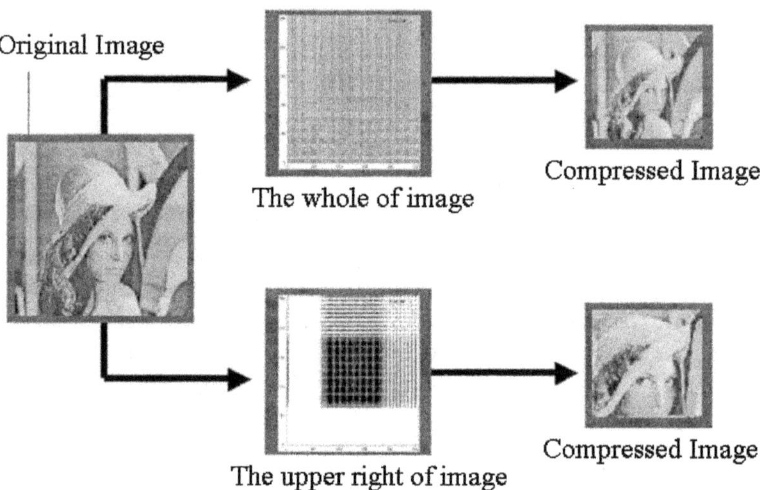

Fig. 14. Example of image compression with restriction of image region

The restricted image compression can be performed by adjusting the distribution of the fuzzy sets of coders **A** and **B** shown in Fig. 14. This feature may be potentially applied in fields like image retrieval, for the extraction and processing of local features (See [27] for a discussion on global and local features in image retrieval). However, research in this area has not been conducted yet. It is a potential research area for fuzzy set theory researchers, and promising results can be expected by combining ICF and natural language processing techniques.

6 Conclusions

An **I**mage **C**ompression and reconstruction method based on **F**uzzy relational equation (ICF) and some of its capabilities were introduced. Furthermore, a fast image reconstruction method for ICF was proposed. Besides, in order to improve the quality of reconstructed image, a new image reconstruction method for ICF was also proposed.

In image compression and reconstruction experiments using 20 images (extracted from SIDBA), we confirmed that the decrease of the image reconstruction time to 1/132.02 and 1/382.29 under compression rates of 0.0156 and 0.0625, respectively. Furthermore, it was confirmed that the quality of the reconstructed image obtained by the proposed method is better than that of the conventional one from the viewpoint of PSNR (Peak Signal to Noise Ratio).

Comparing to other methods, ICF has additional features like the ability to deal with multi resolution images and image compression restricted to an image region. Therefore, hopefully ICF presents the potential to its competitors, not only in terms of the coding performance, but also in the applications.

References

1. Antonini M., Barlaud M., Mathieu P.,and Daubchies I., *Image Coding Using Wavelet Transform*, IEEE Transactions on Image Processing, vol. 1, no. 2, 1992, pp. 205-220.
2. Averbuch A., Lazar D., and Israeli M., *Image Compression Using Wavelet Transform and Multiresolution Decomposition*, IEEE Transactions on Image Processing, vol. 5, no. 1, 1996, pp. 4-15.
3. Banri M., Bartolini F., and Piva A., *Improved Wavelet-Based Watermarking Through Pixel-Wise Masking*, IEEE Transactions on Image Processing, vol. 10, no. 5, 2001, pp. 783-791.
4. Banri M., Podilchuk C.I., Fartolini F., and Delp E.J., *Watermarking Embedding : Hiding a Signal Within a Cover Image*, IEEE Communications Magazine, vol. 39, no. 8, 2001, pp. 102-108.
5. DiNola A., Pedrycz W., and Sessa S., *On Measures of Fuzziness of Solutions of Fuzzy Relation Equation with Generalized Connectives*, Journal of Mathematical Analysis and Applications, 106, 1985, pp. 443-453.
6. DiNola A., Sessa S., Pedrycz W., and Sanchez E., *Fuzzy Relation Equation and Their Applications to Knowledge Engineering*, Kluwer Academic Publishers, 1989.
7. Drewniak J., *Fuzzy Relational Equation and Inequality*, Fuzzy Sets and Systems, vol. 14, no. 3, 1984, pp. 237-247.
8. Hartung F., and Ramme F., *Digital Rights Management and Watermarking of Multimedia Content for M-Commerce Applications*, IEEE Communications Magazine, vol. 38, no. 11, 2000, pp. 78-84.
9. Hirota K., and Pedrycz W., *Fuzzy Relational Compression*, IEEE Transactions on Systems, Man, and Cybernetics, vol. 29, no. 3, 1999, pp. 407-415.
10. Karayiannis N.B., Pai Pin-I, *Fuzzy Vector Quantization Algorithms and Their Application in Image Compression*, IEEE Transactions on Image Processing, vol. 4, no. 9, 1995, pp. 1193-1201.
11. Kurita T., Otsu N., Sato T., *A face recognition method using higher order local autocorrelation and multivariate analysis*, Proc. of Int. Conf. on Pattern Recognition, Aug.30-Sep.3, The Hague, vol.II, 1992, pp. 213-216.
12. Liang K.C., and Jay Kuo C.C., *WaveGuide: A Joint Wavelet-Based Image Representation and Description System*, IEEE Transactions on Image Processing, vol. 8, no. 11, 1999, pp. 1619-1629.

13. Lin C.Y., Wu M., Bloom J.A., Cox I.J., Miller M.L., and Lui Y.M., *Rotation, Scale, and Translation Resilient Watermarking for Images*, IEEE Transactions on Image Processing, vol. 10, no. 5, 2001, pp. 767-782.
14. Maclaughlin J.A., and Raviv J., *Nth-order autocorrelations in pattern recognition*, Information and Control, vol. 12, 1968, pp. 121-12.
15. Huang C.M., and Harris R.W., *A Comparison of Several Vector Quantization Codebook Generation Approaches*, IEEE Transactions on Image Processing, vol. 2, no. 1, 1993, pp. 108-112.
16. Linde Y., Buzo A., and Gray Y.M., "An Algorithm for Vector Quantizer Design," *IEEE Transactions on Communications*, vol. 28, no. 1, 1980, pp. 84-95.
17. Mallat S.G., *A Theory for Multiresolution Signal Decomposition : The Wavelet Representation*, IEEE Transactions on Pattern Analysis and Machine Intelligence, vol. 11, no. 7, 1989, pp. 674-693.
18. Miyakoshi M., and Shinbo M., *Solutions of Composite Fuzzy Relational Equations with Triangular Norms*, Fuzzy Sets and Systems, vol. 16, no. 1, 1985, pp. 53-64.
19. Nasrabadi N.M., and King R.A., *Image Coding Using Vector Quantization : A Review*, IEEE Transactions on Communications, vol. 36, no. 8, 1988, pp. 957-971.
20. Nobuhara H., Pedrycz W., and Hirota K., *Fast Solving Method of Fuzzy Relational Equation and Its Application to Lossy Image Compression/Reconstruction*, IEEE Transactions on Fuzzy Systems, vol. 8, no. 3, 2000, pp. 325-334.
21. Nobuhara H., Takama Y., and Hirota K. *Image Compression/Reconstruction Based on Various Types of Fuzzy Relational Equations*, The Transaction of The Institute of Electrical Engineers of Japan (in Japanese), vol. 121, no. 6, 2001, pp. 1102-1113.
22. Pedrycz W., *Fuzzy Relational Equations with Generalized Connectives and Their Applications*, Fuzzy Sets and Systems, vol. 10, 1983, pp. 185-201.
23. Pedrycz W., *On Generalized Fuzzy Relational Equations and Their Applications*, Journal of Mathematical Analysis and Applications, 107, 1985, pp. 510-536.
24. Pedrycz W., *Processing in Relational Structures: Fuzzy Relational Equations*, Fuzzy Sets and Systems, vol. 40, no. 1, 1991, pp. 77-106.
25. Podilchuk C.I., and Delp E.J., *Digital Watermarking: Algorithms and Applications*, IEEE Signal Processing Magazine, vol. 18, no. 4, 2001, pp. 33-46.
26. Shanbehzadeh J., Moghadam A.M.E., and Mahmoudi F., *Image Indexing and Retrieval Techniques : Past, Present, and Next*, Proc. of SPIE, vol. 3972, 2000, pp. 461-470.
27. Stejic Z., Iyoda E.M., Takama Y., and Hirota K., *Content-Based Image Retrieval using Local Similarity Patterns defined by Interactive Genetic Algorithm* , Late-Breaking Papers of the Genetic and Evolutionary Computation Conference (GECCO-2001), SAN FRANCISCO, CA, USA, July 2001, pp. 390-397.
28. Voloshynovskiy S., Pereira S., Iquise V., and Pun T., *Attack modeling : towards a second generation watermarking benchmark*, Signal Processing, vol. 81, 2001, pp. 1177-1214.
29. Wallace G.K., *The JPEG still picture compression standard*, Communication ACM, vol. 34, no. 4, 1991, pp. 30-44.
30. Yager R.R., *On a General Class of Fuzzy Connectives*, Fuzzy Sets and Systems, vol. 4, no. 3, pp. 235-242, 1980.

31. Yu D., Liu Y., Mu Y., and Yang S., *Integrated System for Image Storage, Retrieval and Transmission Using Wavelet Transform*, Proc. of SPIE, vol. 3656, 1999, pp. 448-457.
32. Zhu B., Ramsey M., and Chen H., *Creating a Large-Scale Content-Based Air-photo Image Digital Library*, IEEE Transactions on Image Processing, vol. 9, no. 1, 2000, pp. 163-167.

Chapter 15

Avoidance of Highlights through ILFOs in Automated Visual Inspection

Aureli Soria-Frisch

Fraunhofer IPK
Dept. Security and Inspection Technologies
Pascalstr. 8-9, 10587 Berlin, Germany
email: aureli.soria_frisch@ipk.fhg.de

Summary. The avoidance of so-called highlights is a very important practical problem in Computer Vision systems. This kind of "noise" is caused by the incidence of light on a specular surface. Images of scenes including such surfaces present blind areas, the highlights, which hinder their employment in Computer Vision applications. This problem especially affects automated visual inspection systems of high reflective materials. In spite of its importance from a practical point of view, very few research efforts have been undertaken in this context in order to solve it.

The recently developed Intelligent Localized Fusion paradigm, which is based on the fuzzy integral as image fusion operator, offers a systematic solution for the avoidance of highlights in the context of automated visual inspection systems. The paper considers the theoretical aspects of the paradigm together with some guidelines for a successful implementation. Among them is worth mentioning the utilization of classical and interactive genetic algorithms for the parameterization of the fuzzy integral. Finally some examples of real applications demonstrate the goodness of the paradigm.

1 Introduction

When acquiring images presenting specular surfaces with a camera device the areas in the camera detector symmetric in the vertical axis to the lighting source are saturated. As a consequence the acquired image presents some areas with no pictorial information but the maximal grayvalue, which receive the name of highlights .

The theoretical classification of this problem is not well defined. Traditionally noise has been considered to be binary, white or gaussian distributed. The image degrading effect formerly described is not explicitly considered in the classical literature on image processing [3] [6]. However such a "noise" is a very important issue from a practical point of view. The image processing practitioner is supposed to cope with the problem by working on the optimality of the illumination conditions. The efforts are then concentrated on finding the right type of lamp and illumination angle. In case of high reflective surfaces as those of e.g. polished metals, glasses, precious crystals or plastic bundles the problem remains in most of the cases unsolved. This problem is especially relevant in the application field of automated visual inspection.

Image highlights are characterized by the lack of structural information. Thus the failing information should be generated in order to suppress these detector saturated areas on the final image. One way to achieve this goal is the acquisition with varying conditions of an image series of the underlying object. These images can be fused, in order to use the areas of each source where the information of interest is not corrupted. A similar approach was presented in [9] for the inspection of firearm bullets. In this case the fusion operation was based on the Regularization Theory, where an energy function of the input and output images was minimized. The inspection of objects with a higher reflectivity than those requires the utilization of more flexible fusion strategies. Moreover the presented methodology increases the interpretability of the fusion procedure because of the usage of fuzzy operators.

Intelligent Localized Fusion (ILF) is the name of a new paradigm for fuzzy aggregation in image processing, which can be successfully used in the pre-processing of high reflective material images. The paradigm is based on the employment of fuzzy integrals as fusion operators, which become so-called Intelligent Localized Fusion Operators (ILFOs). The fuzzy integral generalizes most of the Soft-Computing related aggregation operators and thus constitutes a flexible mean for fuzzy fusion in multisensory systems [10]. The fuzzy integral is computed with respect to a fuzzy measure, which is used for the characterization of the *a priori* importance of the information sources. The coefficients of this measure are normally established as a global parameter for the images being fused. On the other hand the ILF paradigm proposes a localized definition of the fuzzy measure.

2 Intelligent localized fusion operators (ILFOs)

The Intelligent Localized Fusion paradigm is based on the usage of the fuzzy integral as fusion operator at the pixel level. The fuzzy integral is a fuzzy fusion operator introduced by Sugeno [11] in order to make the mathematical integration more flexible. Processes as multi-criteria decision making or subjective evaluation, where human beings show a robust, adapting and compensating behaviour, inspired his research. From this work a new fuzzy fusion operator resulted that generalizes most of the aggregation operators used in the context of Soft-Computing methodologies, e.g. ranking operators and its weighted variants, weighted sum, Ordered Weighted Averaging (OWA) [17] (see Fig. 1).

The utilization of fuzzy aggregation strategies allows overcoming the shortcomings of classical fusion operators [10]. Fuzzy aggregation operators present the following positive features: adaptability, reinforcement capability [16], in-

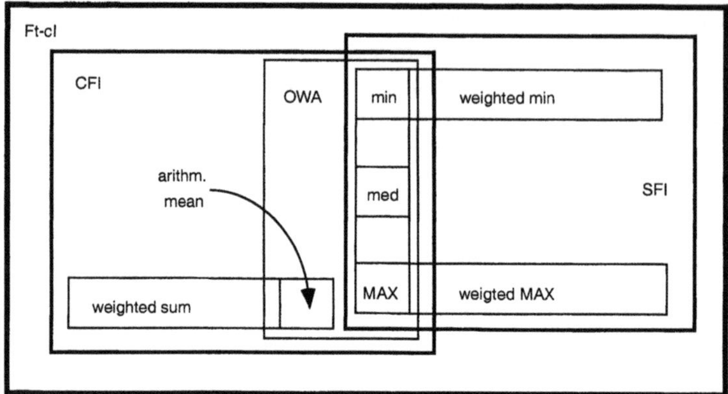

Fig. 1. Generalization relationship between different fuzzy fusion operators. *Ft-cI*: Fuzzy t-conorm Integral. *CFI*: Choquet Fuzzy Integral. *SFI*: Sugeno's Fuzzy Integral. *OWA*: Ordered Weighted Averaging. *med*: median. Modified from a Fig. in [5] (with permission).

clusion of meta-knowledge [16], characterization of the interaction between information sources [4], and tractability of fuzzy information. Specially fuzzy integrals present all mentioned properties, i.e. other fuzzy aggregation operators are not capable of characterizing the interaction between information sources [4].

2.1 Overview on the fuzzy integral and the fuzzy measure theory

The fuzzy integral is a weighted fusion operator that applies on fuzzy information. The expression of the sensor data to be fused, $x_i, \forall i = 1, \ldots, n$, through fuzzy membership functions, which will be denoted here by $h_i(x_i)$, enables the exploitation of the positive features of fuzzy sets for information processing [13].

The theoretical development of the fuzzy integral was based on the establishment of a new type of measures. Since fuzzy sets extended classical sets by the consideration of subjectiveness [18], the definition of new types of measures on these sets were requested. Sugeno presented fuzzy measures as a generalization of classical measures in a similar direction as the classical sets were extended, namely as a mean of reflecting the human capability of subjectively expressing degrees of trust [4]. From a mathematical point of view fuzzy measures generalize probability measures by relaxing its additivity axiom. Thus fuzzy measures are functions on fuzzy sets, $\mu : \mathcal{J}(\mathcal{X}) \rightarrow [0, 1]$, where $\mathcal{X}$ is the set of information sources. Each fuzzy measure coefficient, $\mu(A_j)$, characterizes the *a priori* importance of each subset A_j of $\mathcal{X}$, where $j = 1, \ldots, 2^n - 1$. Fuzzy measures satisfy the following conditions in the discrete case:

I. $\mu\{\emptyset\} = 0; \mu\{\mathcal{X}\} = 1$,
II. $A_j \subset A_k \rightarrow \mu(A_j) \leq \mu(A_k) \, \forall A_j, A_k \in \mathcal{J}(\mathcal{X})$.

In the fuzzy integral the weighting is done through fuzzy measures. They characterize the *a priori* importance of the different criteria supporting the hypothesis evaluated through the fuzzy integral. The idea behind the generalization of classical measures (i.e. probability measures) is that its additivity characterizes independence between criteria, while these can also be redundant or complementary. Thus probability, belief, possibility and fuzzy-λ measures are special cases of general fuzzy measures [14] based on the way the additivity property is fulfilled. Intuitively sub-additive measures like possibility measures are thought to express redundancy, while super-additivity from e.g. belief measures expresses complementarity [4]. Fuzzy-λ measures makes this condition depend on a parameter λ:

$$\mu(A_i \cup A_j) = \mu(A_i) + \mu(A_j) + \lambda\mu(A_i)\mu(A_j). \tag{1}$$

There are basically two types of fuzzy integrals known as Sugeno and Choquet Fuzzy Integral. The Sugeno Fuzzy Integral ($\mathcal{S}_\mu$) is the generalization of other ranking operators as the weighted minimum or the median and thus presents a combination of the fuzzy connectives minimum ($\wedge$) and maximum ($\vee$). On the other hand the Choquet Fuzzy Integral ($\mathcal{C}_\mu$) uses a combination of algebraic product and addition becoming a generalization of operators as the arithmetic mean or the OWAs (see Table 1). The mathematical expressions of these integrals are:

$$\mathcal{S}_\mu[h_1(x_1), \ldots, h_n(x_n)] = \bigvee_{i=1}^{n} [h_{(i)}(x_i) \wedge \mu(A_{(i)})] \tag{2}$$

and

$$\mathcal{C}_\mu[h_1(x_1), \ldots, h_n(x_n)] = \sum_{i=1}^{n} h_{(i)}(x_i) \cdot [\mu(A_{(i)}) - \mu(A_{(i-1)})], \tag{3}$$

where $\mu(A_{(0)}) = \mu(\emptyset)$ and the enclosed subindex states for a sort operation previous to the integration itself, e.g.

$$h_1 \geq h_3 \geq h_2 \rightarrow h_{(1)} = h_1; h_{(2)} = h_3; h_{(3)} = h_2. \tag{4}$$

This operation fixes up the coefficients of the fuzzy measures thence employed in the integration. In the example these would be

$$\mu(A_{(1)}) = \mu(\{x_1\}); \mu(A_{(2)}) = \mu(\{x_1, x_3\}); \mu(A_{(3)}) = \mu(\{x_1, x_2, x_3\}). \tag{5}$$

One important property of these fuzzy integrals for the development of the here presented paradigm is related to their limits. The value of both the Choquet and the Sugeno Fuzzy Integral, represented here as $\mathcal{FI}_\mu$, has its lower and upper bound in the result of the minimum and maximum operators:

$$\bigwedge[x_1,\dots,x_n] \leq \mathcal{FI}_\mu(x_1,\dots,x_n) \leq \bigvee[x_1,\dots,x_n]. \tag{6}$$

The fuzzy t-conorm integral generalizes $\mathcal{S}_\mu$ and $\mathcal{C}_\mu$ by allowing the utilization of optional T- and S-norms [8] (see Fig. 1). This type of integral, which has a larger range than the Choquet and Sugeno integrals, has not been used in image fusion so far.

Table 1. Generalization relationships between the fuzzy integral with respect to a particular fuzzy measure and other fuzzy aggregation operators (*Op.*). *min*: minimum. *max*: maximum. *wmin*: weighted minimum [4]. *med*: median. *wsum*: weighted sum. *OWA*: Ordered Weighted Averaging [17]. N_π states for a necessity measure [14] induced ($\rightarrow$) by a possibility distribution π induced itself by the set of weights W [4].

Op.	Integral	Fuzzy Measure
min	$\mathcal{S}_\mu, \mathcal{C}_\mu$	$\mu(A_j) = 0,\ \forall j/\lVert A_j\rVert < \lVert\mathcal{X}\rVert$
max	$\mathcal{S}_\mu, \mathcal{C}_\mu$	$\mu(A_j) = 1,\ \forall j/\lVert A_j\rVert = 1$
wmin	$\mathcal{S}_\mu$	$\mu = N_\pi, /W = \{w_1,\dots,w_n\} \rightarrow \pi \rightarrow N_\pi$
med	$\mathcal{S}_\mu$	$\mu(A_j) < \mu(A_k) \bigwedge \mu(A_k) = 1,\ \forall j,k/\lVert A_j\rVert < \lVert A_k\rVert \bigwedge \lVert A_k\rVert = n/2$
wsum	$\mathcal{C}_\mu$	$\mu(A_j \cup A_k) = \mu(A_j) + \mu(A_k) = w_j + w_k,\ \forall j,k/\lVert A_j\rVert = \lVert A_k\rVert = 1$
OWA	$\mathcal{C}_\mu$	$\mu(A_j) = \mu(A_k) = w_i,\ \forall j,k/\lVert A_j\rVert = \lVert A_k\rVert$

The generalization of other fuzzy aggregation operators satisfied by the fuzzy integral succeeds through the values of the fuzzy measure coefficients. Due to the mathematical expressions of the fuzzy integrals the computation of their result with respect to a particular fuzzy measure is equivalent to the application of other fuzzy aggregation operators. Table 1 gives an overview of some of these generalization relationships between operators.

2.2 Locally defined fuzzy measures

In the application of the fuzzy integral for image fusion the coefficients of the fuzzy measures are determined globally with respect to the image, i.e. the fuzzy measure coefficients characterize the importance of each image channel and that of their possible coalitions taking the image as a unit.

A novelty concerning this last aspect is presented in the ILF paradigm. Here the used fuzzy measure is not unique, but depends on the region where the fusion is undertaken, i.e. the fuzzy measure changes over the image. Therefore one can talk of "localized" fuzzy measures. The localization of these different fuzzy measures is given in form of a label image. The label image codifies the application area of the different fuzzy measures through a label.

The utilization of these "localized" fuzzy measures allows operating in the same image with different fusion strategies. The "localized" fuzzy measures exploit the generalization property of the fuzzy integral. With regard to the mathematical properties of the fuzzy measures (see Sect. 2.1) the usage of such "localized" measures allows overcoming the monotonicity requirement when looking at them from a global perspective. A fuzzy integral with respect to such locally defined fuzzy measures becomes a so-called Intelligent Localized Fusion Operator, ILFO. These operators are successfully applied in different multisensorial image processing applications [10].

3 Automated determination of fuzzy measures in ILFOs

The utilization of locally defined fuzzy measures implies the proliferation of its number. The number of coefficients to be determined grows with a factor of M, where M is the number of different codes in the label image. If general fuzzy measures are used the number of coefficients becomes $M \cdot 2^n - 2$. Taking a special case of fuzzy measures, as the fuzzy-λ measures (see Eq. (1)), reduces the number of coefficients to be found to $M \cdot n$, but it brings together a loss of flexibility in the fusion result (see Eq. (1)). The automation of the fuzzy integral parameterization can ease the application of the presented methodology for larger values of M or n.

The automated determination of fuzzy measures plays an analogous role in the Fuzzy Measure Theory as the automated computation of fuzzy membership functions in Fuzzy Sets Theory. Currently there is still a lack of efficient standard methods for computing the fuzzy measures, although intensive research work is being targeted on this field [7]. Evolutionary computation methodologies have been successfully implemented with this purpose in different application fields [1] [7] [15] .

The term evolutionary computation groups a set of searching methodologies inspired by the Darwinist theory of evolution. Among them genetic algorithms seek the iterative optimization of a so-called fitness function through the interplay of exploration and exploitation procedures in the search space. The possible solutions of the optimization, which receive the name of individuals, are embedded in a population iteratively saving the best solutions and substituting the worst ones by new individuals. A general introduction on genetic algorithms and its theoretical background can be found in [2].

The general methodology of genetic algorithms with standard crossover, selection and mutation operators [2] applies for the determination of the "localized" fuzzy measures of the ILFOs. Here the individuals code the fuzzy measures with arrays of real numbers, which represent each of their coefficients. The procedure can consider the construction of all the fuzzy measures at once or sequentially repeated M times.

The fulfillment of the monotonicity property of the fuzzy measures (see Sect. 2.1 condition II) obstructs the application of genetic algorithms when constructing general fuzzy measures. A conformance function can be used for lowering the fitness of those individuals not fulfilling the monotonicity relationship between coefficients [1]. Using fuzzy-λ measures solves this question (see Eq. (1)), but again with a loss of flexibility in the fusion result.

Special attention is devoted to the fitness function. The problem for the application on hand is the characterization of the subjective evaluation of the fusion result through a mathematical function. The fused image should present a small number of pixels with maximal grayvalues. This numerical feature together with the minimization of the number of pixels with low grayvalues helps in the achievement of a contrast improved pre-processed image. As a matter of fact a trade-off between contrast and smoothness has to be found, since the smoothness of the final image reflects the absence of artefacts introduced by the image fusion operator and the contrast in the obtained result facilitates the completion of a successful visual inspection. All these factors influence the selection of the fitness function.

4 Highlights filtering through ILFOs

The presented methodologies are implemented for the pre-processing in different automated visual inspection applications. A block diagram of the preprocessing system for the highlight filtering is depicted in Fig. 2. The system is built by an Image Acquisition module (*ImAcq*) and an ILFO, which consists of the modules: Label Image Generator (*LabImGen*), Fuzzy Measure Constructor (*FuzMeCo*), and Fuzzy Integral (*FuzInt*).

The stage previous to the avoidance of highlights through the application of an ILFO is in charge of the image set generation (*ImAcq*). For that purpose different images of the underlying object are taken with a camera from the same position but with different focal distances, shutter times and/or illumination conditions, i.e. lamp type, lighting direction. Therefore the highlights appear in the image set with different structures and/or positions, while the object information remains the same. This operation is not trivial since the high reflective nature of the objects being inspected makes the reflection of the environment on the surface very easy.

Once the image series has been generated the label image for the fusion is computed (*LabImGen*). This is obtained through the application on the input images of standard image processing algorithms that seek a coarse seg-

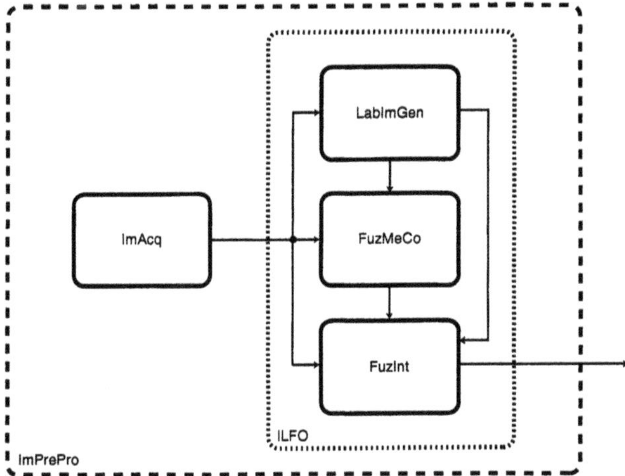

Fig. 2. Block diagram of a pre-processing system for the highlight filtering through the application of ILFOs. ILFO: Intelligent Localized Fusion Operator. *ImPrePro*: Image Pre-Processing. *LabImGen*: Label Image Generator. *FuzMeCo*: Fuzzy Measure Constructor. *FuzInt*: Fuzzy Integral.

mentation. The segmentation is not based as usual on the characterization of all different homogeneous regions on the input image in terms of gray or color value but on the detection of the saturation areas. The ideal case is the utilization of a label image with so many labels as the number of possible subsets on the source set $\mathcal{J}(\mathcal{X})$. Such a label image codes all the different highlight interaction areas: no highlights, highlight in the first image, highlight in the first and second image, etc. However such an extensive characterization should be avoided when the obtained results with more simple label images are satisfactory.

A process for the determination of the fuzzy measures is thence undertaken (*FuzMeCo*). The number of different labels in the label image fixes up how many fuzzy measures have to be constructed. This process is based on the acquired images and can be heuristically realized. In this case it is based on the subjective impression of a user, who iteratively modifies the different coefficients of the fuzzy measures and inspects the obtained result. The search of the coefficients finishes when the user considers that the fusion result is optimal. Such a process can be very tedious. Genetic algorithms can be alternatively used instead of this extensive search.

As already mentioned finding an adequate fitness function for the genetic algorithm is not trivial in case of mathematically assessing a quality image function. Three different fitness functions, which are employed to obtain the results presented in the next section, are exemplary given. The first one uses a weighted sum of three features, namely the contrast of the fused image (characterized by the variance of its grayvalues σ_g^2), and the fuzzy variables "bright" and "dark" (applied on the grayvalues g_i of the fused image pixels):

$$f(x) = 0.8\sigma_g^2 + \sum_{i=0}^{N-1} [0.75 h_{bright}(g_i) + 0.2 h_{dark}(g_i)], \tag{7}$$

where N states for the number of pixels of the final image. The used fuzzy membership functions are depicted in Fig. 3 and present the following mathematical expressions:

$$h_{bright}(g_i) = \begin{cases} 0 & : \quad 0 \le g_i < 200 \\ \frac{g_i - 200}{55} & : \quad 200 \le g_i \le 255 \end{cases} \tag{8}$$

and

$$h_{dark}(g_i) = \begin{cases} 1 - \frac{g_i}{15} & : \quad 0 \le g_i < 15 \\ 0 & : \quad 15 \le g_i \le 255. \end{cases} \tag{9}$$

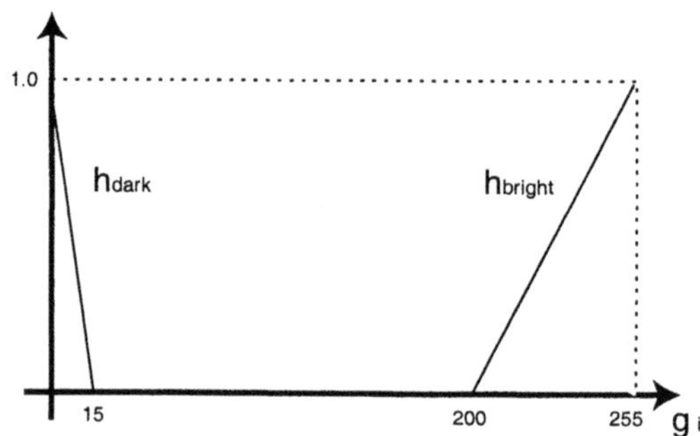

Fig. 3. Fuzzy membership functions for the fuzzification of grayvalue pixel information (g_i) with linguistic terms *dark* and *bright* (used in Eq. (7)).

The second fitness function forces the grayvalue distribution of the fused image to present $\bar{g} = 0.5$ and $\sigma_g^2 = 0.25$. Finally the minimization of the fused image contrast, which can be characterized by σ_g^2, is attained through the third fitness function. These fitness functions and their parameters were heuristically found.

Finally, the fuzzy integral is applied on each pixel of the image set with respect to the fuzzy measures codified in the label image (*FuzInt*). The result is an image, where the highlight areas have been filtered.

5 Automated visual inspection: application report

Two different pre-processing goals are attained in the presented results. On the first hand some of the applications request just a right visualization of the underlying object. In this case the ILFO is the only part of an image enhancement system, whose objective is the highlights suppression. A second type of pre-processing system, embodied generally in fault detection systems, goes further. Since the reflection on the possible faults ease its detection, the highlights in these positions are left untouched, while being suppressed in the rest of the image.

5.1 Pre-processing for visualization

The results presented in the following belong to an automated visual inspection system, whose final goal was the analysis of the chocolate package completeness. Thus a contrast enhanced image with no highlights should be generated. Three different images of the chocolate package are used, where the presence of a bundle produces some highlights to be suppressed (see Fig. 4).

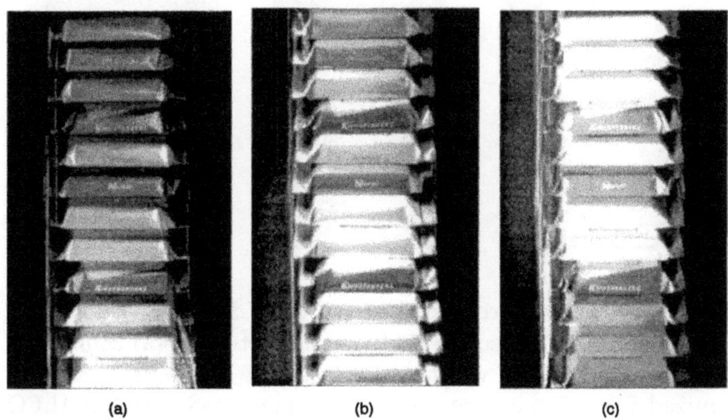

Fig. 4. Input images of the pre-processing stage in a system for the analysis of chocolate package completeness.

Different results, which are obtained with different fusion strategies, are shown in Fig. 5. The utilization of more complex strategies than the minimum increases the luminosity of the final result (compare Fig. 5a with Figs. 5b-d). Furthermore the utilization of an ILFO improves the contrast of the fused image (see Fig. 5f). The label image, Fig. 5e, was built up to an addition of the three input images followed by a multithreshold. Once the label image was computed a genetic algorithm (GA) with the fitness function 7 was applied for the construction of the three fuzzy measures.

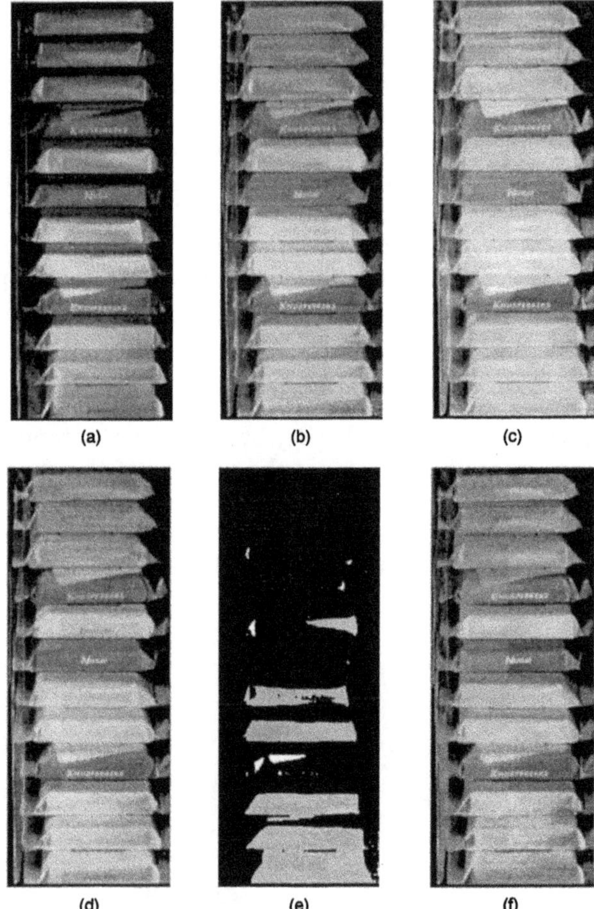

Fig. 5. Results of the pre-processing stage with different fusion strategies in a system for the analysis of chocolate package completeness. (a) Minimum. (b) OWA. (c) OWA with GA optimized weights. (d) Choquet Fuzzy Integral with respect to a GA optimized fuzzy-λ measure. (e) Label image used with the ILFO. (f) ILFO with label image in Fig. (e) and fuzzy-λ measures, whose coefficients were optimized through a genetic algorithm (GA).

A bit more complex is the pre-processing system for the automated visual inspection of automobile headlamp reflectors. The input images of the system are shown in Fig. 6. The obtained results are depicted in Fig. 7. The label image is obtained after the application of a threshold in each of the individual images and the addition of the binary ones (see Fig. 7a). All results were obtained after the construction of the corresponding fuzzy measures through a GA with 40 individuals of population size, 60 generations, 0.8 of crossover probability, and 0.001 of mutation probability.

Fig. 6. Input images of the pre-processing stage in a system for the detection of structural faults on automobile headlamp reflectors.

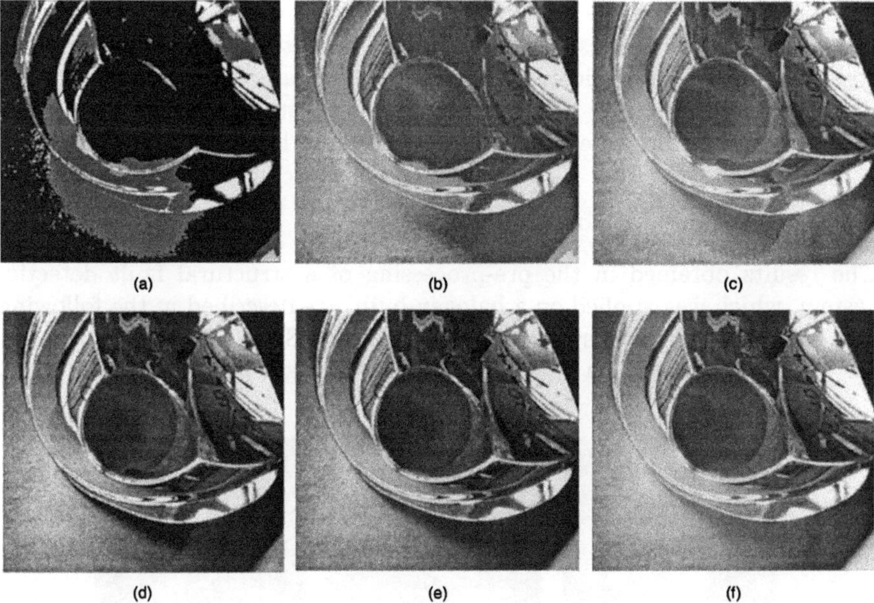

Fig. 7. Results of the pre-processing stage, where the highlights avoidance is undertaken through the application of ILFOs, in a system for the detection of structural faults on automobile headlamp reflectors. (a) Label image for results. (b) Sugeno Fuzzy Integral based ILFO with GA optimized fuzzy-*lambda* measures (fitness favored a mean grayvalue of 0.5 and variance of 0.25). (c) Choquet Fuzzy Integral based ILFO with GA optimized fuzzy-*lambda* measures (fitness Eq. (7) in text). (d) Choquet Fuzzy Integral based ILFO with GA optimized fuzzy-*lambda* measures (minimization of contrast function). (e) Choquet Fuzzy Integral based ILFO with fuzzy-*lambda* measures, which were heuristically found up to fuzzy measures from Fig. (d). (f) Choquet Fuzzy Integral based ILFO with general fuzzy measures, which were heuristically found up to GA optimized fuzzy-*lambda* measures (fitness favored a mean grayvalue of 0.5 and variance of 0.25).

The results show the performance of different fuzzy measure construction strategies. The Sugeno Fuzzy Integral is not adequate for the solution of the application on hand when the fuzzy measures are determined with a GA (see Fig. 7b). The optimization was done by favoring those images with mean grayvalue 0.5 and variance of 0.25. Some artefacts, areas with a grayvalue equal to one of the coefficients of the fuzzy measure, appear very easily when applying this operator. Some pseudo-edges appear when using the Choquet Fuzzy Integral based ILFO (see Fig. 7c), where the fitness function of the GA for the fuzzy-λ construction is Eq. (7). The utilization of another fitness function with the same operator, where the contrast of the image should be minimized, does not change this situation (see Fig. 7d). However an heuristic and slight modification of the results obtained by the GA improves the ILFO performance as can be observed on Fig. 7e. It is worth mentioning that the modification changes the type of the used fuzzy measures from λ measures to general fuzzy measures (see Sect. 3). The same strategy was used in order to obtain the image depicted in Fig. 7f. In this case the GA minimizes the fitness function given by Eq. (7). This image constitutes a very good result in terms of contrast, absence of pseudo-edges and highlight suppression.

5.2 Pre-processing for fault detection

The results obtained in the pre-processing of a structural fault detection system, which was applied on a halogen bulb, are described in the following. The input images of the system are shown in Fig. 8. The fault to be detected can be distinguished as a triangular area in the right part of the bulb neck.

Fig. 8. Input images of the pre-processing stage in a system for the detection of structural faults on halogen bulbs.

A comparison of the results obtained through a weighted minimum (Fig. 9a) and an OWA (Fig. 9b) shows again that the utilization of a more complex operator (OWA in front of weighted minimum) improves the result of the fusion in terms of presence of structural information. Furthermore the usage of an ILFO (Fig. 9d) can ease the posterior detection of the fault, since other highlights than this are suppressed. The saturation area in the center of the bulb body is due to an environment reflection and can not be eliminated because of its presence in all three input images (see Fig. 8).

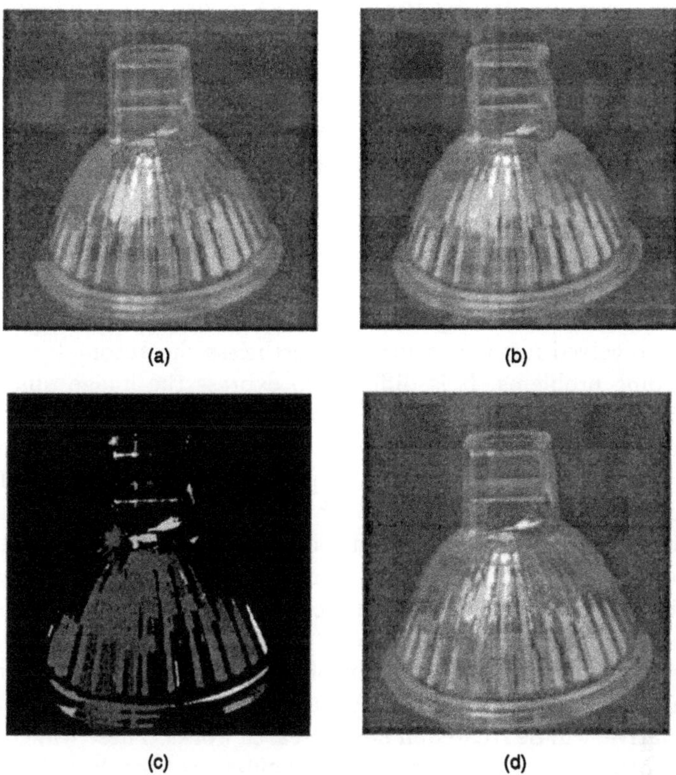

Fig. 9. Results of the pre-processing stage with different fusion strategies in a system for the detection of structural faults on halogen bulbs. (a) Weighted minimum. (b) OWA. (c) Label image used with the ILFO. (d) Choquet Fuzzy Integral based ILFO with label image in Fig. (c) and fuzzy-λ measures, which were heuristically determined.

6 Conclusions

The methodology of the ILF paradigm was presented. It offers a systematic approach for the avoidance of highlights in the inspection of high reflective materials. The approach is based on the generation of an image set of the underlying object acquired under different conditions. These images are thence fused through a so-called ILFO.

ILFOs are based on the utilization of different types of the fuzzy integral, a fuzzy fusion operator which generalizes most of used Soft-Computing related aggregation methodologies. Both the here employed Sugeno and Choquet Fuzzy Integrals constitute a flexible mean of binding information, whose development was inspired by the subjective multi-criteria integration. The fuzzy integral uses so-called fuzzy measures for the assessment of the *a priori* importance of the information sources. In the context of image fusion the presented paradigm proposes a local definition of the fuzzy measures, which is different from its usual global definition in terms of the images being fused. Therefore a label image is used for codifying the fuzzy measure to be used in each image area. This fact allows the exploitation of the generalization property of the fuzzy integral and thus the utilization of different fusion strategies in the same image with just one operator.

Besides heuristic search, genetic algorithms can be used for the construction of the involved fuzzy measures. Nevertheless the automated assessment presents some problems. It is difficult to express the image quality, which is in principle a subjective feature involving e.g. contrast and smoothness, through a mathematical fitness function. Specially the apparition of pseudo-edges on images obtained through Choquet Fuzzy Integral based ILFOs and of artefacts on images obtained through Sugeno Fuzzy Integral based ILFOs have to be taken into consideration. Some of the results described in this paper present these problems. They can be corrected as shown by a posterior modification of the fuzzy measure coefficients found through GAs. Such a combined application of automated (for a first estimation) and heuristic search deliver optimal results in terms of image quality. Therefore interactive evolutionary computation, which offers a methodological approach similar to this alternative, will be treated in next research works. These will consider the analysis of other fitness functions for the avoidance of the described problems as well.

The ILF paradigm was successfully used in different cases taken from automated visual inspection systems. Both the visualization and the facilitation of the posterior fault detection can be attained in a pre-processing system based on the here elucidated paradigm. Thus this methodology constitutes a promising technology for the implementation of automated inspection of high reflective objects.

Aknowledgments. The author wants to thank: Jing Zhou, for the completion of some of the here presented results, Daniel Kotow, for porting part of the functionality of GALib to Python, and Mario Köppen, for his helpful comments on the manuscript. The software for this work used the GAlib genetic algorithm package, written by Matthew Wall at the Massachusetts Institute of Technology, and Python as prototyping language (http://www.python.org).

References

1. T. Chen, J. Wang, G. Tzeng, *Identification of General Fuzzy Measures by Genetic Algorithms Based on Partial Information*, in: IEEE Trans. on Systems, Man and Cybernetics Part B, 30 (4), 2000, pp. 517-528.
2. D. Goldberg, *Genetic Algorithms in Search, Optimization and Machine Learning*, Addison Wesley Pub. Co., 1989.
3. R.C. Gonzalez, R.E.Woods, *Digital Image Processing*, Addison-Wesley Pub. Co., 1992.
4. M. Grabisch, H.T. Nguyen and E.A. Walker, *Fundamentals of Uncertainty Calculi with Applications to Fuzzy Inference*, Kluwer Ac. Pub., 1995.
5. M. Grabisch, *Fuzzy Measures and Integrals for Decision Making and Pattern Recognition*, in: Fuzzy Structures: Current Trends, TATRA MOUNTAINS Mathematical Pub., 1997, pp. 7-34.
6. B. Jähne, *Digital Image Processing: Concepts, Algorithms, and Scientific Applications*, Springer-Verlag, 1997.
7. G.J. Klir, Z. Wang and D. Harmanec,*Constructing Fuzzy Measures in Expert systems*, in: Fuzzy Sets and Systems, 92, 1997, pp. 251-264.
8. T. Murofushi, M. Sugeno, *Fuzzy t-conorm integral with respect to fuzzy measures: Generalization of Sugeno and Choquet integral*, in: Fuzzy Sets and Systems, 42, 1991, pp. 57-71.
9. F. Puente León, J. Beyerer *Datenfusion zur Gewinnung hochwertiger Bilder in der automatischen Sichtprüfung*, in: Automatisierungstechnik, 45, 1997, pp. 480-489 (in German).
10. A. Soria-Frisch, *Soft Data Fusion in Image Processing* , to appear in: Soft Computing in Industry -Recent Applications, R. Roy *et al.* eds., Springer-Verlag, Berlin, 2002.
11. M. Sugeno, *Theory of fuzzy integral and its applications*, Ph.D. thesis, Tokyo Institute of Technology, 1974.
12. H. Takagi, *Interactive evolutionary computation: Fusion of the capabilities of EC optimization and human evaluation*, in: Proceedings of the IEEE, 89 (9), 2001, pp. 1275-1296.
13. H.R. Tizhoosh, *Fuzzy Bildverarbeitung*, Springer-Verlag, Heidelberg, 1998 (in German).
14. Z. Wang and G.J. Klir, *Fuzzy Measure Theory*, Plenum Press, 1992.
15. Z. Wang, K. Leung, J. Wang, *A genetic algorithm for the determining nonadditive set functions in information fusion*, in: Fuzzy Sets and Systems, 102, 1999, pp. 463-469.
16. R.R. Yager and A. Kelman, *Fusion of Fuzzy Information With Considerations for Compatibility, Partial Aggregation, and Reinforcement*, in: Int. J. of Approximate Reasoning, 15, 1996, pp. 93-122.
17. R.R. Yager and J. Kacprzyk eds., *The Ordered Weighted Averaging Operators: Theory and Applications*, Kluwer Ac. Pub., 1997.
18. L.A. Zadeh, *Fuzzy Sets*, in: Information and Control, 8, 1965, pp. 338-353.

Appendix

Color Images

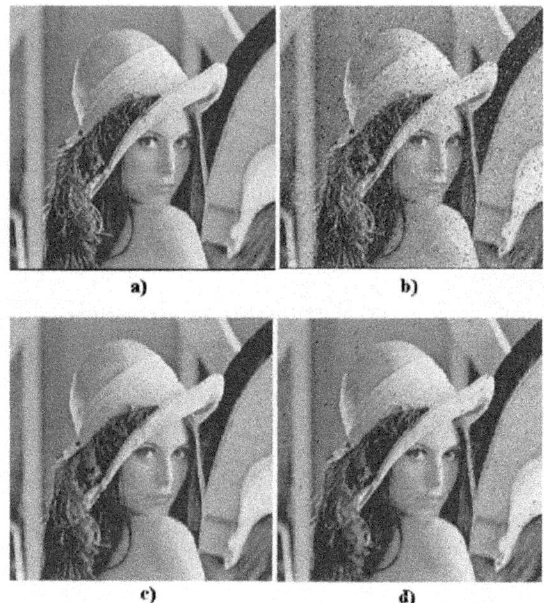

Fig. 1. Chapter 1, Figure 10

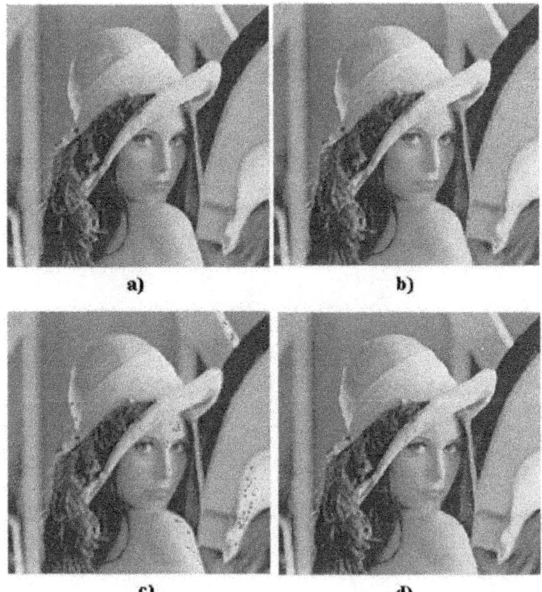

Fig. 2. Chapter 1, Figure 11

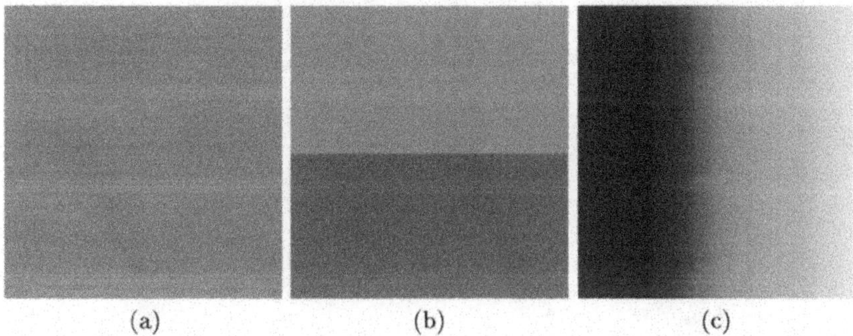

Fig. 3. Chapter 4, Figure 2

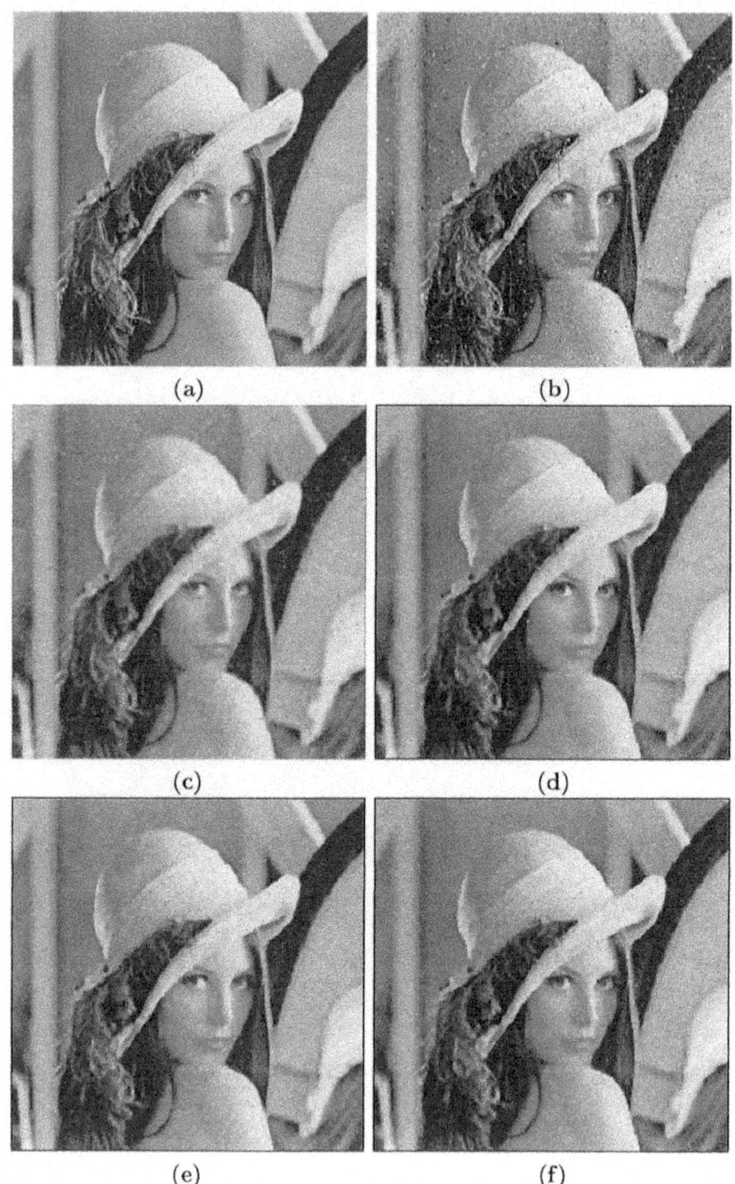

Fig. 4. Chapter 4, Figure 9

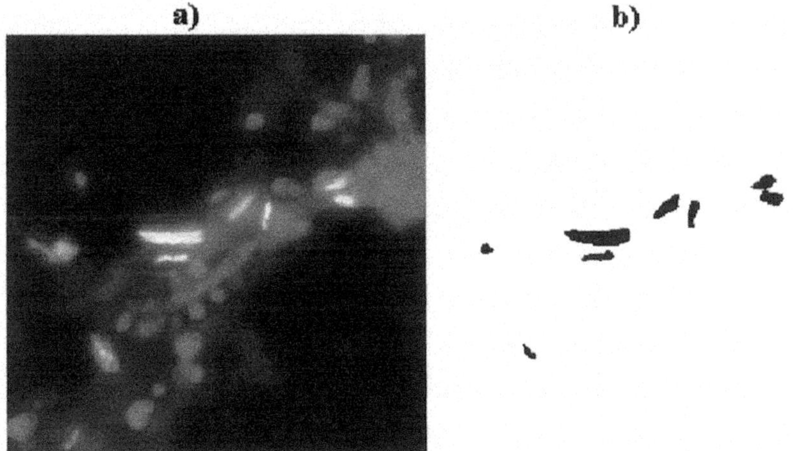

Fig. 5. Chapter 6, Figure 14

Fig. 6. Chapter 7, Figure 10 and 15

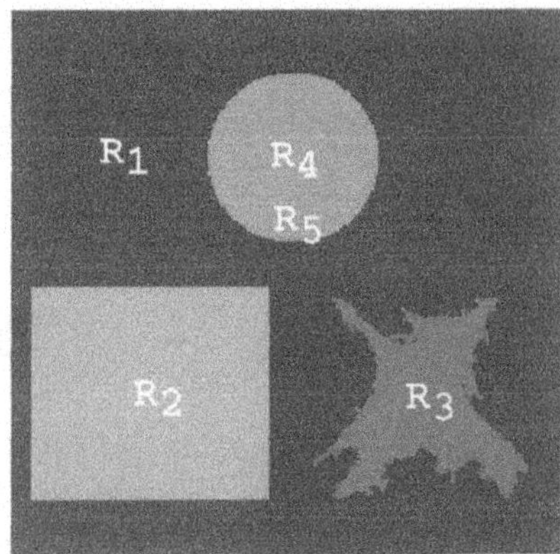

Fig. 7. Chapter 7, Figure 16

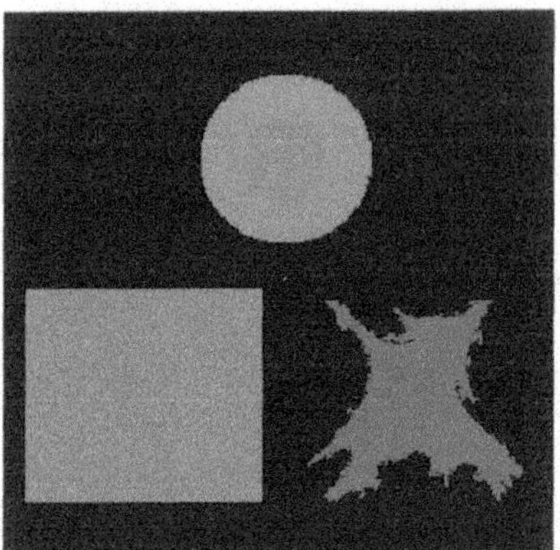

Fig. 8. Chapter 7, Figure 18

Fig. 9. Chapter 7, Figure 21

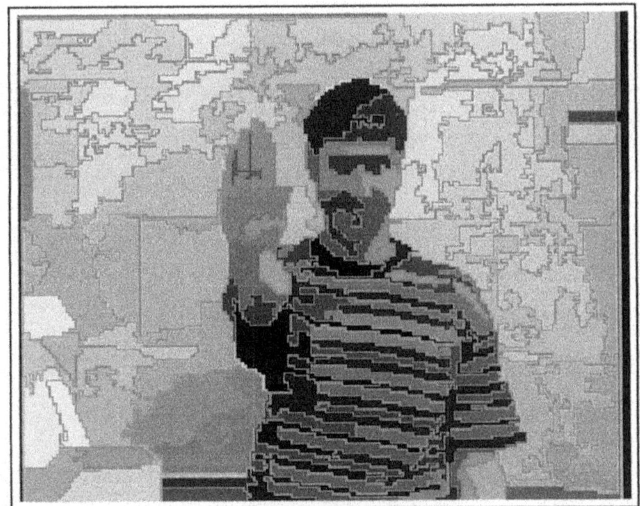

Color Segmentation

Motion Segmentation

Depth Segmentation

Previous Fusion

Fusion Result

Fig. 10. Chapter 9, Figure 5

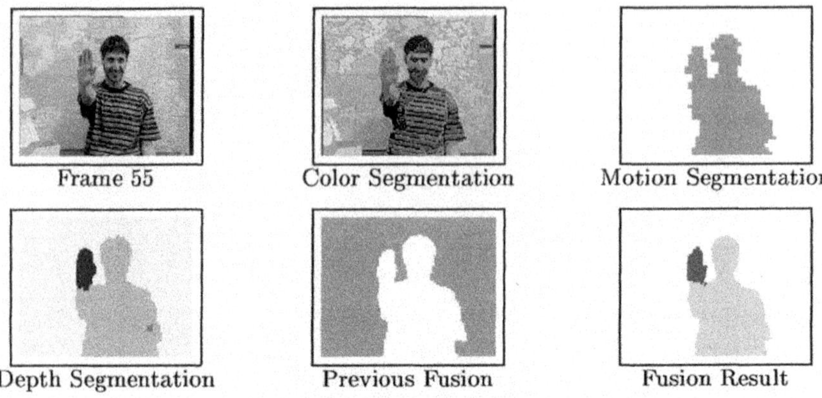

Fig. 11. Chapter 9 : Sequence 2, Frame 55

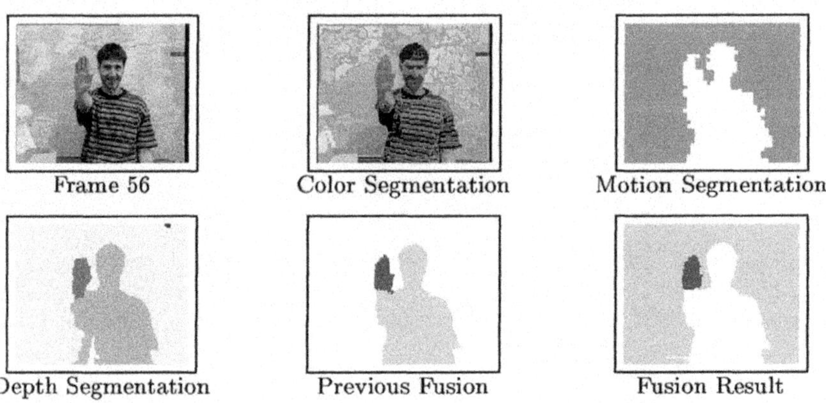

Fig. 12. Chapter 9 : Sequence 2, Frame 56

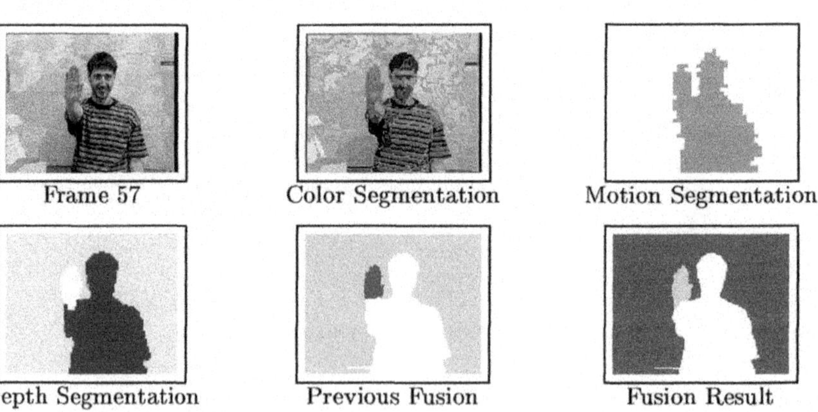

Fig. 13. Chapter 9 : Sequence 2, Frame 57

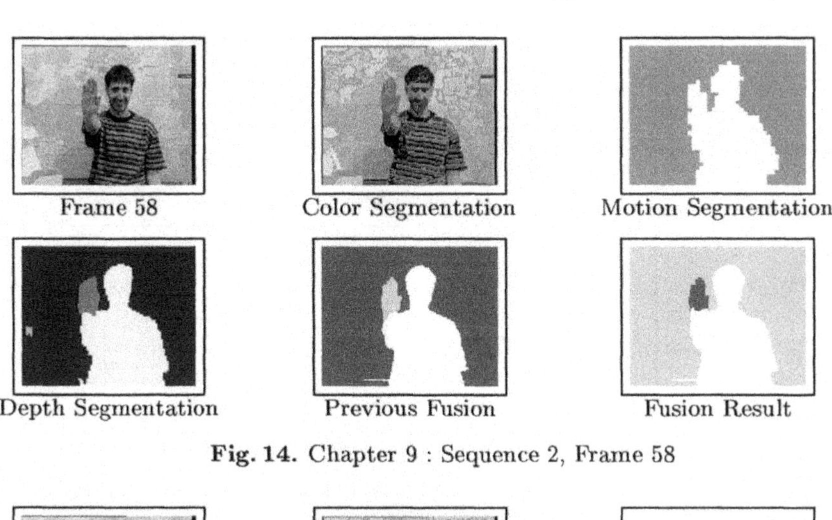

Fig. 14. Chapter 9 : Sequence 2, Frame 58

Frame 59 Color Segmentation Motion Segmentation

Depth Segmentation Previous Fusion Fusion Result

Fig. 15. Chapter 9 : Sequence 2, Frame 59

Fig. 16. Chapter 9 : Sequence 2, Frame 60

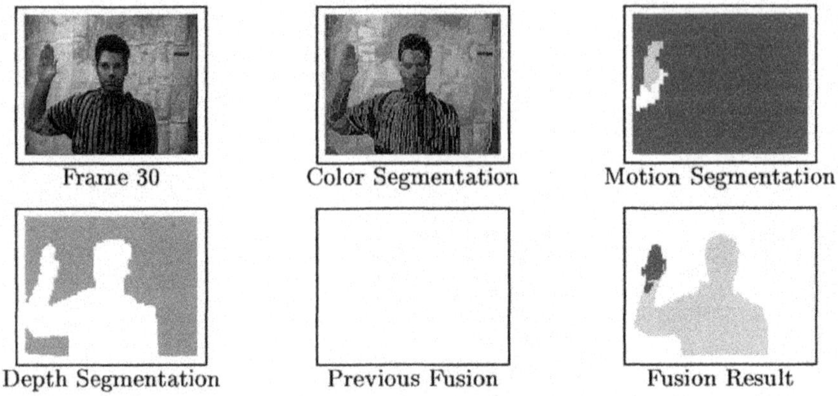

Frame 30 | Color Segmentation | Motion Segmentation

Depth Segmentation | Previous Fusion | Fusion Result

Fig. 17. Chapter 9 : Sequence 1, Frame 30

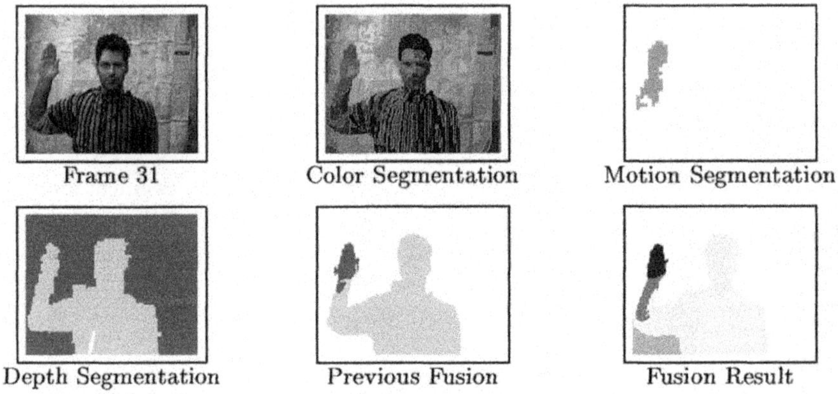

Frame 31 | Color Segmentation | Motion Segmentation

Depth Segmentation | Previous Fusion | Fusion Result

Fig. 18. Chapter 9 : Sequence 1, Frame 31

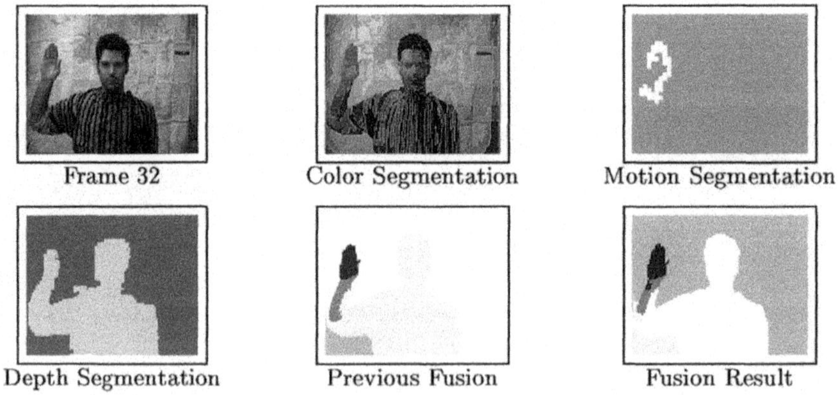

Frame 32 | Color Segmentation | Motion Segmentation

Depth Segmentation | Previous Fusion | Fusion Result

Fig. 19. Chapter 9 : Sequence 1, Frame 32

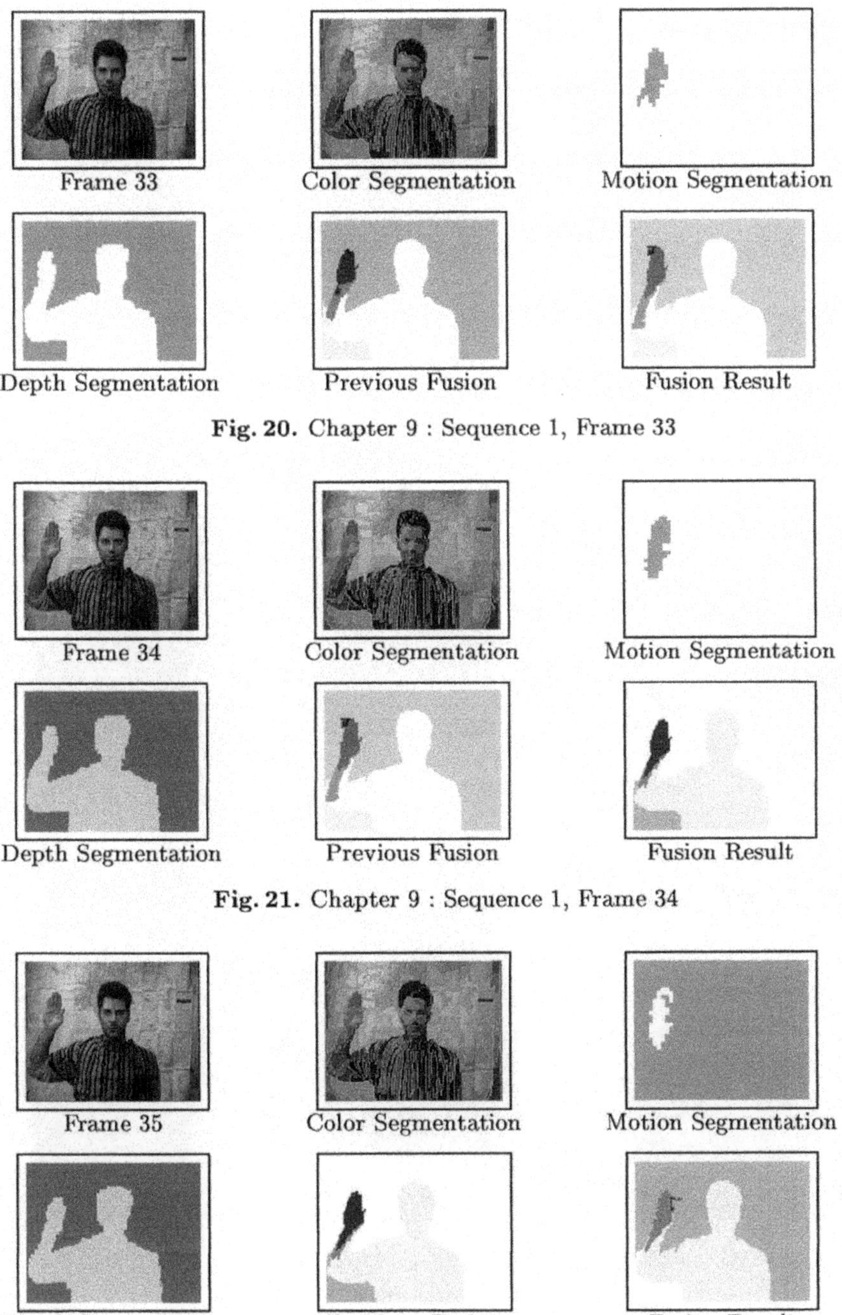

Fig. 20. Chapter 9 : Sequence 1, Frame 33

Fig. 21. Chapter 9 : Sequence 1, Frame 34

Fig. 22. Chapter 9 : Sequence 1, Frame 35

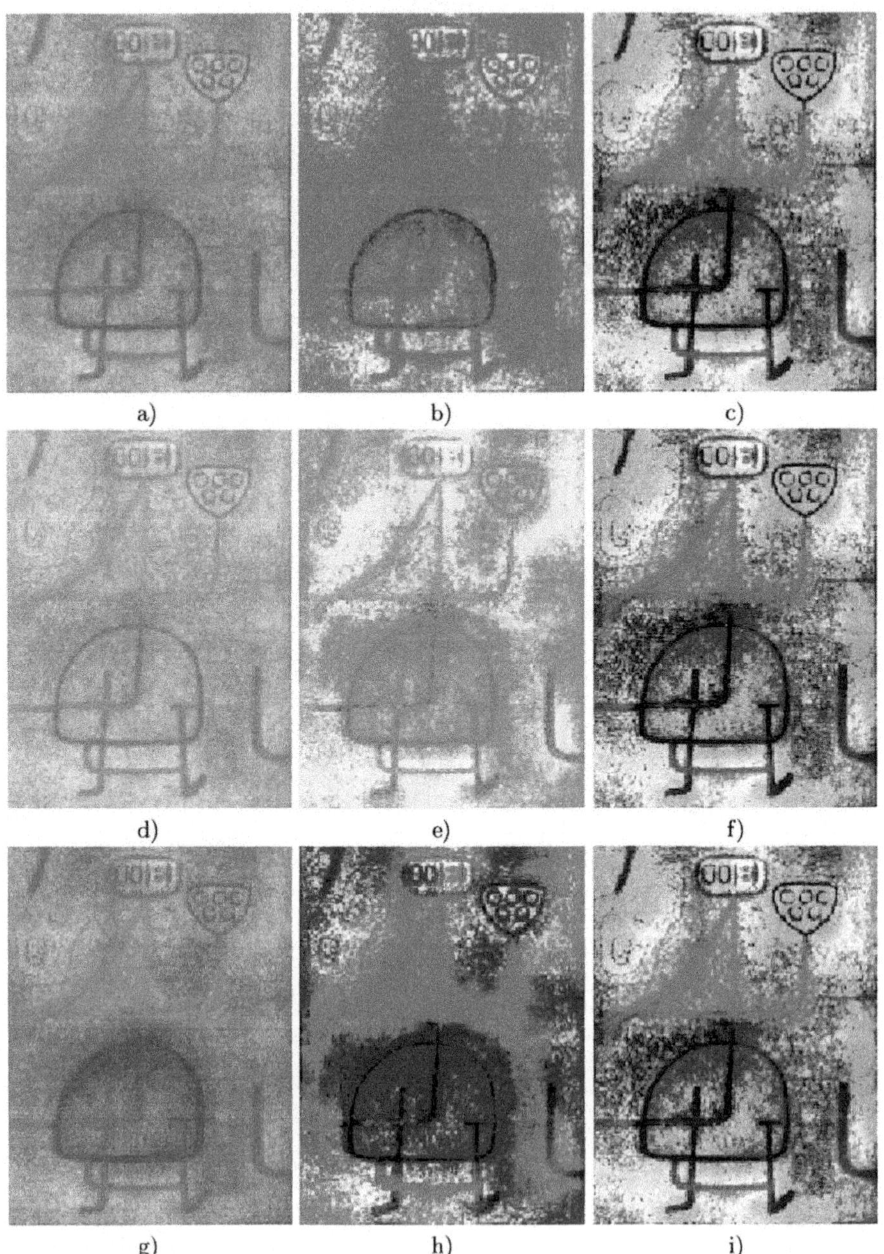

Fig. 23. Chapter 10, Figure 4

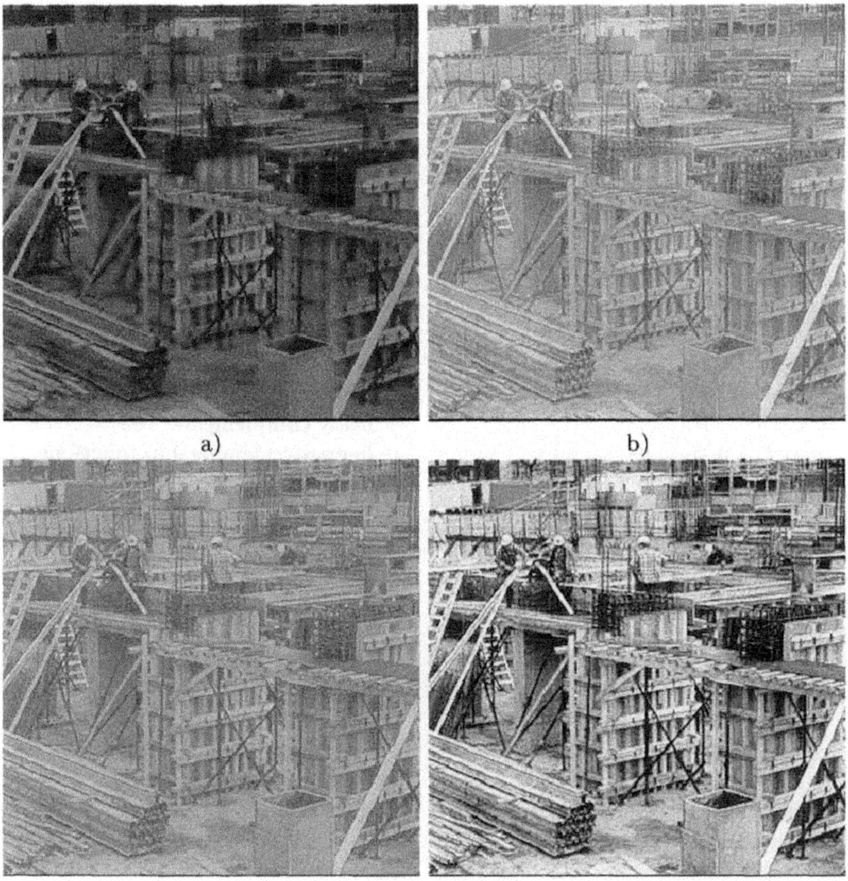

a) b)

c) d)

Fig. 24. Chapter 10, Figure 5

(a) (b) (c)

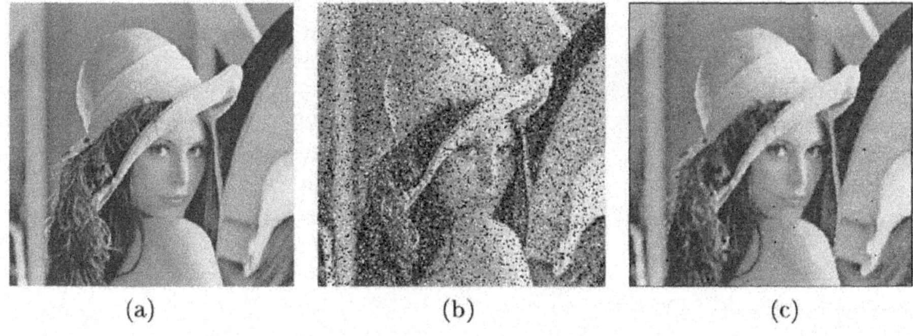

Fig. 25. Chapter 12, Figure 2

Index

www.ingramcontent.com/pod-product-compliance
Ingram Content Group UK Ltd.
Pitfield, Milton Keynes, MK11 3LW, UK
UKHW021858190726
13853UKWH00003B/1323

* 9 7 8 3 6 4 2 5 3 5 1 8 5 *